148 Graph Directed Markov Systems

R. Daniel Mauldin and Mariusz Urbański

Graph Directed Markov Systems

Geometry and dynamics of limit sets

CAMBRIDGE
UNIVERSITY PRESS

PUBLISHED BY THE PRESS SYNDICATE OF THE UNIVERSITY OF CAMBRIDGE
The Pitt Building, Trumpington Street, Cambridge, United Kingdom

CAMBRIDGE UNIVERSITY PRESS
The Edinburgh Building, Cambridge CB2 2RU, UK
40 West 20th Street, New York, NY 10011-4211, USA
477 Williamstown Road, Port Melbourne, VIC 3207, Australia
Ruiz de Alarcón 13, 28014 Madrid, Spain
Dock House, The Waterfront, Cape Town 8001, South Africa

http://www.cambridge.org

First published 2003

Printed in the United Kingdom at the University Press, Cambridge

Typeface Computer Modern 10/13 pt *System* LaTeX 2ε [TB]

A catalogue record for this book is available from the British Library

ISBN 0 521 82538 5 hardback

Contents

Introduction *page* vii
1 **Preliminaries** 1
2 **Symbolic Dynamics** 4
2.1 Topological pressure and variational principles 5
2.2 Gibbs states, equilibrium states and potentials 12
2.3 Perron–Frobenius operator 26
2.4 Ionescu-Tulcea and Marinescu inequality 31
2.5 Stochastic laws 40
2.6 Analytic properties of pressure and the Perron–Frobenius operator 43
2.7 The existence of eigenmeasures of the Conjugate Perron–Frobenius operator and of Gibbs states 48
3 **Hölder Families of Functions and *F*-Conformal Measures** 54
3.1 Summable Hölder families 54
3.2 *F*-conformal measures 57
4 **Conformal Graph Directed Markov Systems** 62
4.1 Some properties of conformal maps in $\mathbb{R}^d$ with $d \geq 2$ 62
4.2 Conformal measures; Hausdorff and box dimensions 71
4.3 Strongly regular, hereditarily regular and irregular systems 87
4.4 Dimensions of measures 90
4.5 Hausdorff, packing and Lebesgue measures 94
4.6 Porosity of limit sets 103
4.7 The associated iterated function system 107
4.8 Refined geometry, *F*-conformal measures versus Hausdorff measures 109
4.9 Multifractal analysis 123

5 Examples of GDMSs 136
5.1 Examples of GDMSs in other fields of mathematics 136
5.2 Examples with special geometric features 139
6 Conformal Iterated Function Systems 144
6.1 The Radon-Nikodym derivative $\rho = \frac{d\mu}{dm}$ 144
6.2 Rate of approximation of the Hausdorff dimension by finite subsystems 153
6.3 Uniform perfectness 156
6.4 Geometric rigidity 160
6.5 Refined geometric rigidity 165
7 Dynamical Rigidity of CIFSs 171
7.1 General results 171
7.2 One-dimensional systems 176
7.3 Two-dimensional systems 181
7.4 Rigidity in dimension $d \geq 3$ 195
8 Parabolic Iterated Function Systems 209
8.1 Preliminaries 209
8.2 Topological pressure and associated parameters 212
8.3 Perron–Frobenius operator, semiconformal measures and Hausdorff dimension 218
8.4 The associated hyperbolic system. Conformal and invariant measures 222
8.5 Examples 234
9 Parabolic Systems: Hausdorff and Packing Measures 238
9.1 Preliminaries 238
9.2 The Case $d \geq 3$ 240
9.3 The plane case, $d = 2$ 246
9.4 Proofs of the main theorems 255
Appendix 1 Ergodic theory 262
Appendix 2 Geometric measure theory 264
Glossary of Notation 269
Bibliography 272
Index 280

Introduction

The geometric and dynamic theory of the limit set generated by the iteration of finitely many similarity maps satisfying the open set condition has been well developed for some time now. Over the past several years, the authors have in turn developed a technically more complicated geometric and dynamic theory of the limit set generated by the iteration of infinitely many uniformly contracting conformal maps, a (hyperbolic) conformal iterated function system. This theory allows one to analyze many more limit sets, for example sets of continued fractions with restricted entries. We recall and extend this theory in the later chapters. The main focus of this book is the exploration of the geometric and dynamic properties of a far reaching generalization of a conformal iterated function system called a graph directed Markov system (GDMS). These systems are very robust in that they apply to many settings that do not fit into the scheme of conformal iterated systems. While the basic theory is laid out here and we touch on many natural questions arising in its context, we emphasize that there are many issues and current research topics which we do not cover: for examples, the detailed analysis of the structure of harmonic measures of limit sets provided in [UZd], the examination of the doubling property of conformal measures performed in [MU6], the extensive study of generalized polynomial like mappings (see [U7] and [SU]), the multifractal analysis of geometrically finite Kleinian groups (see [KS]), and the connection to quantization dimension from engineering (see [LM] and [GL]). There are many research problems in this active area that remain unsolved.

Our book is organized as follows. In the very short first chapter we describe the basic setting for GDMSs. Essentially, one iterates a family of uniformly contracting maps which are indexed by the directed edges in a multigraph which may have infinitely many vertices and infinitely many edges. (This includes the finite graph directed systems of similarities introduced in [MW2, EM] and expounded by Edgar in [E].) One generates points in the limit set by performing an infinite directed walk through the graph. This leads to a natural map from the coding space or space of infinite walks through the graph to the points of the limit set.

Chapter 2 forms a self-contained unit and can be read independently of the rest of the book. Here we develop the symbolic dynamics and thermodynamic formalism for subshifts of finite type with infinitely many symbols. Thus, we are given an incidence matrix A and we consider the space E of all infinite sequences of symbols such that A has value 1 on all pairs of consecutive terms and the shift map on E. This formalism includes the action on the coding space described in the first chapter. One of our main goals is to develop the theory and properties of various measures and functions on the symbol space. To a function f on the symbol space is associated its topological pressure with respect to the shift map. Of course, this is a standard thing to do, but care must be taken because we have infinitely many symbols and the space E is not compact. In the first section we determine some conditions under which the topological pressure may be approximated by the more usual pressures when the system is restricted to subsystems on finitely many symbols. Next, we determine conditions under which there exists an invariant Gibbs state for the functions f. We present some results of ourselves and of Sarig which state that for a reasonable class of functions f which we call acceptable, there is an invariant Gibbs state for f if and only if the matrix A is finitely irreducible. We also determine some conditions under which f has a unique invariant equilibrium state. In the third section we develop the properties of the transfer or Perron–Frobenius operator associated to f. In order to fully analyze our system, we provide some results from functional analysis in the fourth section. We prove an exponential decay of correlations, a central limit theorem and a generalized law of the iterated logarithm in section 5. In section 6, we vary the function f with a complex parameter t. We show that various operators are then holomorphic, which implies under appropriate assumptions real analyticity of the pressure function. In section 7, we show that for certain functions f, the associated conjugate Perron–Frobenius operator has a Borel probability measure as an eigenmeasure. This allows us to conclude that if A is finitely primitive then there is a Gibbs state for f.

In Chapter 3, by using the tools developed in Chapter 2, we put the thermodynamic formalism for infinite subshifts of finite type into the context of GDMSs. We begin with F, a Hölder family of weight functions associated to the GDMS. By using an associated topological pressure and Perron–Frobenius operator, we determine conditions under which there is an F-conformal measure. This very general definition of F-conformal measure not only generalizes the usual notion of conformal measure, but forms a basic tool for later use in obtaining the geometric properties of the limit sets.

In Chapter 4 we deal in detail with geometric and fractal properties of conformal GDMSs. We begin by proving various kinds of distortion properties of conformal maps in $\mathbb{R}^d$ with $d \geq 2$. We then deal in this chapter with the various basic notions of dimension for the limit set: Hausdorff, upper and lower Minkowski or box (ball) counting dimension and packing dimension and the corresponding Hausdorff and packing measures. We deal with the Hausdorff and packing dimension of various natural measures supported on the limit set and some geometric properties of the limit set, e.g. porosity. Finally, we obtain the multifractal analysis of various conformal measures supported on the limit set. We emphasize that it is in this chapter that we must transfer many results from the abstract coding space to the limit set. As a point may have more than one code, this leads to several delicate issues in geometric measure theory. Therefore, the roles of distortion properties of our system of maps and the geometric properties of our seed sets, e.g. a relatively reasonable boundary, "the cone condition," play a crucial role in obtaining the necessary estimates of our analysis.

Chapter 5 is devoted to various illustrative examples including Kleinian groups of Schottky type, expanding repellers and a number of one-dimensional systems with prescribed geometric features.

In Chapter 6 we start to present the special case of a GDMS, a conformal iterated function system (CIFS). We study the real-analytic extension of the Radon-Nikodym derivative of an invariant measure with respect to a conformal measure, the classical example being Gauss' measure for the shift map on continued fractions. We estimate the rate at which the Hausdorff dimension of the limit sets generated by the finite subsystems of a CIFS approximates the dimension of the limit set. We determine conditions under which the limit set is uniformly perfect. In Section 4 of this chapter we begin the discussion of geometric rigidity by dealing with the limits set of a CIFS whose closure is connected. In essence the rigidity in this section means that in case $d \geq 3$ either the

Hausdorff dimension of the limit set is larger than 1 or else in case $d \geq 3$ the limit set is a subset of circle or line and in case $d = 2$ it is a subset of an analytic Jordan curve. In the next section we improve on this rigidity by showing that if $d \geq 3$ then essentially either the Hausdorff dimension of the limit set J exceeds the topological dimension k of the closure of J or else the closure of J is a proper compact subset of either a geometric sphere or an affine subspace of dimension k.

In Chapter 7 we deal with dynamical rigidity stemming from the work of Sullivan (see [Su3]) on conformal expanding repellers in the complex plane. We ask the fundamental question when two topologically conjugate infinite iterated function systems are conjugate in a smoother fashion. The answer is that such conjugacy extends to a conformal conjugacy on some neighborhoods of limit sets if and only if it is Lipschitz continuous. This turns out to equivalently mean that this conjugacy exchanges measure classes of appropriate conformal measures or that the multipliers of corresponding fixed points of all compositions of our generators coincide.

In Chapter 8 we study PIFS, parabolic iterated function systems. These are systems that are almost conformal except that we allow finitely many of the maps to have a parabolic or neutral fixed point instead of being uniformly contracting. A prime example of such a system is given by the system of three conformal maps in the plane whose limit set is the residual set in Apollonian packing. We analyze these systems by showing how to associate a conformal iterated function system to the parabolic system and how the properties of the parabolic limit set and measures may be derived from this associated conformal system. It is interesting that the parabolic system may consist of finitely many maps, but the associated conformal system is infinite. By moving from the finite to the infinite, the analysis becomes easier.

In Chapter 9 we provide a detailed quantative analysis of the dynamical behavior of parabolic maps (in dimension $d \geq 2$) around parabolic points and we apply it to provide a complete characterization of conformal measures of finite parabolic systems in terms of Hausdorff and packing measures. This simultaneously provides the answer to the question about necessary and sufficient conditions for these two geometric measures to be finite and positive.

In first section of the Appendix we collect some basic concepts and theorems from ergodic theory and in the second section contains a compressed exposition of some topics from geometric measure theory which are of interest here.

We have also provided two indexes, one for terminology and the other for special symbolic notation, and some references. We thank Cambridge University Press for the enormous help provided in bringing this project to fruition. We thank Larry Lindsay for his corrections to parts of the manuscript. Of course, we bear responsibility for all errors and omissions and ask foregiveness of all whom we have overlooked in our credits. Finally, we wish to thank the National Science Foundation for its support for our research during the preparation of this book.

1
Preliminaries

Graph directed Markov systems are based upon a directed multigraph and an associated incidence matrix, (V, E, i, t, A). The multigraph consists of a finite set V of vertices and a countable (either finite or infinite) set of directed edges E and two functions $i, t : E \to V$. For each edge e, $i(e)$ is the initial vertex of the edge e and $t(e)$ is the terminal vertex of e. The edge goes from $i(e)$ to $t(e)$. Also, a function $A : E \times E \to \{0, 1\}$ is given, called an incidence matrix. The matrix A is an edge incidence matrix. It determines which edges may follow a given edge. So, the matrix has the property that if $A_{uv} = 1$, then $t(u) = i(v)$. We will consider finite and infinite walks through the vertex set consistent with the incidence matrix. Thus, we define the set of infinite admissible words

$$E_A^\infty = \{\omega \in E^\infty : A_{\omega_i \omega_{i+1}} = 1 \text{ for all } i \geq 1\},$$

by E_A^n we denote the set of all subwords of E_A^∞ of length $n \geq 1$, and by E_A^* we denote the set of all finite subwords of E_A^∞. We will drop the subscript A when the matrix is clear from context. We will consider the left shift map $\sigma : E^\infty \to E^\infty$ defined by dropping the first entry of ω. Sometimes we also consider this shift as defined on words of finite length. Given $\omega \in E^*$ by $|\omega|$ we denote the length of the word ω, i.e./ the unique n such that $\omega \in E^n$. If $\omega \in E^\infty$ and $n \geq 1$, then

$$\omega|_n = \omega_1 \dots \omega_n.$$

A *Graph Directed Markov System* (GDMS) consists of a directed multigraph and incidence matrix together with a set of non-empty compact metric spaces $\{X_v\}_{v \in V}$, a number s, $0 < s < 1$, and for every $e \in E$, a 1-to-1 contraction $\phi_e : X_{t(e)} \to X_{i(e)}$ with a Lipschitz constant $\leq s$. Briefly, the set

$$S = \{\phi_e : X_{t(e)} \to X_{i(e)}\}_{e \in E}$$

is called a GDMS. The main object of interest in this book will be the limit set of the system S and objects associated to this set. We now describe the limit set. For each $\omega \in E_A^*$, say $\omega \in E_A^n$, we consider the map coded by ω:

$$\phi_\omega = \phi_{\omega_1} \circ \cdots \circ \phi_{\omega_n} : X_{t(\omega_n)} \to X_{i(\omega_1)}.$$

For $\omega \in E_A^\infty$, the sets $\{\phi_{\omega|_n}(X_{t(\omega_n)})\}_{n\geq 1}$ form a descending sequence of non-empty compact sets and therefore $\bigcap_{n\geq 1} \phi_{\omega|_n}(X_{t(\omega_n)}) \neq \emptyset$. Since for every $n \geq 1$, $\mathrm{diam}(\phi_{\omega|_n}(X_{t(\omega_n)})) \leq s^n \mathrm{diam}(X_{t(\omega_n)}) \leq s^n \max\{\mathrm{diam}\ (X_v) : v \in V\}$, we conclude that the intersection

$$\bigcap_{n\geq 1} \phi_{\omega|_n}(X_{t(\omega_n)})$$

is a singleton and we denote its only element by $\pi(\omega)$. In this way we have defined the *coding map* π:

$$\pi : E_A^\infty \to \bigoplus_{v\in V} X_v$$

from E^∞ to $\bigoplus_{v\in V} X_v$, the disjoint union of the compact sets X_v. The set

$$J = J_S = \pi(E_A^\infty)$$

will be called the *limit set* of the GDMS S. We will also deal with the sets coded by words starting with a given vertex v,

$$J_v = \pi(\{\omega \in E^\infty : i(\omega_1) = v\}).$$

Obviously $J_S = \bigoplus_{v\in V} J_v$. From now on we will assume that

$$\forall a \in E\, \exists b \in E \quad A_{ab} = 1. \tag{1.1}$$

We extend the functions i, t to E^* by putting

$$i(\omega) = i(\omega_1) \ \text{ and } \ t(\omega) = t(\omega_{|\omega|}).$$

For each $v \in V$, let $S_v(\infty)$ be the set of limit points of all sequences $\{x_n\}_{n=1}^\infty$, where $x_n \in \phi_{e_n}(X_{t(e_n)})$, for some mutually distinct edges e_n with $i(e_n) = v$. Put

$$S(\infty) = \bigcup_{v\in V} S_v(\infty).$$

We shall prove the following.

Lemma 1.0.1 *If* $\lim_{e\in E} \mathrm{diam}(\phi_e(X_{t(e)})) = 0$, *then for every* $v \in V$

$$\overline{J_v} = J_v \cup S_v(\infty) \cup \bigcup_{\omega\in E_v^*} \phi_\omega(S_{t(\omega)}(\infty)),$$

where $E_v^* = \{\omega \in E^* : i(\omega) = v\}$.

Proof. It follows from the assumption of our lemma and (1.1) that $S_v(\infty) \subset \overline{J_v}$. Thus, if $\omega \in E_v^*$, then

$$\phi_\omega(S_{t(\omega)}(\infty)) \subset \phi_\omega(\overline{J_{t(\omega)}}) \subset \overline{\phi_\omega(J_{t(\omega)})} \subset \overline{J_v}.$$

Hence, the inclusion

$$J_v \cup S_v(\infty) \cup \bigcup_{\omega\in E_v^*} \phi_\omega(S_{t(\omega)}(\infty)) \subset \overline{J_v}$$

is proved. Let $E_v^\infty = \{\omega \in E^\infty : i(\omega_1) = v\}$ and $E_v = \{e \in E : i(e) = v\}$. In order to prove the opposite inclusion to that just given, fix $x \in \overline{J_v}$. Then there exists a sequence $\{\omega^{(n)}\}_{n=1}^\infty$ of points in E_v^∞ such that $x = \lim_{n\to\infty} \pi(\omega^{(n)})$. If the sequence of the first coordinates of the words $\omega^{(n)}$ is infinite, then $x \in S_v(\infty)$ and we are done. So, suppose that the set of the first coordinates is finite. If the set of second coordinates is infinite, then there exists $e_1 \in E_v$ and $y \in S_{t(e_1)}(\infty)$ such that $x = \phi_{e_1}(y)$ and we are done in this case too. So, suppose that the set of second coordinates is also finite, but the set of the third coordinates is infinite. Then there exist $e_1 \in E_v$, $e_2 \in E$ such that $A_{e_1e_2} = 1$ and $y \in S_{t(e_2)}(\infty)$ such that $x = \phi_{e_1e_2}(y)$ and we are done. If this procedure halts after finitely many steps, say n, our proof is complete since then $x \in \phi_{e_1e_2\ldots e_n}(S_{t(e_n)}(\infty))$, where $e_1e_2\ldots e_n \in E_v^*$. Otherwise, we will produce a word $\omega \in E_v^\infty$ such that $\mathrm{dist}(x, \phi_{\omega|_n}(X_{t(\omega_n)}))$ tends to zero which implies that $x = \pi(\omega) \in J_v$. □

We end this short introductory chapter by distinguishing two special subclasses of GDMSs. We call a GDMS simply a *graph directed system*(GDS) if $A_{e_1e_2} = 1$ if and only if $t(e_1) = i(e_2)$. If, moreover, the set of vertices V is a singleton, then the GDS is called an *iterated function system*. The GDSs with finitely many edges have been introduced in [MW2] (see also [EM]). The infinite (the set of edges is infinite) iterated function systems have been introduced in [MU1] (see also [MU2]).

2
Symbolic Dynamics

This chapter is of abstract character. By this we mean that we consider a 0-1 incidence matrix $A : I \times I \to \{0, 1\}$ where I is a countable alphabet and

$$E^\infty = \{\omega \in I^\infty : A_{\omega_i \omega_{i+1}} = 1 \text{ for all } i \geq 1\}$$

is the space of all A-admissible infinite sequences with terms in I. Thus, in this chapter we are considering a directed graph with vertices the elements of I and a directed edge from i to j if and only if $A_{ij} = 1$. So, in this chapter A determines which vertices may follow a given vertex. The notation E^*, $|\omega|$, $\omega|_n$ and the shift map $\sigma : E^\infty \to E^\infty$ have the same meaning as in Chapter 1. We do not consider graph directed Markov systems here although one may think of the alphabet I as the set of edges of a multigraph. Let us remark that we use the notation E^∞ instead of I_A^∞ in order not to have so many subscripts. Let us fix some more notation and definitions. Given $\omega, \tau \in I^\infty$, we define $\omega \wedge \tau \in I^\infty \cup I^*$ to be the longest initial block common to both ω and τ. For each $\alpha > 0$, we define a *metric* d_α, on I^∞, by setting $d_\alpha(\omega, \tau) = \mathrm{e}^{-\alpha|\omega \wedge \tau|}$. These metrics are all equivalent and induce the same topology and Borel sets. A function is uniformly continuous with respect to one of these metrics if and only if it is uniformly continuous with respect to all. Also, a function is Hölder with respect to one of these metrics if and only if it is Hölder with respect to all; of course the Hölder order depends on the metric. If no metric is specifically mentioned, we take it to be d_1.

In this chapter we present various aspects of the thermodynamic formalism of a continuous potential on a shift space generated by a countable alphabet. Our approach has been developed in several papers culminating in [MU3] and stems from that of Ruelle [Ru] and Bowen [B1], cf. also [Wa] and [PU]. The case of a countable shift has also been considered (see e.g. [Gu], [GS], [PP], [Sar] and [Zar]). Our definition of topological pressure is more traditional than that proposed in these works. It fits better with our geometric applications and needs no compactifications of the shift space. In particular we are able to construct Gibbs states and equilibrium states of unbounded potentials.

2.1 Topological pressure and variational principles

The incidence matrix A is said to be *irreducible* if for all $i, j \in I$ there exists a path $\omega \in E^*$ such that $\omega_1 = i$ and $\omega_{|\omega|} = j$. This is equivalent to saying that the directed graph is *strongly connected*: for any two elements a, b of I there is a finite path starting at a and ending at b. This is also equivalent to saying that the left shift map, σ, on $E^\infty = I_A^\infty$ is *topologically mixing*: for any two non-empty open subsets U, V of I_A^∞ there is a non-negative integer n such that $\sigma^n(U) \cap V \neq \emptyset$. We say A is *primitive* if there exists $p \geq 1$ such that all the entries of A^p are positive, or in other words, for all $i, j \in I$ there exists a path $\omega \in E^p$ such that $\omega_1 = i$ and $\omega_{|\omega|} = j$. The matrix A is said to be *finitely irreducible* if there exists a finite set $\Lambda \subset E^*$ such that for all $i, j \in I$ there exists a path $\omega \in \Lambda$ for which $i\omega j \in E^*$. We note the following fact. If A is irreducible, then A is finitely irreducible if and only if there is a finite set of letters F such that for every $a \in I$, there are $p, q \in F$ such that $A_{ap} = A_{qa} = 1$. Finally, A is said to be *finitely primitive* if there exists a finite set $\Lambda \subset E^*$ consisting of words of the same length such that for all $i, j \in I$ there exists a path $\omega \in \Lambda$ for which $i\omega j \in E^*$. Notice that a finitely irreducible matrix does not have to be primitive nor conversely. Notice also that the set Λ (associated either with a finitely irreducible or finitely primitive matrix) can be taken to be empty provided E^∞ consists of all infinite words from I. Given a set $F \subset I$, we put

$$E_F^\infty = \{\omega \in E^\infty : \omega_i \in F \ \text{ for all } i \geq 1\}.$$

A sequence $\{a_n\}_{n=1}^\infty$ consisting of real numbers is said to be subadditive if $a_{n+m} \leq a_n + a_n$ for all $m, n \geq 1$. For the sake of completness we provide the proof of the following well-known elementary fact.

Lemma 2.1.1 *If a sequence $\{a_n\}_{n=1}^\infty$ is subadditive, then $\lim_{n\to\infty} a_n$ exists and is equal to $\inf_n a_n/n$. The limit could be $-\infty$, but if the a_n's are bounded below, then the limit is nonnegative.*

Proof. Fix $m \geq 1$. Each $n \geq 1$ can be expressed as $n = km + i$ with $0 \leq i < m$. Then

$$\frac{a_n}{n} = \frac{a_{i+km}}{i+km} \leq \frac{a_i}{km} + \frac{a_{km}}{km} \leq \frac{a_i}{km} + \frac{ka_m}{km} = \frac{a_i}{km} + \frac{a_m}{m}$$

If $n \to \infty$ then also $k \to \infty$ and therefore $\limsup_{n\to\infty} \frac{a_n}{n} \leq \frac{a_m}{m}$. Thus $\limsup_{n\to\infty} \frac{a_n}{n} \leq \inf \frac{a_m}{m}$. Now the inequality $\inf \frac{a_m}{m} \leq \liminf_{n\to\infty} \frac{a_n}{n}$ finishes the proof. □

Given a function $f : E_F^\infty \to I\!R$ we define the standard *nth partition function* by

$$Z_n(F, f) = \sum_{\omega \in E^n \cap F^n} \exp\Big(\sup_{\tau \in [\omega \cap F]} \sum_{j=0}^{n-1} f(\sigma^j(\tau))\Big),$$

where $[\omega \cap F] = \{\tau \in E_F^\infty : \tau|_{|\omega|} = \omega\}$. If $F = I$, we simply write $[\omega]$ for $[\omega \cap F]$. We will need the following.

Lemma 2.1.2 *The sequence $n \mapsto \log Z_n(F, f)$ is subadditive.*

Proof. We need to show that the sequence $n \mapsto Z_n(F, f)$ is submultiplicative, i.e. that $Z_{m+n}(F, f) \leq Z_m(F, f)Z_n(F, f)$ for all $m, n \geq 1$. And indeed,

$$\begin{aligned}
&Z_{m+n}(F, f) \\
&= \sum_{\omega \in E^{m+n} \cap F^{m+n}} \exp\Big(\sup_{\tau \in [\omega \cap F]} \sum_{j=0}^{mn-1} f(\sigma^j(\tau))\Big) \\
&= \sum_{\omega \in E^{m+n} \cap F^{m+n}} \exp\Big(\sup_{\tau \in [\omega \cap F]} \sum_{j=0}^{m-1} f(\sigma^j(\tau)) + \sum_{j=0}^{n-1} f(\sigma^j(\sigma^m(\tau)))\Big) \\
&\leq \sum_{\omega \in E^{m+n} \cap F^{m+n}} \exp\Big(\sup_{\tau \in [\omega \cap F]} \sum_{j=0}^{m-1} f(\sigma^j(\tau)) + \sup_{\tau \in [\omega \cap F]} \sum_{j=0}^{n-1} f(\sigma^j(\sigma^m(\tau)))\Big) \\
&\leq \sum_{\omega \in E^m \cap F^m} \sum_{\rho \in E^n \cap F^n} \exp\Big(\sup_{\tau \in [\omega \cap F]} \sum_{j=0}^{m-1} f(\sigma^j(\tau)) + \sup_{\gamma \in [\rho \cap F]} \sum_{j=0}^{n-1} f(\sigma^j(\gamma))\Big) \\
&= \sum_{\omega \in E^m \cap F^m} \exp\Big(\sup_{\tau \in [\omega \cap F]} \sum_{j=0}^{m-1} f(\sigma^j(\tau))\Big) \cdot \sum_{\rho \in E^n \cap F^n} \exp\Big(\sup_{\gamma \in [\rho \cap F]} \sum_{j=0}^{m-1} f(\sigma^j(\gamma))\Big) \\
&= Z_m(F, f)Z_n(F, f).
\end{aligned}$$

□

We can now define the *topological pressure of* f with respect to the shift map $\sigma : E_F^\infty \to E_F^\infty$ to be

$$\mathrm{P}_F(f) = \lim_{n\to\infty} \frac{1}{n} \log Z_n(F,f) = \inf\{\frac{1}{n} \log Z_n(F,f)\}. \qquad (2.1)$$

If $F = I$, we suppress the subscript F and write simply $\mathrm{P}(f)$ for $\mathrm{P}_I(f)$ and $Z_n(f)$ for $Z_n(I,f)$.
We let $S_n f$ be the nth partial *orbit sum* of f with respect to σ:

$$S_n f = \sum_{j=0}^{n-1} f \circ \sigma^j.$$

First, we shall provide a characterization of topological pressure expressed in the style of a *Poincaré exponent.*

Theorem 2.1.3 *For every continuous function* $f : E_F^\infty \to I\!R$, *we have*

$$\mathrm{P}_F(f) = \inf \left\{ t \in I\!R : \sum_{\omega \in F^* \cap E^*} \exp\bigl(\sup(S_{|\omega|} f|_{[\omega]})\bigr) e^{-t|\omega|} < \infty \right\}.$$

Proof. Fix $t > \mathrm{P}_F(f)$. By the definition of pressure there exists $n_t \geq 1$ such that for every $n \geq n_t$

$$\log \sum_{\omega \in F^n \cap E^n} \exp\bigl(\sup(S_n f|_{[\omega \cap F]})\bigr) < \left(\mathrm{P}_F(f) + \frac{t - \mathrm{P}_F(f)}{2}\right) n$$

and therefore

$$\sum_{\omega \in F^n} \exp\bigl(\sup(S_n f|_{[\omega \cap F]})\bigr) \mathrm{e}^{-tn} \leq \exp\left(\frac{\mathrm{P}_F(f) - t}{2} n\right).$$

Consequently,

$$\sum_{n\geq 0} \sum_{\omega \in F^n \cap E^n} \exp\bigl(\sup(S_n f|_{[\omega \cap F]})\bigr) \mathrm{e}^{-tn} < \infty.$$

Suppose in turn that $t < \mathrm{P}_F(f)$. By the definition of pressure, if n is large enough,

$$\left(\mathrm{P}_F(f) + \frac{t - \mathrm{P}_F(f)}{2}\right) n \leq \log \sum_{\omega \in F^n} \exp\bigl(\sup(S_n f|_{[\omega \cap F]})\bigr)$$

and therefore,

$$\exp\left(\frac{\mathrm{P}_F(f) - t}{2} n\right) \leq \sum_{\omega \in F^n} \exp\bigl(\sup(S_n f|_{[\omega \cap F]})\bigr) \mathrm{e}^{-tn}.$$

Consequently,

$$\sum_{n\geq 0}\sum_{\omega\in F^n} \exp\big(\sup(S_nf|_{[\omega\cap F]})\big)\mathrm{e}^{-tn} = \infty.$$

□

There are several things concerning pressure which may differ radically from the case when the alphabet is finite. However, there is a reasonably wide class of functions introduced in [MU3] for which the pressure function is fairly well behaved.

Definition 2.1.4 *(see [MU3]) A function $f : E^\infty \to I\!R$ is acceptable provided it is uniformly continuous and*

$$\mathrm{osc}(f) := \sup_{i\in I}\{\sup(f|_{[i]}) - \inf(f|_{[i]})\} < \infty.$$

Note that an acceptable function need not be bounded. We shall prove the following.

Theorem 2.1.5 *If $f : E^\infty \to I\!R$ is acceptable and A is finitely irreducible, then*

$$\mathrm{P}(f) = \sup\{\mathrm{P}_F(f)\},$$

where the supremum is taken over all finite subsets F of I.

Proof. The inequality $\mathrm{P}(f) \geq \sup\{P_F(f)\}$ is obvious. Let Λ witness that A is finitely irreducible. To prove the converse suppose first that $\mathrm{P}(f) < \infty$. Put $q = \#\Lambda$ and $p = \max\{|\omega| : \omega \in \Lambda\}$ and $T = \min\left\{\inf \sum_{j=0}^{|\omega|-1} f\circ\sigma^j|_{[\omega]} : \omega\in\Lambda\right\}$, where $[\omega] = \{\tau \in E^\infty : \tau|_{|\omega|} = \omega\}$. Fix $\epsilon > 0$. By the acceptability of f, there exists $l \geq 1$ such that $|f(\omega) - f(\tau)| < \epsilon$, if $\omega|_l = \tau|_l$ and $M = \mathrm{osc}(f) < \infty$. Now, fix $k \geq l$. By subadditivity, $\frac{1}{k}\log Z_k(f) \geq \mathrm{P}(f)$. Notice that there exists a finite set $F \subset I$ such that

$$\frac{1}{k}\log Z_k(F, f) > \mathrm{P}(f) - \epsilon. \tag{2.2}$$

We may assume that F contains Λ. Put

$$\overline{f} = \sum_{j=0}^{k-1} f\circ\sigma^j.$$

Now, for every element $\tau = \tau_1, \tau_2, \ldots, \tau_n \in F^k \cap E^k \times \cdots \times F^k \cap E^k$ (n factors) one can choose elements $\alpha_1, \alpha_2, \ldots, \alpha_{n-1} \in \Lambda$ such that $\overline{\tau} = \tau_1\alpha_1\tau_2\alpha_2\ldots\tau_{n-1}\alpha_{n-1}\tau_n \in E^*$. Notice that the function $\tau \mapsto \overline{\tau}$ is at most q^{n-1}-to-1 (in fact u^{n-1}-to-1, where u is the number of lengths of words composing Λ. Then for every $n \geq 1$,

$$
\begin{aligned}
q^{n-1} & \sum_{i=kn}^{kn+p(n-1)} Z_i(F,f) \\
& \geq \sum_{\tau \in (F^k \cap E^k)^n} \exp\left(\sup_{[\overline{\tau} \cap F]} \sum_{j=0}^{|\overline{\tau}|} f \circ \sigma^j\right) \\
& \geq \sum_{\tau \in (F^k \cap E^k)^n} \exp\left(\inf_{[\overline{\tau}]} \sum_{j=0}^{|\overline{\tau}|} f \circ \sigma^j\right) \\
& \geq \sum_{\tau \in (F^k \cap E^k)^n} \exp\left(\sum_{i=1}^{n} \inf \overline{f}|_{[\tau_i]} + T(n-1)\right) \\
& = \exp(T(n-1)) \sum_{\tau \in (F^k \cap E^k)^n} \exp \sum_{i=1}^{n} \inf \overline{f}|_{[\tau_i]} \\
& \geq \exp(T(n-1)) \sum_{\tau \in (F^k \cap E^k)^n} \exp\left(\sum_{i=1}^{n} (\sup \overline{f}|_{[\tau_i]} - (k-l)\epsilon - Ml)\right) \\
& = \exp(T(n-1) - (k-l)\epsilon n - Mln) \sum_{\tau \in (F^k \cap E^k)^n} \exp \sum_{i=1}^{n} \sup \overline{f}|_{[\tau_i]} \\
& = \mathrm{e}^{-T} \exp\big(n(T - (k-l)\epsilon - Ml)\big) \left(\sum_{\tau \in (F^k \cap E^k)} \exp(\sup \overline{f}|_{[\tau]})\right)^n .
\end{aligned}
$$

Hence, there exists $kn \leq i_n \leq (k+p)n$ such that

$$
Z_{i_n}(F,f) \geq \frac{1}{pn} \mathrm{e}^{-T} \exp\big(n(T - (k-l)\epsilon - Ml - \log q)\big) Z_k(F,f)^n
$$

and therefore, using (2.2), we obtain

$$
\begin{aligned}
\mathrm{P}_F(f) = \lim_{n\to\infty} \frac{1}{i_n} \log Z_{i_n}(F,f) & \geq \frac{-|T|}{k} - \epsilon + \frac{l\epsilon}{k+p} - \frac{Ml + \log p}{k} \\
& + \mathrm{P}(f) - 2\epsilon \geq \mathrm{P}(f) - 7\epsilon
\end{aligned}
$$

provided that k is large enough. Thus, letting $\epsilon \searrow 0$, the theorem follows. The case $\mathrm{P}(f) = \infty$ can be treated similarly. □

We say a σ-invariant Borel probability measure $\tilde{\mu}$ on E^∞ is *finitely supported* provided there exists a finite set $F \subset I$ such that $\tilde{\mu}(E_F^\infty) = 1$. The well-known *variational principle* for finitely supported measures (see

[B1], [Ru], comp. [Wa] and [PU]) tells us that for every finite set $F \subset I$

$$\mathrm{P}_F(f) = \sup\{\mathrm{h}_{\tilde{\mu}}(\sigma) + \int f d\tilde{\mu}\},$$

where the supremum is taken over all σ-invariant ergodic Borel probability measures $\tilde{\mu}$ with $\tilde{\mu}(F^\infty) = 1$ and $\mathrm{h}_{\tilde{\mu}}(\sigma)$ is the *entropy of* $\tilde{\mu}$ with respect to σ. Applying Theorem 2.1.5, we therefore obtain the following.

Theorem 2.1.6 *(1st variational principle) If A is finitely irreducible and if $f : E^\infty \to I\!R$ is acceptable, then*

$$\mathrm{P}(f) = \sup\{(\mathrm{h}_{\tilde{\mu}}(\sigma) + \int f d\tilde{\mu}\},$$

where the supremum is taken over all σ-invariant ergodic Borel probability measures $\tilde{\mu}$ which are finitely supported.

For $n \geq 1$, let α^n be the *standard partition* of E^∞ into cylinders of length n:

$$\alpha^n = \{[\omega] : |\omega| = n\}.$$

If $n = 1$, we write also α for α^1. If β is a countable measurable partition of E^∞ and $\tilde{\mu}$ is a probability measure, then the *entropy of* $\tilde{\mu}$ *with respect to the partition* β is $\mathrm{H}_{\tilde{\mu}}(\beta) = -\sum_{B\in\beta} \tilde{\mu}(B) \log \tilde{\mu}(B)$. Our next theorem is the following.

Theorem 2.1.7 *(2nd variational principle) If $f : E^\infty \to I\!R$ is a continuous function and $\tilde{\mu}$ is a σ-invariant Borel probability measure on E^∞ such that $\int f d\tilde{\mu} > -\infty$, then*

$$\mathrm{h}_{\tilde{\mu}}(\sigma) + \int f d\tilde{\mu} \leq \mathrm{P}(f).$$

In addition, if $\mathrm{P}(f) < \infty$, *then there exists* $q \geq 1$ *such that* $\mathrm{H}_{\tilde{\mu}}(\alpha^q) < \infty$.

Proof. If $\mathrm{P}(f) = +\infty$, there is nothing to prove. So, suppose that $\mathrm{P}(f) < \infty$. Then there exists $q \geq 1$ such that $Z_n(f) < \infty$ for every $n \geq q$. Also, for every $n \geq 1$, we have

$$\sum_{|\omega|=n} \tilde{\mu}([\omega]) \sup(S_n f|_{[\omega]}) \geq \int S_n f d\tilde{\mu} = n \int f d\tilde{\mu} > -\infty.$$

Therefore, using the concavity of the function $h(x) = -x\log x$, we obtain for every $n \geq q$,

$$\begin{aligned}
&\mathrm{H}_{\tilde{\mu}}(\alpha^n) + \int S_n f\, d\tilde{\mu} \\
&\leq \sum_{|\omega|=n} \tilde{\mu}([\omega])(\sup S_n f|_{[\omega]} - \log \tilde{\mu}([\omega])) \\
&= Z_n(f) \sum_{|\omega|=n} Z_n(f)^{-1} \mathrm{e}^{\sup S_n f|_{[\omega]}} h\big(\tilde{\mu}([\omega])\mathrm{e}^{-\sup S_n f|_{[\omega]}}\big) \\
&\leq Z_n(f) h\left(\sum_{|\omega|=n} Z_n(f)^{-1} \mathrm{e}^{\sup S_n f|_{[\omega]}} \tilde{\mu}([\omega]) \mathrm{e}^{-\sup S_n f|_{[\omega]}} \right) \\
&= Z_n(f) h(Z_n(f)^{-1}) \\
&= \log\big(\sum_{|\omega|=n} \exp(\sup S_n f|_{[\omega]})\big) = \log Z_n(f).
\end{aligned}$$

Therefore, $\mathrm{H}_{\tilde{\mu}}(\alpha^n) \leq \log Z_n(f) + n \int (-f) d\tilde{\mu} < \infty$ for every $n \geq q$, and since in addition α^q is a generator, we obtain

$$\begin{aligned}
\mathrm{h}_{\tilde{\mu}}(\sigma) + \int f\, d\tilde{\mu} &\leq \liminf_{n\to\infty} \left(\frac{1}{n} \left(\mathrm{H}_{\tilde{\mu}}(\alpha^n) + \int S_n f\, d\tilde{\mu} \right) \right) \\
&\leq \lim_{n\to\infty} \frac{1}{n} \log Z_n(f) = \mathrm{P}(f).
\end{aligned}$$

□

As an immediate consequence of Theorem 2.1.6 and Theorem 2.1.7, we find the following.

Theorem 2.1.8 *(3rd variational principle) Suppose the incidence matrix A is finitely irreducible. If $f : E^\infty \to \mathbb{R}$ is acceptable, then*

$$\mathrm{P}(f) = \sup\{\mathrm{h}_{\tilde{\mu}}(\sigma) + \int f\, d\tilde{\mu}\},$$

where the supremum is taken over all σ-invariant ergodic Borel probability measures $\tilde{\mu}$ such that $\int f\, d\tilde{\mu} > -\infty$.

We end this section with the following useful technical fact.

Proposition 2.1.9 *If the incidence matrix is finitely irreducible and the function f is acceptable, then $\mathrm{P}(f) < \infty$ if and only if $Z_1(f) < \infty$.*

Proof. Let $\Lambda \subset E^*$ be a set of words which witnesses the finite irreducibility of the incidence matrix A. Let $s = \max\{|\alpha| : \alpha \in \Lambda\}$ and let

$$M = \min\left\{\inf_{[\alpha]}\left\{\sum_{j=0}^{|\alpha|-1} f\circ\sigma^j\right\} : \alpha\in\Lambda\right\}.$$

For $n \geq 1$ and $\omega \in I^n$, let $\overline{\omega} = \omega_1\alpha_1\omega_2\alpha_2\ldots\omega_{n-1}\alpha_{n-1}\omega_n$, where all $\alpha_1,\ldots,\alpha_{n-1}$ are appropriately taken from Λ so that $\overline{\omega}\in E^*$. Thus, $n \leq |\overline{\omega}| \leq n + s(n-1)$. Similarly as in the proof of Theorem 2.1.5, the function $\omega \mapsto \overline{\omega}$ is at most q^{n-1}-to-1, where $q = \#\Lambda$. Since f is acceptable, we therefore get

$$\begin{aligned}
q^{n-1}\sum_{i=n}^{n+s(n-1)} Z_i(f) &= q^{n-1}\sum_{i=n}^{n+s(n-1)}\sum_{\omega\in E^i}\exp\left(\sup_{[\omega]}\left\{\sum_{j=0}^{i-1} f\circ\sigma^j\right\}\right)\\
&\geq \sum_{\omega\in I^n}\exp\left(\sup_{[\overline{\omega}]}\left\{\sum_{j=0}^{|\overline{\omega}|-1} f\circ\sigma^j\right\}\right)\\
&\geq \sum_{\omega\in I^n}\exp\left(\sum_{j=1}^{n}\inf\big(f|_{[\omega_j]}\big) + M(n-1)\right)\\
&\geq \mathrm{e}^{M(n-1)}\sum_{\omega\in I^n}\exp\left(\sum_{j=1}^{n}\sup\big(f|_{[\omega_j]}\big) - \operatorname{osc}(f)n\right)\\
&= \exp\big(-M + (M-\operatorname{osc}(f))n\big)\left(\sum_{e\in I}\exp\big(\sup(f|_{[e]})\big)\right)^n\\
&= \exp\big(-M + (M-\operatorname{osc}(f))n\big)Z_1(f)^n.
\end{aligned}$$

From this it follows that if $\mathrm{P}(f) < \infty$, then also $Z_1(f) < \infty$. The opposite implication is obvious since $Z_n(f) \leq Z_1(f)^n$. □

2.2 Gibbs states, equilibrium states and potentials

If $f : E^\infty \to I\!R$ is a continuous function, then a Borel probability measure $\tilde{m}$ on E^∞ is called a *Gibbs state* for f (cf. [B1], [HMU], [PU], [Ru], [Wa] and [U1]) if there exist constants $Q \geq 1$ and $\mathrm{P}_{\tilde{m}}$ such that for every $\omega \in E^*$ and every $\tau \in [\omega]$

$$Q^{-1} \leq \frac{\tilde{m}([\omega])}{\exp\big(S_{|\omega|}f(\tau) - \mathrm{P}_{\tilde{m}}|\omega|\big)} \leq Q. \tag{2.3}$$

If additionally $\tilde{m}$ is shift-invariant, it is then called an *invariant Gibbs state.*

Remark 2.2.1 *Notice that the sum $S_{|\omega|}f(\tau)$ in (2.3) can be replaced by $\sup(S_{|\omega|}f|_{[\omega]})$ or by $\inf(S_{|\omega|}f|_{[\omega]})$. Also, notice that if $\tilde{m}$ is a Gibbs state and if $\tilde{\mu}$ and $\tilde{m}$ are boundedly equivalent (meaning there is some $K \geq 1$ such that $K^{-1} \leq \tilde{\mu}([\omega])/\tilde{m}([\omega]) \leq K$ for all $\omega \in E^*$), then $\tilde{\mu}$ is a Gibbs state for the potential f. We will use these facts from time to time.*

We start with the following.

Proposition 2.2.2

(a) For every Gibbs state $\tilde{m}$, $\mathrm{P}_{\tilde{m}} = \mathrm{P}(f)$.

(b) Any two Gibbs states for the function f are equivalent, with Radon-Nikodym derivatives bounded away from zero and infinity.

Proof. We shall first prove (a). Towards this end fix $n \geq 1$ and, using Remark 2.2.1, sum (2.3) over all words $\omega \in E^n$. Since $\sum_{|\omega|=n} \tilde{m}([\omega]) = 1$, we therefore get

$$Q^{-1}\mathrm{e}^{-\mathrm{P}_{\tilde{m}}n} \sum_{|\omega|=n} \exp\bigl(\sup S_n f|_{[\omega]}\bigr) \leq 1 \leq Q\mathrm{e}^{-\mathrm{P}_{\tilde{m}}n} \sum_{|\omega|=n} \exp\bigl(\sup S_n f|_{[\omega]}\bigr).$$

Applying logarithms to all three terms of this formula, dividing all the terms by n and taking the limit as $n \to \infty$, we obtain $-\mathrm{P}_{\tilde{m}} + \mathrm{P}(f) \leq 0 \leq -\mathrm{P}_{\tilde{m}} + \mathrm{P}(f)$, which means that $\mathrm{P}_{\tilde{m}} = \mathrm{P}(f)$. The proof of item (a) is thus complete.

In order to prove part (b) suppose that m and ν are two Gibbs states of the function f. Notice now that part (a) implies the existence of a constant $T \geq 1$ such that

$$T^{-1} \leq \frac{\nu([\omega])}{m([\omega])} \leq T$$

for all words $\omega \in E^*$. Straightforward reasoning gives now that ν and m are equivalent and $T^{-1} \leq \frac{d\nu}{dm} \leq T$. □

As an immediate consequence of (2.3) and Remark 2.2.1 we get the following.

Proposition 2.2.3 *Any uniformly continuous function $f : E^\infty \to \mathbb{R}$ that has a Gibbs state is acceptable.*

For $\omega \in E^*$ and $n \geq 1$, let

$$E_n^\omega = \{\tau \in E^n : A_{\tau_n \omega_1} = 1\} \text{ and } E_*^\omega = \{\tau \in E^* : A_{\tau_{|\tau|}\omega_1} = 1\}.$$

We shall prove the following result concerning uniqueness and some stochastic properties of Gibbs states. Stronger stochastic properties will be proved in Section 2.5

Theorem 2.2.4 *If an acceptable function f has a Gibbs state and the incidence matrix A is finitely irreducible, then f has a unique invariant Gibbs state. The invariant Gibbs state is ergodic. Moreover, if A is finitely primitive, the invariant Gibbs state is completely ergodic.*

Proof. Let $\tilde{m}$ be a Gibbs state for f. Fixing $\omega \in E^*$, using (2.3), Remark 2.2.1 and Proposition 2.2.2(a) we get for every $n \geq 1$

$$\begin{aligned}
&\tilde{m}(\sigma^{-n}([\omega])) \\
&\quad = \sum_{\tau \in E_n^\omega} \tilde{m}([\tau\omega]) \leq \sum_{\tau \in E_n^\omega} Q \exp\big(\sup(S_{n+|\omega|} f|_{[\tau\omega]}) - \mathrm{P}(f)(n + |\omega|)\big) \\
&\quad \leq \sum_{\tau \in E_n^\omega} Q \exp\big(\sup(S_n f|_{[\tau]}) - \mathrm{P}(f)n\big) \exp\big(\sup(S_{|\omega|} f|_{[\omega]}) - \mathrm{P}(f)|\omega|\big) \\
&\quad \leq \sum_{\tau \in E_n^\omega} QQ\tilde{m}([\tau])Q\tilde{m}([\omega]) \leq Q^3 \tilde{m}([\omega]).
\end{aligned} \tag{2.4}$$

Let the finite set of words Λ witness the finite irreducibility of the incidence matrix A and let p be the maximal length of a word in Λ. Since f is acceptable,

$$T = \min\{\inf(S_\alpha |f|_{[\alpha]}) - \mathrm{P}(f)|\alpha| : \alpha \in \Lambda\} > -\infty.$$

For each $\tau, \omega \in E^*$, let $\alpha = \alpha(\tau, \omega) \in \Lambda$ be such that $\tau\alpha\omega \in E^*$. Then, we have for all $\omega \in E^*$ and all n

$$\begin{aligned}
\sum_{i=n}^{n+p} \tilde{m}(\sigma^{-i}([\omega])) &= \sum_{i=n}^{n+p} \sum_{\tau \in E_i^\omega} \tilde{m}([\tau\omega]) \geq \sum_{\tau \in E^n} \tilde{m}([\tau\alpha(\tau,\omega)\omega]) \\
&\geq \sum_{\tau \in E^n} Q^{-1} \exp\big(\inf(S_{|\tau|+|\alpha(\tau,\omega)|+|\omega|} f|_{[\tau\alpha\omega]}) \\
&\quad - \mathrm{P}(f)(|\tau| + |\alpha(\tau,\omega)| + |\omega|)\big) \\
&\geq Q^{-1} \sum_{\tau \in E^n} \exp\big(\inf(S_n f|_{[\tau]}) - \mathrm{P}(f)(n) \\
&\quad + \inf(S_{|\alpha(\tau,\omega)|} |f|_{[\alpha(\tau,\omega)]}) - \mathrm{P}(f)|\alpha(\tau,\omega)|) \\
&\quad + \inf(S_{|\omega|} f|_{[\omega]}) - \mathrm{P}(f)|\omega|\big)
\end{aligned} \tag{2.5}$$

$$\begin{aligned}
&\geq Q^{-1}e^T \exp\big(\inf(S_{|\omega|}f|_{[\omega]}) - \mathrm{P}(f)|\omega|\big) \sum_{\tau\in E^n} \\
&\times \exp\big(\inf(S_n f|_{[\tau]}) - \mathrm{P}(f)(n)\big) \\
&\geq Q^{-2}e^T \tilde{m}([\omega]) \sum_{\tau\in E^n} \exp\big(\inf(S_n f|_{[\tau]}) - \mathrm{P}(f)(n)\big) \\
&\geq Q^{-2}e^T \tilde{m}([\omega]) Q^{-1} \sum_{\tau\in E^n|} \tilde{m}([\tau]) \\
&= Q^{-3}e^T \tilde{m}([\omega]).
\end{aligned}$$

Let L be a Banach limit defined on the Banach space of all bounded sequences of real numbers. It is not difficult to check that the formula $\tilde{\mu}(A) = L\big((\tilde{m}(\sigma^{-n}(A)))_{n\geq 0}\big)$ defines a finite, non-zero, invariant, finitely additive measure on Borel sets of E^∞ satisfying

$$\frac{Q^{-3}e^T}{p}\tilde{m}(A) \leq \tilde{\mu}(A) \leq Q^3\tilde{m}(A). \tag{2.6}$$

Since $\tilde{m}$ is a countably additive measure, we deduce that $\tilde{\mu}$ is also countably additive.

Let us prove the ergodicity of $\tilde{\mu}$ or, equivalently, of $\tilde{m}$. Let $\omega \in E^n$. For each $\tau \in E^*$, we find:

$$\begin{aligned}
\sum_{i=n}^{n+p} \tilde{m}(\sigma^{-i}([\tau]) \cap [\omega]) &\geq \tilde{m}([\omega\alpha(\omega,\tau)\tau]) \\
&\geq Q^{-3}e^T\tilde{m}([\tau])\tilde{m}([\omega]).
\end{aligned} \tag{2.7}$$

Take now an arbitrary Borel set $A \subset E^\infty$. Fix $\epsilon > 0$. Since the nested family of sets $\{[\tau] : \tau \in E^*\}$ generates the Borel σ-algebra on E^∞, for every $n \geq 0$ and every $\omega \in E^n$ we can find a subfamily Z of E^* consisting of mutually incomparable words such that $A \subset \bigcup\{[\tau] : \tau \in Z\}$ and for $n \leq i \leq n+p$,

$$\sum_{\tau\in Z} \tilde{m}(\sigma^{-i}([\tau]) \cap [\omega]) \leq \tilde{m}([\omega] \cap \sigma^{-i}(A)) + \epsilon/p.$$

Then, using (2.7) we get

$$\begin{aligned}
\epsilon + \sum_{i=n}^{n+p} \tilde{m}([\omega] \cap \sigma^{-i}(A)) + \epsilon &\geq \sum_{i=n}^{n+p}\sum_{\tau\in Z} \tilde{m}([\omega] \cap \sigma^{-i}(\tau)) \\
&\geq \sum_{\tau\in Z} Q^{-3}e^T\tilde{m}([\tau])\tilde{m}([\omega]) \\
&\geq Q^{-3}e^T\tilde{m}(A)\tilde{m}([\omega]).
\end{aligned} \tag{2.8}$$

Hence, letting $\epsilon \searrow 0$, we get

$$\sum_{i=n}^{n+p} \tilde{m}([\omega] \cap \sigma^{-i}(A)) \geq Q^{-3}\mathrm{e}^T \tilde{m}(A)\tilde{m}([\omega]).$$

From this inequality we find

$$\begin{aligned}\sum_{i=n}^{n+p} \tilde{m}(\sigma^{-i}(E^\infty \setminus B) \cap [\omega]) &= \sum_{i=n}^{n+p} \tilde{m}([\omega] \setminus \sigma^{-i}(B) \cap [\omega]) \\ &= \sum_{i=n}^{n+p} \tilde{m}([\omega]) - \tilde{m}(\sigma^{-i}(B) \cap [\omega]) \\ &\leq [p+1 - Q^{-3}\mathrm{e}^T \tilde{m}(B)]\tilde{m}([\omega]).\end{aligned}$$

Thus, for every Borel set $B \subset E^\infty$, for every $n \geq 0$, and for every $\omega \in E^n$ we have

$$\sum_{i=n}^{n+p} \tilde{m}(\sigma^{-i}(B) \cap [\omega]) \leq (p+1 - Q^{-3}\mathrm{e}^T(1 - \tilde{m}(B)))\tilde{m}([\omega]). \tag{2.9}$$

In order to conclude the proof of the ergodicity of σ, suppose that $\sigma^{-1}(B) = B$ with $0 < \tilde{m}(B) < 1$. Put $\gamma = 1 - Q^{-3}\mathrm{e}^T(1 - \tilde{m}(B))/(p+1)$. Note that $0 < \gamma < 1$. In view of (2.9), for every $\omega \in E^*$ we get $\tilde{m}(B \cap [\omega]) = \tilde{m}(\sigma^{-i}(B) \cap [\omega]) \leq \gamma \tilde{m}([\omega])$. Take now $\eta > 1$ so small that $\gamma\eta < 1$ and choose a subfamily R of E^* consisting of mutually incomparable words and such that $B \subset \bigcup\{[\omega] : \omega \in R\}$ and $\tilde{m}(\bigcup\{[\omega] : \omega \in R\}) \leq \eta \tilde{m}(B)$. Then $\tilde{m}(B) = \sum_{\omega \in R} \tilde{m}(B \cap [\omega]) \leq \sum_{\omega \in R} \gamma \tilde{m}([\omega]) = \gamma \tilde{m}(\bigcup\{[\omega] : \omega \in R\}) \leq \gamma\eta\tilde{m}(B) < \tilde{m}(B)$. This contradiction finishes the proof of the existence part.

The uniqueness of the invariant Gibbs state follows immediately from the ergodicity of any invariant Gibbs state and Proposition 2.2.2(b).

Finally, let us prove the complete ergodicity of $\tilde{\mu}$ or, equivalently, of $\tilde{m}$ in case A is finitely primitive. Essentially, we repeat the argument just given. Let Λ be a finite set of words all of length q which witnesses the finite primitiveness of A. Fix $r \geq 1$. Let $\omega \in E^n$. For each $\tau \in E^*$, we find the following improvement of (2.6).

$$\begin{aligned}\tilde{m}(\sigma^{-(n+qr)}([\tau]) \cap [\omega]) &\geq \sum_{\alpha \in \Lambda^r \cap E^{qr} : A_{\omega_n \alpha_1} = A_{\alpha_{qr}\tau_1} = 1} \tilde{m}([\omega\alpha\tau]) \\ &\geq Q^{-3}\mathrm{e}^{rT}\tilde{m}([\tau])\tilde{m}([\omega]).\end{aligned} \tag{2.10}$$

Take now an arbitrary Borel set $A \subset E^\infty$. Fix $\epsilon > 0$. Since the nested family of sets $\{[\tau] : \tau \in E^*\}$ generates the Borel σ-algebra on E^∞, for every $n \geq 0$ and every $\omega \in E^n$ we can find a subfamily Z of E^* consisting of mutually incomparable words such that $A \subset \bigcup\{[\tau] : \tau \in Z\}$ and

$$\sum_{\tau \in Z} \tilde{m}(\sigma^{-(n+qr)}([\tau]) \cap [\omega]) \leq \tilde{m}\big([\omega] \cap \sigma^{-(n+qr)}(A)\big) + \epsilon.$$

Then, using (2.10) we get

$$\begin{aligned} \epsilon + \tilde{m}\big([\omega] \cap \sigma^{-(n+qr)}(A)\big) &\geq \sum_{\tau \in Z} Q^{-3} e^{rT} \tilde{m}([\tau]) \tilde{m}([\omega]) \\ &\geq Q^{-3} e^{rT} \tilde{m}(A) \tilde{m}([\omega]). \end{aligned}$$

Hence, letting $\epsilon \searrow 0$, we get

$$\tilde{m}\big([\omega] \cap \sigma^{-(n+qr)}(A)\big) \geq \tilde{Q}(r) \tilde{m}(A) \tilde{m}([\omega]),$$

where $\tilde{Q}(r) = Q^{-3} \exp(rT)$. Note that it follows from this last inequality that $\tilde{Q} = \tilde{Q}(r) \leq 1$. Also, from this inequality we find $\tilde{m}\big(\sigma^{-(n+qr)}(E^\infty \setminus B) \cap [\omega]\big) = \tilde{m}\big([\omega] \setminus \sigma^{-(n+qr)}(B) \cap [\omega]\big) = \tilde{m}([\omega]) - \tilde{m}\big(\sigma^{-(n+qr)}(B) \cap [\omega]\big) \leq (1 - \tilde{Q}\tilde{m}(B))\tilde{m}([\omega])$. Thus, for every Borel set $B \subset E^\infty$, for every $n \geq 0$, and for every $\omega \in I^n$ we have

$$\tilde{m}(\sigma^{-(n+qr)}(B) \cap [\omega]) \leq \big(1 - \tilde{Q}(1 - \tilde{m}(B))\big)\tilde{m}([\omega]). \tag{2.11}$$

In order to conclude the proof of the complete ergodicity of σ suppose that $\sigma^{-r}(B) = B$ with $0 < \tilde{m}(B) < 1$. Put $\gamma = 1 - \tilde{Q}(1 - \tilde{m}(B))$. Note that $0 < \gamma < 1$. In view of (2.11), for every $\omega \in (E^r)^*$ we get $\tilde{m}(B \cap [\omega]) = \tilde{m}(\sigma^{-(|\omega|+qr)}(B) \cap [\omega]) \leq \gamma \tilde{m}([\omega])$. Take now $\eta > 1$ so small that $\gamma\eta < 1$ and choose a subfamily R of $(E^r)^*$ consisting of mutually incomparable words and such that $B \subset \bigcup\{[\omega] : \omega \in R\}$ and $\tilde{m}\big(\bigcup\{[\omega] : \omega \in R\}\big) \leq \eta \tilde{m}(B)$. Then $\tilde{m}(B) \leq \sum_{\omega \in R} \tilde{m}(B \cap [\omega]) \leq \sum_{\omega \in R} \gamma \tilde{m}([\omega]) = \gamma \tilde{m}\big(\bigcup\{[\omega] : \omega \in R\}\big) \leq \gamma\eta \tilde{m}(B) < \tilde{m}(B)$. This contradiction finishes the proof of the complete ergodicity of $\tilde{m}$. □

There is a sort of converse to part of the preceding theorem. We need the following lemma first. The next two results are due to Sarig [Sar].

Lemma 2.2.5 *Suppose the incidence matrix A is irreducible and $\tilde{m}$ is an invariant Gibbs state for the acceptable function f. There is a positive constant K such that for every $p \in I$,*

$$K \leq \min\left\{ \sum_{a \in I : A_{pa} = 1} \exp \sup f|_{[a]}, \sum_{a \in I : A_{ap} = 1} \exp \sup f|_{[a]} \right\}$$

Proof. Let $\tilde{m}$ be an invariant Gibbs state for f. Suppose $p, q \in I$, $A_{p,q} = 1$, and $\tau \in [pq]$. Since $\tilde{m}$ is a Gibbs state for f, we have $|f(x) - f(y)| \leq 2\log Q$, for $x, y \in [a], a \in I$. This implies for every $a \in I$, $\exp\inf f|_{[a]} \leq \exp f(x) \leq \exp\sup f|_{[a]}$, for all $x \in E^\infty$ with $x_1 = a$. Also,

$$Q^{-1}\exp(f(\tau) + f(\sigma(\tau)) - 2P) \leq \tilde{m}([pq]) \leq Q\exp(f(\tau) + f(\sigma(\tau)) - 2P).$$

Thus,

$$Q^{-1}\mathrm{e}^{-2P}\exp f(\tau)\exp f(\sigma(\tau)) \leq \tilde{m}([pq]) \leq Q\mathrm{e}^{-2P}\exp f(\tau)\exp f(\sigma(\tau)).$$

Therefore,

$$\begin{aligned} Q^{-1}\mathrm{e}^{-2P}\exp\inf f|_{[p]}\exp\inf f|_{[q]} &\leq \tilde{m}([pq]) \\ &\leq Q^1\mathrm{e}^{-2P}\exp\sup f|_{[p]}\exp\sup f|_{[q]}. \end{aligned}$$

Now,

$$\tilde{m}([p]) = \sum_{a\in I:A_{pa}=1} m([pa]) \leq Q^1\mathrm{e}^{-2P}\exp\sup f|_{[p]} \sum_{a\in I:A_{pa}=1} \exp\sup f|_{[a]}.$$

Since $\exp\sup f|_[p] \leq Q\mathrm{e}^P\tilde{m}([p])$, we find

$$Q^{-2}\mathrm{e}^P \leq \sum_{a\in I:A_{pa}=1} \exp\sup f|_{[a]}.$$

If, in addition, $\tilde{m}$ is invariant, $\tilde{m}([p]) = \tilde{m}(\sigma^{-1}([p])) = \sum_{a\in I:A_{ap}=1} \tilde{m}([ap])$. But,

$$\begin{aligned} Q^{-1}\mathrm{e}^{-P}\sup f|_{[p]} &\leq \tilde{m}(\sigma^{-1}([p])) \\ &\leq Q^5\mathrm{e}^{-2P} \sum_{a\in I:A_{ap}=1} \exp\sup f|_{[a]}\exp\sup f|_{[p]}. \end{aligned}$$

Therefore,

$$Q^{-2}\mathrm{e}^P \leq \sum_{a\in I:A_{ap}=1} \exp\sup f|_{[a]}.$$

□

Theorem 2.2.6 *Let the incidence matrix A be irreducible. If an acceptable function f has an invariant Gibbs state, then the incidence matrix A is finitely irreducible.*

Proof. It suffices to show there is a finite set of letters F such that for every letter p there are $a, b \in F$ such that $A_{ap} = A_{pb} = 1$. Enumerate the elements of $I = \{a_n\}_{n=1}^\infty$. Since $\sum_{a\in I}\exp\sup f|_{[a]} < \infty$, there is some k_0 such that $\sum_{j>k_0}\exp\sup f|_{[a_j]} \leq K$ where K comes from lemma 2.2.5.

Let $F = \{a_i : \; i \le k_0\}$. Now, it follows from the preceding lemma that every letter is followed by some element of F and every letter is preceded by some element of F. Since A is irreducible, it follows that A is finitely irreducible. □

Recall that for $\omega, \tau \in I^\infty$, we defined $\omega \wedge \tau \in I^\infty \cup I^*$ to be the longest initial block common to both ω and τ. We say that a function $f : E^\infty \to I\!R$ is *Hölder continuous with an exponent* $\alpha > 0$ if

$$V_\alpha(f) := \sup_{n\ge 1}\{V_{\alpha,n}(f)\} < \infty,$$

where

$$V_{\alpha,n}(f) = \sup\{|f(\omega) - f(\tau)|\mathrm{e}^{\alpha(n-1)} : \omega, \tau \in E^\infty \text{ and } |\omega \wedge \tau| \ge n\}.$$

Note that if g is Hölder continuous of order α and $\theta, \psi \in E^\infty$, then

$$V_\alpha(g)d_\alpha(\theta,\psi) = V_\alpha(g)\mathrm{e}^{-\alpha|\theta\wedge\psi|} \ge \mathrm{e}^{-\alpha}|g(\theta) - g(\psi)|.$$

Also, note that each Hölder continuous function is acceptable. We say that two functions $f, g : E^\infty \to I\!R$ are *cohomologous* in a class $\mathcal{H}$ if there exists a function $u : E^\infty \to I\!R$ in the class $\mathcal{H}$ such that

$$g - f = u - u \circ \sigma.$$

We shall now provide a list of necessary and sufficient conditions for two Hölder continuous functions to have the same invariant Gibbs states. The proof is analogous to the proof of Theorem 1.28 in [B1] (see also [HMU]).

Theorem 2.2.7 *Suppose that $f, g : E^\infty \to I\!R$ are two Hölder continuous functions that have invariant Gibbs states $\tilde\mu_f$ and $\tilde\mu_g$ respectively. Suppose also that the incidence matrix A is finitely irreducible. Then the following conditions are equivalent:*

(1) $\tilde\mu_f = \tilde\mu_g$.

(2) There exists a constant R such that for each $n \ge 1$, if $\sigma^n(\omega) = \omega$, then

$$S_nf(\omega) - S_ng(\omega) = nR.$$

(3) The difference $g - f$ is cohomologous to a constant R in the class of bounded Hölder continuous functions.

(4) The difference $g - f$ is cohomologous to a constant in the class of bounded continuous functions.

(5) *There exist constants S and T such that for every $\omega \in E^\infty$ and every $n \geq 1$*

$$|S_n f(\omega) - S_n g(\omega) - Sn| \leq T.$$

If these conditions are satisfied then $R = S = \mathrm{P}(f) - \mathrm{P}(g)$.

Proof. (1) $\Rightarrow$ (2). It follows from (2.3) that

$$Q^{-2} \leq \frac{\exp\big(S_k f(\omega) - \mathrm{P}(f)k\big)}{\exp\big(S_k g(\omega) - \mathrm{P}(g)k\big)} \leq Q^2$$

for every $\omega \in E^\infty$ and every $k \geq 1$. Suppose that $\sigma^n(\omega) = \omega$. Then for every $k = ln$, $l \geq 1$, and every h, $S_{ln}h(\omega) = lS_n h(\omega)$ and so

$$Q^{-2} \leq \exp\big(l[(S_n f(\omega) - S_n g(\omega)) - (\mathrm{P}(f) - \mathrm{P}(g))n]\big) \leq Q^2.$$

Hence, there exists a constant $T \geq 0$ such that for all positive integers l,

$$l|S_n f(\omega) - S_n g(\omega) - (\mathrm{P}(f) - \mathrm{P}(g))n| \leq T.$$

Therefore, letting $l \nearrow \infty$, we conclude that $S_n f(\omega) - S_n g(\omega) = (\mathrm{P}(f) - \mathrm{P}(g))n$. Thus, putting $R = \mathrm{P}(f) - \mathrm{P}(g)$ completes the proof of the implication (1) $\Rightarrow$ (2).

(2) $\Rightarrow$ (3). Define

$$\eta = f - g - R.$$

Since the incidence matrix A is irreducible, there exists a point $\tau \in E^\infty$ transitive for the shift map $\sigma : E^\infty \to E^\infty$. Put

$$\Gamma = \{\sigma^k(\tau) : k \geq 1\}$$

and define the function $u : \Gamma \to I\!R$ by setting

$$u(\sigma^k(\tau)) = \sum_{j=0}^{k-1} \eta(\sigma^j(\tau)).$$

Note that the function u is well defined since the points $\sigma^k(\tau)$, $k \geq 1$, are mutually distinct. Taking the minimum of exponents we may assume that both functions f and g are Hölder continuous with the same order β. Let Λ witness the finite irreducibility of the incidence matrix A. Let $|\Lambda| = \sup\{|\alpha| : \alpha \in \Lambda\}$ and $S = \sup\{|S_{|\alpha|}\eta| : \alpha \in \Lambda\}$. Fix $k \geq 1$, some point $\alpha \in \Lambda$ such that $\tau_k \alpha \tau_1 \in E^*$, and consider the periodic point

$\omega = (\tau|_k\alpha)^\infty$. Then by our assumption

$$\begin{aligned}|u(\sigma^k(\tau))| &= \Bigg|\sum_{j=0}^{k-1}\big(\eta(\sigma^j(\tau)) - \big(f(\sigma^j(\omega)) - g(\sigma^j(\omega))\big) + Rk \\ &\quad + \sum_{j=0}^{|\alpha|-1}\big(g(\sigma^{k+j}\omega) - f(\sigma^{k+j}\omega)\big) + R|\alpha|\Bigg| \\ &= \Bigg|\sum_{j=0}^{k-1}\big((f(\sigma^j(\tau)) - f(\sigma^j(\omega))) - (g(\sigma^j(\tau)) \\ &\quad - g(\sigma^j(\omega))\big) - S_{|\alpha|}\eta(\sigma^k(\omega))\Bigg| \qquad (2.12) \\ &\le \sum_{j=0}^{k-1}|f(\sigma^j(\tau)) - f(\sigma^j(\omega))| + \sum_{j=0}^{k-1}|g(\sigma^j(\tau)) - g(\sigma^j(\omega))| \\ &\quad + |S_{|\alpha|}\eta(\sigma^k(\omega))| \\ &\le \sum_{j=0}^{k-1}V_\beta(f)\mathrm{e}^{-\beta(k-j)} + \sum_{j=0}^{k-1}V_\beta(g)\mathrm{e}^{-\beta(k-j)} + S \\ &\le (V_\beta(f) + V_\beta(g))\frac{\mathrm{e}^{-\beta}}{1-\mathrm{e}^{-\beta}} + S < \infty.\end{aligned}$$

Assume now $\sigma^k(\tau)|_r = \sigma^l(\tau)|_r$ for some $k < l$ and some $r \ge 1$. Let $\omega = \tau|_k(\sigma^k(\tau)|_{l-k})^\infty \in E^\infty$. By our assumption $\sum_{j=k}^{l-1}\eta(\sigma^j(\omega)) = 0$. Hence,

$$\begin{aligned}&|u(\sigma^l(\tau)) - u(\sigma^k(\tau))| \\ &\quad = \Bigg|\sum_{j=k}^{l-1}\eta(\sigma^j(\tau))\Bigg| = \Bigg|\sum_{j=k}^{l-1}\eta(\sigma^j(\tau)) - \eta(\sigma^j(\omega))\Bigg| \\ &\quad \le \sum_{j=k}^{l-1}\big(|f(\sigma^j(\tau)) - f(\sigma^j(\omega))| + |g(\sigma^j(\tau)) - g(\sigma^j(\omega))|\big) \qquad (2.13) \\ &\quad \le \sum_{j=k}^{l-1}(V_\beta(f) + V_\beta(g))\mathrm{e}^{-\beta(r+l-j-1)} \\ &\quad \le \mathrm{e}^{-\beta r}(V_\beta(f) + V_\beta(g))\sum_{j=0}^{\infty}\mathrm{e}^{-\beta j} = \frac{V_\beta(f) + V_\beta(g)}{1-\mathrm{e}^{-\beta}}\mathrm{e}^{-\beta r}\end{aligned}$$

In particular, it follows from (2.13) that u is uniformly continuous on Γ. Since Γ is a dense subset of E^∞, we therefore conclude that u has a unique continuous extension on E^∞. Moreover, it follows from (2.12) and (2.13) that u is bounded and Hölder continuous. Since $g(\gamma)-f(\gamma)-R = u(\gamma) - u(\sigma(\gamma))$ for all $\gamma \in \Gamma$, this holds for all γ. The proof of the implication (2) $\Rightarrow$ (3) is therefore complete.

The implications (3) $\Rightarrow$ (4) and (4) $\Rightarrow$ (5) are obvious.

(5) $\Rightarrow$ (1). It follows from (5) and (2.3) that for every $\omega \in E^*$, say $\omega \in E^n$,

$$Q^{-2}\mathrm{e}^{-T}\exp\big((S+\mathrm{P}(g)-\mathrm{P}(f))n\big) \le \frac{\tilde\mu_f([\omega])}{\tilde\mu_g([\omega])} \le Q^2\mathrm{e}^{T}\exp\big((S+\mathrm{P}(g)-\mathrm{P}(f))n\big). \tag{2.14}$$

Suppose that $S \ne \mathrm{P}(f)-\mathrm{P}(g)$. Without loss of generality we may assume that $S < \mathrm{P}(f)-\mathrm{P}(g)$. But then it would follow from (2.14) that for every $n \ge 1$

$$1 = \tilde\mu_f(E^\infty) = \sum_{|\omega|=n} \tilde\mu_f([\omega]) \le Q^2\mathrm{e}^{T}\exp\big((S+\mathrm{P}(g)-\mathrm{P}(f))n\big)$$

which yields a contradiction for $n \ge 1$ large enough. Hence $S = \mathrm{P}(f)-\mathrm{P}(g)$. But then (2.14) implies that the measures $\tilde\mu_f$ and $\tilde\mu_g$ are equivalent. Since in view of Theorem 2.2.4 these measures are ergodic, they must coincide. The proof of the implication (5) $\Rightarrow$ (1) and simultaneously the entire proof of Theorem 2.2.7 is complete. □

A *potential* is simply a continuous function $f : E^\infty \to I\!R$. We call a σ-invariant probability measure $\tilde\mu$ an *equilibrium state of the potential* f if $\int -f d\mu < +\infty$ and

$$\mathrm{h}_{\tilde\mu}(\sigma) + \int f d\tilde\mu = \mathrm{P}(f). \tag{2.15}$$

We end this section with the following two results.

Lemma 2.2.8 *Suppose that the incidence matrix A is finitely irreducible and that an acceptable function $f : E^\infty \to I\!R$ has a Gibbs state. Denote by $\tilde\mu_f$ its unique invariant Gibbs state (see Theorem 2.2.4). Then the following three conditions are equivalent:*

(a) $\int_{E^\infty} -f d\tilde\mu_f < \infty$.
(b) $\sum_{i\in I} \inf(-f|_{[i]})\exp(\inf f|_{[i]}) < \infty$.
(c) $\mathrm{H}_{\tilde\mu_f}(\alpha) < \infty$, *where* $\alpha = \{[i] : i \in I\}$ *is the partition of* E^∞ *into initial cylinders of length 1.*

Proof. (a) ⇒ (b). Suppose that $\int -f d\tilde{\mu}_f < \infty$. This means that $\sum_{i\in I}\int_{[i]} -f d\tilde{\mu}_f < \infty$ and consequently

$$\begin{aligned}\infty > \sum_{i\in I}\inf(-f|_{[i]})\tilde{\mu}_f([i]) &\geq Q^{-1}\sum_{i\in I}\inf(-f|_{[i]})\exp\big(\inf f|_{[i]} - \mathrm{P}(f)\big)\\ &= Q^{-1}\mathrm{e}^{-\mathrm{P}(f)}\sum_{i\in I}\inf(-f|_{[i]})\exp\big(\inf f|_{[i]}\big).\end{aligned}$$

(b) ⇒ (c). Assume that $\sum_{i\in I}\inf(-f|_{[i]})\exp\big(\inf(f|_{[i]})\big) < \infty$. We shall show that $\mathrm{H}_{\tilde{\mu}_f}(\alpha) < \infty$. By definition,

$$\mathrm{H}_{\tilde{\mu}_f}(\alpha) = \sum_{i\in I} -\tilde{\mu}_f([i])\log\tilde{\mu}_f([i]) \leq \sum_{i\in I} -\tilde{\mu}_f([i])(\inf(f|_{[i]}) - \mathrm{P}(f) - \log Q).$$

Since $\sum_{i\in I}\tilde{\mu}_f([i])(\mathrm{P}(f)+\log Q) < \infty$, it suffices to show that $\sum_{i\in I} -\tilde{\mu}_f([i])\inf(f|_{[i]}) < \infty$. And indeed,

$$\begin{aligned}\sum_{i\in I} -\tilde{\mu}_f([i])\inf(f|_{[i]}) &= \sum_{i\in I}\tilde{\mu}_f([i])\sup(-f|_{[i]})\\ &\leq \sum_{i\in I}\tilde{\mu}_f([i])\big(\inf(-f|_{[i]}) + \mathrm{osc}(f)\big).\end{aligned}$$

Since $\sum_{i\in I}\tilde{\mu}_f([i])\mathrm{osc}(f) = \mathrm{osc}(f)$, it is enough to show that

$$\sum_{i\in I}\tilde{\mu}_f([i])\inf(-f|_{[i]}) < \infty.$$

Since $\tilde{\mu}_f$ is a probability measure, $\lim_{i\to\infty}\tilde{\mu}_f([i]) = 0$. Therefore, it follows from (2.3) that $\lim_{i\to\infty}\big(\sup(f|_{[i]}) - \mathrm{P}(f)\big) = -\infty$. Thus, for all i sufficiently large, say $i \geq k$, $\sup(f|_{[i]}) < 0$. Hence, for all $i \geq k$, $\inf(-f|_{[i]}) = -\sup(f|_{[i]}) > 0$. So, using (2.3) again, we get

$$\begin{aligned}\sum_{i\geq k}\tilde{\mu}_f([i])\inf(-f|_{[i]}) &\leq \sum_{i\geq k} Q\exp\big(\inf(f|_{[i]}) - \mathrm{P}(f)\big)\inf(-f|_{[i]})\\ &= Q\mathrm{e}^{-\mathrm{P}(f)}\sum_{i\geq k}\exp\big(\inf(f|_{[i]})\big)\inf(-f|_{[i]})\end{aligned}$$

which is finite due to our assumption. Finally, we find $\mathrm{H}_{\tilde{\mu}_f}(\alpha) < \infty$.

(c) ⇒ (a). Suppose that $\mathrm{H}_{\tilde{\mu}_f}(\alpha) < \infty$. We need to show that $\int -f d\tilde{\mu}_f < \infty$. We have

$$\begin{aligned}\infty > \mathrm{H}_{\tilde{\mu}_f}(\alpha) &= \sum_{i\in I} -\tilde{\mu}_f([i])\log\big(\tilde{\mu}_f([i])\big)\\ &\geq \sum_{i\in I} -\tilde{\mu}_f([i])\big(\inf(f|_{[i]}) - \mathrm{P}(f) + \log Q\big).\end{aligned}$$

Hence, $\sum_{i\in I} -\tilde{\mu}_f([i])\inf(f|_{[i]}) < \infty$ and therefore

$$\int -f d\tilde{\mu}_f = \sum_{i\in I}\int_{[i]} -f d\tilde{\mu}_f \le \sum_{i\in I}\sup(-f|_{[i]})\tilde{\mu}_f([i])$$
$$= \sum_{i\in I} -\inf(f|_{[i]})\tilde{\mu}_f([i]) < \infty.$$

□

The next theorem shows that the assumption $\int -f d\tilde{\mu}_f < \infty$ is sufficient for the appropriate Gibbs state to be a unique equilibrium state.

Theorem 2.2.9 *Suppose that the incidence matrix A is finitely irreducible. Suppose that $f : E^\infty \to I\!R$ is a Hölder continuous function with a Gibbs state and that $\int -f d\tilde{\mu}_f < \infty$, where $\tilde{\mu}_f$ is the unique invariant Gibbs state for the potential f (see Theorem 2.2.4). Then $\tilde{\mu}_f$ is the unique equilibrium state for the potential f.*

Proof. In order to show that $\tilde{\mu}_f$ is an equilibrium state of the potential f consider $\alpha = \{[i] : i \in I\}$, the partition of E^∞ into initial cylinders of length 1. By Lemma 2.2.8, $\mathrm{H}_{\tilde{\mu}_f}(\alpha) < \infty$. Applying the Shannon–McMillan–Breiman theorem, Birkhoff's ergodic theorem, and (2.3), we get for $\tilde{\mu}_f$-a.e. $\omega \in E^\infty$

$$\begin{aligned}\mathrm{h}_{\tilde{\mu}_f}(\sigma) \ge \mathrm{h}_{\tilde{\mu}_f}(\sigma,\alpha) &= \lim_{n\to\infty} -\frac{1}{n}\log\tilde{\mu}_f([\omega|_n])\\ &\ge \lim_{n\to\infty} -\frac{1}{n}(\log Q + S_nf(\omega) - \mathrm{P}(f)n)\\ &= \lim_{n\to\infty}\frac{-1}{n}S_nf(\omega) + \mathrm{P}(f) = \int -f d\tilde{\mu}_f + \mathrm{P}(f)\end{aligned}$$

which, in view of Theorem 2.1.7, implies that $\tilde{\mu}_f$ is an equilibrium state for the potential f.

To prove uniqueness of the equilibrium state we suppose that $\tilde{\nu}$ is an equilibrium state for the potential $f : E^\infty \to I\!R$ and $\tilde{\nu} \ne \tilde{\mu}_f$. Applying the ergodic decomposition theorem, we may assume that $\tilde{\nu}$ is ergodic. Then, using (2.3), we have for every $n \ge 1$

$$0 = n(\mathrm{h}_{\tilde{\nu}}(\sigma) + \int (f - \mathrm{P}(f))d\tilde{\nu}) \le \mathrm{H}_{\tilde{\nu}}(\alpha^n) + \int (S_nf - \mathrm{P}(f)n)d\tilde{\nu}$$
$$= -\sum_{|\omega|=n}\tilde{\nu}([\omega])\left(\log\tilde{\nu}([\omega]) - \frac{1}{\tilde{\nu}([\omega])}\int_{[\omega]}(S_nf - \mathrm{P}(f)n)d\tilde{\nu}\right)$$

$$\leq -\sum_{|\omega|=n} \tilde{\nu}([\omega])\left(\log \tilde{\nu}([\omega]) - (S_nf(\tau_\omega) - \mathrm{P}(f)n)\right)$$

for a suitable $\tau_\omega \in [\omega]$

$$= -\sum_{|\omega|=n} \tilde{\nu}([\omega])\left(\log[\tilde{\nu}([\omega])\exp(\mathrm{P}(f)n - S_nf(\tau_\omega))]\right)$$

$$\leq -\sum_{|\omega|=n} \tilde{\nu}([\omega])\left(\log[\tilde{\nu}([\omega])(\mu_f([\omega])Q)^{-1}]\right)$$

$$= \log Q - \sum_{|\omega|=n} \tilde{\nu}([\omega]) \log\left(\frac{\tilde{\nu}([\omega])}{\tilde{\mu}_f([\omega])}\right).$$

Therefore, in order to conclude the proof, it suffices to show that

$$\lim_{n\to\infty}\left(-\sum_{|\omega|=n} \tilde{\nu}([\omega]) \log\left(\frac{\tilde{\nu}([\omega])}{\tilde{\mu}_f([\omega])}\right)\right) = -\infty.$$

Since both measures $\tilde{\nu}$ and $\tilde{\mu}_f$ are ergodic and $\tilde{\nu} \neq \tilde{\mu}_f$, the measures $\tilde{\nu}$ and $\tilde{\mu}_f$ must be mutually singular. In particular,

$$\lim_{n\to\infty} \tilde{\nu}\left(\left\{\omega \in E^\infty : \frac{\tilde{\nu}([\omega|_n])}{\tilde{\mu}_f([\omega|_n])} \leq S\right\}\right) = 0$$

for every $S > 0$. For every $j \in \boldsymbol{Z}$ and every $n \geq 1$, set

$$F_{n,j} = \left\{\omega \in E^\infty : \mathrm{e}^{-j} \leq \frac{\tilde{\nu}([\omega|_n])}{\tilde{\mu}_f([\omega|_n])} < \mathrm{e}^{-j+1}\right\}.$$

Then

$$\tilde{\nu}(F_{n,j}) = \int_{F_{n,j}} \frac{\tilde{\nu}([\omega|_n])}{\tilde{\mu}_f([\omega|_n])} d\tilde{\mu}_f(\omega) \leq \mathrm{e}^{-j+1}\tilde{\mu}_f(F_{n,j}) \leq \mathrm{e}^{-j+1}.$$

Notice

$$-\sum_{|\omega|=n} \tilde{\nu}([\omega]) \log\left(\frac{\tilde{\nu}([\omega])}{\tilde{\mu}_f([\omega])}\right) = -\int \log\left(\frac{\tilde{\nu}([\omega|_n])}{\tilde{\mu}_f([\omega|_n])}\right) d\tilde{\nu}(\omega) \leq \sum_{j\in\boldsymbol{Z}} j\tilde{\nu}(F_{n,j}).$$

Now, for each $k = -1, -2, -3, \dots$ we have

$$\begin{aligned}
-\sum_{|\omega|=n} \tilde{\nu}([\omega]) \log\left(\frac{\tilde{\nu}([\omega])}{\tilde{\mu}_f([\omega])}\right) &\leq k\sum_{j\leq k} \tilde{\nu}(F_{n,j}) + \sum_{j\geq 1} j\mathrm{e}^{-j+1} \\
&= k\tilde{\nu}\left(\left\{\omega \in E^\infty : \frac{\tilde{\nu}([\omega|_n])}{\tilde{\mu}_f([\omega|_n])} \geq \mathrm{e}^{-k}\right\}\right) \\
&\quad + \sum_{j\geq 1} j\mathrm{e}^{-j+1}.
\end{aligned}$$

Thus, we have for each negative integer k,

$$\limsup_{n\to\infty}\left(-\sum_{|\omega|=n}\tilde{\nu}([\omega])\log\left(\frac{\tilde{\nu}([\omega])}{\tilde{\mu}_f([\omega])}\right)\right)\le k+\sum_{j\ge 1} j\mathrm{e}^{-j+1}.$$

Now, letting k go to $-\infty$ completes the proof. □

2.3 Perron–Frobenius operator

In this section we collect some basic properties of the Perron–Frobenius operator. This operator is ultimately applied (see Theorem 2.7.3 and Corollary 2.7.4) to prove the existence of Gibbs states. It is also used to demonstrate several stochastic laws of Gibbs states and the real analyticity of topological pressure. The key facts in this section making possible further work are Theorem 2.3.4 and Theorem 2.3.5. We start with the following technical result usually referred to as a *bounded distortion lemma.*

Lemma 2.3.1 *If $g : E^\infty \to \mathbb{C}$ and $V_\alpha(g) < \infty$, then for all $n \ge 1$, for all $\omega, \tau \in E^\infty$ and all $\rho \in E^n$ with $A_{\rho_n\omega_1} = A_{\rho_n\tau_1} = 1$, we have*

$$|S_n g(\rho\omega) - S_n g(\rho\tau)| \le \frac{V_\alpha(g)}{\mathrm{e}^\alpha - 1} d_\alpha(\omega,\tau).$$

Proof. We have

$$\begin{aligned}
|S_n g(\rho\omega) - S_n g(\rho\tau)| &\le \sum_{i=0}^{n-1} |g(\sigma^i(\rho\omega)) - g(\sigma^i(\rho\tau))| \\
&\le \sum_{i=0}^{n-1} \mathrm{e}^\alpha V_\alpha(g) d_\alpha(\sigma^i(\rho\omega), \sigma^i(\rho\tau)) \\
&\le \mathrm{e}^\alpha V_\alpha(g) \sum_{i=0}^{n-1} \mathrm{e}^{-\alpha(n-i)} d_\alpha(\omega,\tau) \\
&= V_\alpha(g)\frac{1}{1-\mathrm{e}^{-\alpha}} d_\alpha(\omega,\tau) \frac{V_\alpha(g)}{\mathrm{e}^\alpha - 1} d_\alpha(\omega,\tau).
\end{aligned}$$

□

We set

$$T(g) = \exp\left(\frac{V_\alpha(g)}{\mathrm{e}^\alpha - 1}\right),$$

We suppress the dependence on α. From now on throughout this section $f : E^\infty \to \mathbb{R}$ is assumed to be a bounded Hölder continuous function with an exponent $\beta > 0$ and is assumed to satisfy

$$\sum_{e \in I} \exp(\sup(f|_{[e]})) < \infty. \tag{2.16}$$

Functions f satisfying this condition will be called in the sequel *summable.* We note that if f has a Gibbs state, then f is summable. This requirement allows us to define the *Perron–Frobenius operator* $\mathcal{L}_f : C_b(E^\infty) \to C_b(E^\infty)$, acting on the space of bounded continuous functions $C_b(E^\infty)$ provided with $||\cdot||_0$, the uniform norm, as follows:

$$\mathcal{L}_f(g)(\omega) = \sum_{e \in I: A_{e\omega_1}=1} \exp(f(e\omega))g(e\omega).$$

Then $||\mathcal{L}_f||_0 \leq \sum_{e\in I} \exp(\sup(f|_{[e]})) < \infty$ and for every $n \geq 1$,

$$\mathcal{L}_f^n(g)(\omega) = \sum_{\tau \in E^n: A_{\tau_n \omega_1}=1} \exp\bigl(S_n f(\tau\omega)\bigr)g(\tau\omega).$$

The conjugate operator $\mathcal{L}_f^*$ acting on the space $C_b^*(E^\infty)$ has the following form:

$$\mathcal{L}_f^*(\mu)(g) = \mu(\mathcal{L}_f(g)) = \int \mathcal{L}_f(g)d\mu.$$

From now on throughout this section we also assume $\tilde{m}$ is a probability measure defined on the Borel subsets of E^∞ which is an eigenmeasure of the conjugate operator $\mathcal{L}_f^* : C_b^*(E^\infty) \to C_b^*(E^\infty)$. The corresponding eigenvalue is denoted by λ. Since $\mathcal{L}_f$ is a positive operator, $\lambda \geq 0$. Obviously $\mathcal{L}_f^{*n}(\tilde{m}) = \lambda^n \tilde{m}$. The integral version of this equality takes on the following form

$$\int \sum_{\tau \in E^n: A_{\tau_n \omega_1}=1} \exp\bigl(S_n f(\tau\omega)\bigr)g(\tau\omega)d\tilde{m}(\omega) = \lambda^n \int g d\tilde{m}, \tag{2.17}$$

for every function $g \in C_b(E^\infty)$. In fact this equality extends to the space of all bounded Borel functions on E^∞. In particular, taking $\omega \in E^*$, say $\omega \in E^n$, a Borel set $B \subset E^\infty$ such that $A_{\omega_n \tau_1} = 1$, for every $\tau \in B$, and $g = 1\!\!1_{\omega B}$, we obtain from (2.17)

$$\begin{aligned}
\lambda^n \tilde{m}(\omega B) &= \int \sum_{\tau \in E^n: A_{\tau_n \rho_1}=1} \exp\bigl(S_n f(\tau\rho)\bigr) 1\!\!1_{\omega B}(\tau\rho) d\tilde{m}(\rho) \\
&= \int_{\{\rho \in B: A_{\omega_n \rho_1}=1\}} \exp\bigl(S_n f(\omega\rho)\bigr)d\tilde{m}(\rho) \qquad (2.18) \\
&= \int_B \exp\bigl(S_n f(\omega\rho)\bigr)d\tilde{m}(\rho)
\end{aligned}$$

Remark 2.3.2 *Note that if (2.18) holds, then by representing a Borel set $B \subset E^\infty$ as a union $\bigcup_{\omega \in E^n}[\omega B_\omega]$, where $B_\omega = \{\alpha \in E^\infty : A_{\omega_n \alpha_1} = 1 \text{ and } \omega\alpha \in B\}$, a straightforward calculation based on (2.18) demonstrates that (2.17) is satisfied for the characteristic function $1\!\!1_B$ of the set B. Next, it follows from standard approximation arguments that (2.17) is satisfied for all $\tilde{m}$-integrable functions g. Finally, we note that $\tilde{m}$ is an eigenmeasure of the conjugate operator $\mathcal{L}_f^*$ if and only if formula (2.18) is satisfied.*

Theorem 2.3.3 *If the incidence matrix is finitely irreducible, then the eigenmeasure $\tilde{m}$ is a Gibbs state for f. In addition, $\lambda = e^{P(f)}$.*

Proof. It immediately follows from (2.18) and Lemma 2.3.1 that for every $\omega \in E^*$ and every $\tau \in [\omega]$

$$\tilde{m}([\omega]) \le \lambda^{-n} T(f) \exp\bigl(S_n f(\tau)\bigr) = T(f) \exp\bigl(S_n f(\tau) - n \log \lambda\bigr), \quad (2.19)$$

where $n = |\omega|$. On the other hand, let Λ be a minimal set which witnesses the finite irreducibility of A. For every $\alpha \in \Lambda$, let

$$E_\alpha = \{\tau \in E^\infty : \omega\alpha\tau \in E^\infty\}.$$

By the definition of Λ, $\bigcup_{\alpha \in \Lambda} E_\alpha = E^\infty$. Hence, there exists $\gamma \in \Lambda$ such that $\tilde{m}(E_\gamma) \ge (\#\Lambda)^{-1}$. Writing $p = |\gamma|$ we therefore have

$$\begin{aligned}
\tilde{m}([\omega]) \ge \tilde{m}([\omega\gamma]) &= \lambda^{-(n+p)} \int_{\rho \in E^\infty : A_{\gamma_p \rho_1} = 1} \exp\bigl(S_{n+p} f(\omega\gamma\rho)\bigr) d\tilde{m}(\rho) \\
&= \lambda^{-(n+p)} \int_{\rho \in E^\infty : A_{\gamma_p \rho_1} = 1} \exp\bigl(S_n f(\omega\gamma\rho)\bigr) \exp\bigl(S_p f(\gamma\rho)\bigr) d\tilde{m}(\rho) \\
&\ge \lambda^{-n} \exp\bigl(\min\{\inf(S_{|\alpha|} f|_{[\alpha]}) : \alpha \in \Lambda\} - p \log \lambda\bigr) \\
&\times \int_{\rho \in E^\infty : A_{\gamma_p \rho_1} = 1} \exp\bigl(S_n f(\omega\gamma\rho)\bigr) d\tilde{m}(\rho) \\
&= C \lambda^{-n} \int_{E_\gamma} \exp\bigl(S_n f(\omega\gamma\rho)\bigr) d\tilde{m}(\rho) \\
&\ge C T(f)^{-1} \lambda^{-n} \tilde{m}(E_\gamma) \exp\bigl(S_n f(\tau)\bigr) \\
&\ge C T(f)^{-1} (\#\Lambda)^{-1} \exp\bigl(S_n f(\tau) - n \log \lambda\bigr),
\end{aligned} \quad (2.20)$$

where $C = \exp\bigl(\min\{\inf(S_{|\alpha|} f|_{[\alpha]}) : \alpha \in \Lambda\} - p \log \lambda\bigr)$. Thus $\tilde{m}$ is a Gibbs state for f. The equality $\lambda = e^{P(f)}$ follows now immediately from Proposition 2.2.2. □

In most of our work concerning the Perron–Frobenius operator and its applications we will deal with the following normalized version of it.

$$\mathcal{L}_0 = e^{-P(f)}\mathcal{L}_f.$$

The first result concerning the *normalized Perron–Frobenius operator* is the following.

Theorem 2.3.4 *If a function $f : E^\infty \to I\!R$ has a Gibbs state with ratio bounding constant Q, then for every $n \geq 1$ and every $\omega \in I^n$*

$$\mathcal{L}_0^n(1\!\!1)(\omega) \leq Q.$$

Thus, for all $g \in C_b(E^\infty), ||\mathcal{L}_0^n(g)||_0 \leq Q||g||_0$.

Proof. Let ν be a Gibbs measure for f. In view of Lemma 2.3.1 and the definition of Gibbs states we get

$$\begin{aligned}\mathcal{L}_0^n(1\!\!1)(\omega) &= \sum_{\tau\in\sigma^{-n}(\omega)} \exp\big(S_n f(\tau) - P(f)n\big) \\ &\leq \sum_{\tau\in\sigma^{-n}(\omega)} Q\nu([\tau|_n]) \leq Q\nu(\sigma^{-n}([\omega])) \leq Q.\end{aligned}$$

□

We emphasize that in Theorem 2.3.4 we assumed only the existence of a Gibbs state and not an eigenmeasure of the conjugate Perron–Frobenius operator. We shall now prove the following.

Theorem 2.3.5 *Suppose Λ is a finite set of words witnessing the finite irreducibility of the incidence matrix. Let M be the maximal length of a word in Λ. Then there exists a constant $R > 0$ such that*

$$\sum_{j=n}^{n+M} \mathcal{L}_0^j(1\!\!1)(\omega) \geq R,$$

for all $n \geq 1$ and all $\omega \in E^\infty$.

Proof. It follows from (2.17) and Theorem 2.3.3 that $\int \mathcal{L}_0^k(1\!\!1)d\tilde{m} = 1$ for all $k \geq 1$. So, for every $k \geq 1$, there exists $\gamma(k) \in E^\infty$ such that $\mathcal{L}_0^k(1\!\!1)(\gamma(k)) \geq 1$. Since Λ is finite and f is Hölder continuous, we have

$$N = \min\{\exp\big(\inf(S_{|\alpha|}f|_{[\alpha]}) - |\alpha|P(f)\big) : \alpha \in \Lambda\} > 0.$$

Let $n \geq 1$ and let $\tau \in E^\infty$. Thus,

$$\sum_{j=n}^{n+M} \mathcal{L}_0^j(1\!\!1)(\tau) = \sum_{j=n}^{n+M} \sum_{\omega \in E^j : A_{\omega_j \tau_1}=1} \exp\big(S_j f(\omega\tau) - \mathrm{P}(f)j\big).$$

For each $\omega \in E^n$ choose $\alpha(\omega, \tau) \in \Lambda$ such that $\omega\alpha(\omega, \tau)\tau$ is admissible. Thus,

$$\begin{aligned}
\sum_{j=n}^{n+M} \mathcal{L}_0^j(1\!\!1)(\tau) &\geq \sum_{\omega \in E^n} \exp\big(S_{n+|\alpha(\omega,\tau)|} f(\omega\alpha(\omega,\tau)\tau) - \mathrm{P}(f)(n + |\alpha(\omega,\tau)|)\big) \\
&\geq \sum_{\omega \in E^n} \exp\big(S_n f(\omega\alpha(\omega,\tau)\tau) - \mathrm{P}(f)n\big) \\
&\quad \times \exp\big(S_{|\alpha(\omega,\tau)|} f(\alpha(\omega,\tau)\tau) - \mathrm{P}(f)|\alpha(\omega,\tau)|\big) \\
&\geq N \sum_{\omega \in E^n} \exp\big(S_n f(\omega\alpha(\omega,\tau)\tau) - \mathrm{P}(f)n\big).
\end{aligned}$$

Noting that if $A_{\omega_n \gamma(n)_1} = 1$ then

$$\exp\big(S_n f(\omega\alpha(\omega,\tau)\tau) - \mathrm{P}(f)n\big) \geq (Tf)^{-1} \exp\big(S_n f(\omega\gamma(n)) - \mathrm{P}(f)n\big),$$

we have

$$\begin{aligned}
\sum_{j=n}^{n+M} \mathcal{L}_0^j(1\!\!1)(\tau) &\geq N(Tf)^{-1} \sum_{\omega \in E^n : A_{\omega_n \gamma(n)_1}=1} \exp\big(S_n f(\omega\gamma(n)) - \mathrm{P}(f)n\big) \\
&\geq N(Tf)^{-1} \mathcal{L}_0^n(1\!\!1)(\gamma(n)) \geq NT(f)^{-1}.
\end{aligned}$$

□

Remark: *if in theorem 2.3.5 we assume A is finitely primitive then there is a constant $R > 0$ such that $\mathrm{L}_0^n(1\!\!1)(\omega) \geq R$ for all n and ω.*

Theorem 2.3.6 *If the incidence matrix is finitely irreducible, then the conjugate operator $\mathcal{L}_0^*$ fixes at most one Borel probability measure.*

Proof. Suppose that $\tilde{m}$ and $\tilde{m}_1$ are such two fixed points. In view of Proposition 2.2.2(b) and Theorem 2.3.3, the measures $\tilde{m}$ and $\tilde{m}_1$ are equivalent. Consider the Radon-Nikodym derivitive $\rho = \frac{d\tilde{m}_1}{d\tilde{m}}$. Temporarily fix

$$\omega \in E^*, \quad say \quad \omega \in E^n.$$

It then follows from (2.18) and Theorem 2.3.3 that

$$\begin{aligned}
\tilde{m}([\omega]) &= \int_{\tau\in E^\infty: A_{\omega_n\tau_1}=1} \exp\big(S_nf(\omega\tau) - \mathrm{P}(f)n\big)d\tilde{m}(\tau) \\
&= \int_{\tau\in E^\infty: A_{\omega_n\tau_1}=1} \exp\big(S_{n-1}f(\sigma(\omega\tau)) - \mathrm{P}(f)(n-1)\big) \\
&\quad\times \exp\big(f(\omega\tau) - \mathrm{P}(f)\big)d\tilde{m}(\tau) \\
&= \int_{\tau\in E^\infty: A_{(\sigma(\omega))_{n-1}\tau_1}=1} \exp\big(S_{n-1}f(\sigma(\omega\tau)) - \mathrm{P}(f)(n-1)\big) \\
&\quad\times \exp\big(f(\omega\tau) - \mathrm{P}(f)\big)d\tilde{m}(\tau).
\end{aligned}$$

Thus,

$$\inf\big(\exp(f|_{[\omega]}-\mathrm{P}(f))\big)\tilde{m}([\sigma\omega]) \le \tilde{m}([\omega]) \le \sup\big(\exp(f|_{[\omega]}-\mathrm{P}(f))\big)\tilde{m}([\sigma\omega]).$$

Since $f : E^\infty \to I\!R$ is Hölder continuous, we therefore conclude that for every $\omega \in E^\infty$

$$\lim_{n\to\infty} \frac{\tilde{m}\big([\omega|_n]\big)}{\tilde{m}\big([\sigma(\omega)|_{n-1}]\big)} = \exp\big(f(\omega) - \mathrm{P}(f)\big) \tag{2.21}$$

and the same formula is true with $\tilde{m}$ replaced by $\tilde{m}_1$. Using Theorem 2.3.3 and Theorem 2.2.4, there exists a set of points $\omega \in E^\infty$ with $\tilde{m}$ measure 1 for which the Radon-Nikodym derivatives $\rho(\omega)$ and $\rho(\sigma(\omega))$ are both defined. Let $\omega \in E^\infty$ be such a point. Then from (2.21) and its version for $\tilde{m}_1$ we obtain

$$\begin{aligned}
\rho(\omega) &= \lim_{n\to\infty}\left(\frac{\tilde{m}_1\big([\omega|_n]\big)}{\tilde{m}\big([\omega|_n]\big)}\right) \\
&= \lim_{n\to\infty}\left(\frac{\tilde{m}_1\big([\omega|_n]\big)}{\tilde{m}_1\big([\sigma(\omega)|_{n-1}]\big)}\cdot\frac{\tilde{m}_1\big([\sigma(\omega)|_{n-1}]\big)}{\tilde{m}\big([\sigma(\omega)|_{n-1}]\big)}\cdot\frac{\tilde{m}\big([\sigma(\omega)|_{n-1}]\big)}{\tilde{m}\big([\omega|_n]\big)}\right) \\
&= \exp\big(f(\omega)-\mathrm{P}(f)\big)\rho(\sigma(\omega))\exp\big(\mathrm{P}(f)-f(\omega)\big) = \rho(\sigma(\omega)).
\end{aligned}$$

But according to Theorem 2.2.4, $\sigma : E^\infty \to E^\infty$ is ergodic with respect to a σ-invariant measure equivalent with $\tilde{m}$. We conclude that ρ is $\tilde{m}$-almost everywhere constant. Since $\tilde{m}_1$ and $\tilde{m}$ are both probability measures, $\tilde{m}_1 = \tilde{m}$. □

2.4 Ionescu-Tulcea and Marinescu inequality

In this section we prove in our context the famous Tulcea-Ionescu and Marinescu inequality (see Lemma 2.4.1). It is our aim to obtain a spectral picture and the rate of convergence of the iterates of the Perron–Frobenius operator and, in the next section, some important stochastic laws. Let

$$\mathcal{H}_0 = \{g : E^\infty \to \mathbb{C} : g \text{ is bounded and continuous}\}$$

and for every $\alpha > 0$, let

$$\mathcal{H}_\alpha = \{g \in \mathcal{H}_0 : V_\alpha(g) < \infty\}.$$

The set $\mathcal{H}_\alpha$ becomes a Banach space when endowed with the norm

$$\|g\|_\alpha = \|g\|_0 + V_\alpha(g).$$

The subclass of $\mathcal{H}_\alpha$ consisting of summable functions will be denoted by $\mathcal{H}_\alpha^s$. The main technical result of this section, called the *Ionescu-Tulcea and Marinescu inequality*, is the following.

Lemma 2.4.1 *Let $f : E^\infty \to \mathbb{R}$ be a summable Hölder continuous function with an exponent $\beta > 0$ and suppose f has a Gibbs state. Then the normalized operator $\mathcal{L}_0 : \mathcal{H}_0 \to \mathcal{H}_0$ preserves the space $\mathcal{H}_\beta$. Moreover, there exists a constant $C > 0$ such that for every $n \geq 1$ and every $g \in \mathcal{H}_\beta$,*

$$||\mathcal{L}_0^n(g)||_\beta \leq Qe^{-\beta n}||g||_\beta + C||g||_0.$$

Proof. Let $\tau, \rho \in E^\infty$, $\tau|_k = \rho|_k$ and $\tau_{k+1} \neq \rho_{k+1}$ for some $k \geq 1$. Then for every $n \geq 1$

$$\begin{aligned}
&\mathcal{L}_0^n(g)(\rho) - \mathcal{L}_0^n(g)(\tau) \\
&\quad = \sum_{\omega \in E^n : A_{\omega_n \rho_1} = 1} \exp(S_n f(\omega\rho) - \mathrm{P}(f)n) g(\omega\rho) - \sum_{\omega \in E^n : A_{\omega_n \rho_1} = 1} \\
&\quad \times \exp(S_n f(\omega\tau) - \mathrm{P}(f)n) g(\omega\tau) \\
&\quad = \sum_{\omega \in E^n : A_{\omega_n \rho_1} = 1} \exp(S_n f(\omega\rho) - \mathrm{P}(f)n)\big(g(\omega\rho) - g(\omega\tau)\big) \\
&\quad + \sum_{\omega \in E^n : A_{\omega_n \rho_1} = 1} g(\omega\tau)\big(\exp(S_n f(\omega\rho) - \mathrm{P}(f)n) \\
&\quad - \exp(S_n f(\omega\tau) - \mathrm{P}(f)n)\big).
\end{aligned} \tag{2.22}$$

But $|g(\omega\rho) - g(\omega\tau)| \le V_\beta(g)\mathrm{e}^{-\beta(n+k)}$, and therefore, employing Theorem 2.3.4, we obtain

$$\begin{aligned}\sum_{\omega\in E^n : A_{\omega_n\rho_1}=1} &\exp(S_n f(\omega\rho) - \mathrm{P}(f)n)|g(\omega\rho) - g(\omega\tau)| \\ &\le \mathcal{L}_0^n(1\!\!1)(\rho)V_\beta(g)\mathrm{e}^{-\beta(n+k)} \\ &\le QV_\beta(g)\mathrm{e}^{-\beta(n+k)} \le \mathrm{e}^{-\beta n}Q||g||_\beta d_\beta(\rho,\tau)\end{aligned} \tag{2.23}$$

Now notice that there exists a constant $M \ge 1$ such that $|1 - \mathrm{e}^x| \le M|x|$ for all x with $|x| \le \log(T(f))$. Since by Lemma 2.3.1 $|S_n f(\omega\rho) - S_n f(\omega\tau)| \le d_\beta(\rho,\tau)\log(T(f)) \le \log(T(f))$, we can make the following estimate:

$$\begin{aligned}&|\exp(S_n f(\omega\rho) - \mathrm{P}(f)n) - \exp(S_n f(\omega\tau) - \mathrm{P}(f)n)| \\ &\quad = \exp(S_n f(\omega\rho) - \mathrm{P}(f)n)|1 - \exp(S_n f(\omega\tau) - S_n f(\omega\rho))| \\ &\quad \le M\exp(S_n f(\omega\rho) - \mathrm{P}(f)n)|S_n f(\omega\rho) - S_n f(\omega\tau)| \\ &\quad \le M\exp(S_n f(\omega\rho) - \mathrm{P}(f)n)\log(T(f))d_\beta(\rho,\tau) \\ &\quad = M\log(T(f))\exp(S_n f(\omega\rho) - \mathrm{P}(f)n)d_\beta(\rho,\tau).\end{aligned}$$

So, using Theorem 2.3.4 again, we get

$$\begin{aligned}\sum_{\omega\in E^n : A_{\omega_n\rho_1}=1} &|g(\omega\tau)||\exp(S_n f(\omega\rho) - \mathrm{P}(f)n) - \exp(S_n f(\omega\tau) - \mathrm{P}(f)n)| \\ &\le ||g||_0 M\log(T(f))d_\beta(\rho,\tau)\sum_{\omega\in E^n : A_{\omega_n\rho_1}=1}\exp(S_n f(\omega\rho) - \mathrm{P}(f)n) \\ &= ||g||_0 M\log(T(f))d_\beta(\rho,\tau)\mathcal{L}_0^n(1\!\!1)(\rho) \le MQ\log(T(f))||g||_0 d_\beta(\rho,\tau).\end{aligned}$$

Combining this inequality, (2.23) and (2.22), we get

$$|\mathcal{L}_0^n(g)(\rho) - \mathcal{L}_0^n(g)(\tau)| \le \mathrm{e}^{-\beta n}Q||g||_\beta d_\beta(\rho,\tau) + MQ\log(T(f))||g||_0 d_\beta(\rho,\tau).$$

Combining in turn this and Theorem 2.3.4 we get

$$||\mathcal{L}_0^n(g)||_\beta \le Q\mathrm{e}^{-\beta n}||g||_\beta + Q(M\log(T(f)) + 1)||g||_0.$$

□

Remark 2.4.2 *We remark that the proof of Lemma 2.4.1 used only a "weaker" property of Gibbs states, namely the right-hand side inequality of (2.3).*

If the unit ball in $\mathcal{H}_\beta$ were compact as a subset of the Banach space $\mathcal{H}_0$ with the supremum norm $||\cdot||_0$, we could use now the famous Ionescu-Tulcea and Marinescu theorem (see [ITM]) to establish some useful spectral properties of the Perron–Frobenius operator $\mathcal{L}_0$. But this ball is compact only in the topology of uniform convergence on compact subsets of E^∞ and we need to prove these spectral properties directly as we cannot apply the Ionescu-Tulcea and Marinescu theorem. Next, we introduce and important fixed point ψ for $\mathcal{L}_0$.

Theorem 2.4.3 *Suppose a summable Hölder continuous function $f : E^\infty \to \mathbb{R}$ with an exponent $\beta > 0$ has a Gibbs state and the operator conjugate to the normalized Perron–Frobenius operator $\mathcal{L}_0$ fixes a Borel probability measure $\tilde{m}$. Then the operator $\mathcal{L}_0 : \mathcal{H}_\beta \to \mathcal{H}_\beta$ has a fixed point $\psi \le Q$ such that $\int \psi d\tilde{m} = 1$. If, in addition, the incidence matrix A is finitely primitive then $\psi \ge R$, where R is the constant produced in Theorem 2.3.5.*

Proof. In view of Lemma 2.4.1, $\|\mathcal{L}_0^n(1\!\!1)\|_\beta \le Q + C$ for every $n \ge 0$. Thus,

$$\left\| \frac{1}{n} \sum_{j=0}^{n-1} \mathcal{L}_0^j(1\!\!1) \right\|_\beta \le Q + C \tag{2.24}$$

for every $n \ge 1$. Therefore, by the Ascoli–Arzelá theorem, there exists a strictly increasing sequence of positive integers $\{n_k\}_{k\ge1}$ such that the sequence $\left\{ \frac{1}{n_k} \sum_{j=0}^{n_k-1} \mathcal{L}_0^j(1\!\!1) \right\}_{k\ge1}$ converges uniformly on compact sets of E^∞ to $\psi : E^\infty \to \mathbb{R}$. Obviously, $||\psi||_\beta \le Q + C$ and, in particular $\psi \in \mathcal{H}_\beta$. Since $\tilde{m}$ is a fixed point of the operator conjugate to $\mathcal{L}_0$, $\int \mathcal{L}_0^j(1\!\!1) d\tilde{m} = 1$ for every $j \ge 0$. Consequently, $\int \frac{1}{n} \sum_{j=0}^{n-1} \mathcal{L}_0^j(1\!\!1) d\tilde{m} = 1$ for every $n \ge 1$. So, applying Lebesgue's dominated convergence theorem along with Theorem 2.3.4, we conclude that $\int \psi d\tilde{m} = 1$ and $\psi \le Q$. Assuming in addition that the incidence matrix A is finitely primitive, using Theorem 2.3.5, we simultaneously get $\psi \ge R$. It remains to show that $\mathcal{L}_0(\psi) = \psi$. Indeed, using Theorem 2.3.4, we get for every $k \ge 1$ that

$$\begin{aligned} \left\| \mathcal{L}_0 \left(\frac{1}{n_k} \sum_{j=0}^{n_k-1} \mathcal{L}_0^j(1\!\!1) \right) - \frac{1}{n_k} \sum_{j=0}^{n_k-1} \mathcal{L}_0^j(1\!\!1) \right\|_0 &= \frac{1}{n_k} \left\| \mathcal{L}_0^{n_k}(1\!\!1) - \mathcal{L}_0(1\!\!1) \right|_0 \\ &\le \frac{1}{n_k} Q. \end{aligned}$$

Thus,

$$\mathcal{L}_0\left(\frac{1}{n_k}\sum_{j=0}^{n_k-1}\mathcal{L}_0^j(1\!\!1)\right)-\frac{1}{n_k}\sum_{j=0}^{n_k-1}\mathcal{L}_0^j(1\!\!1)\to 0 \tag{2.25}$$

uniformly. Therefore, in order to conclude the proof it suffices to show that if a sequence $\{g_k\}_{k=1}^{\infty}\subset\mathcal{H}_0$ is uniformly bounded and converges uniformly on compact sets of E^{∞} to a function g, then $\mathcal{L}_0(g_k)$, $k\geq 1$, converges uniformly on compact sets of E^{∞} to $\mathcal{L}_0(g)$. To see this, first notice that $\|g\|_0\leq B$, where B is an upper bound of the sequence $\{g_k\}_{k=1}^{\infty}$. Now, fix $\epsilon>0$. Since f has a Gibbs state, the series $M=\sum_{i\in I}\exp\big(\sup(f|_{[i]})-\mathrm{P}\big)$ converges, where P is the pressure of f and therefore there exists a finite set $I_\epsilon\subset I$ such that

$$\sum_{i\in I\setminus I_\epsilon}2B\exp\big(\sup(f|_{[i]})-\mathrm{P}\big)<\frac{\epsilon}{2}. \tag{2.26}$$

Fix an arbitrary compact set $K\subset E^{\infty}$. Then for every $i\in I$, the set $iK=\{i\omega:\omega\in K\text{ and }A_{i\omega_1}=1\}$ is also compact and so is the set $\bigcup_{i\in I_\epsilon}iK$. Since $\{g_k\}_{k=1}^{\infty}$ converges uniformly on compact sets to g, there exists $q\geq 1$ such that for every $n\geq q$, $\|(g_n-g)|_{\bigcup_{i\in I_\epsilon}iK}\|_0\leq\frac{\epsilon}{2M}$. Applying this, Theorem 2.3.4 and (2.26), we get for every $n\geq q$ and every $\omega\in K$ that

$$\begin{aligned}
&|\mathcal{L}_0(g)(\omega)-\mathcal{L}_0(g_n)(\omega)|=|\mathcal{L}_0(g-g_n)(\omega)|\\
&\quad\leq\sum_{i\in I_\epsilon:A_{i\omega_1}=1}|g_n(i\omega)-g(i\omega)|\exp(f(i\omega)-\mathrm{P})\\
&\quad+\sum_{i\in I\setminus I_\epsilon:A_{i\omega_1}=1}|g_n(i\omega)-g(i\omega)|\exp(f(i\omega)-\mathrm{P})\\
&\quad\leq\sum_{i\in I_\epsilon:A_{i\omega_1}=1}\frac{\epsilon}{2M}\exp(f(i\omega)-\mathrm{P})\\
&\quad+\sum_{i\in I\setminus I_\epsilon:A_{i\omega_1}=1}|g_n(i\omega)-g(i\omega)|\exp(f(i\omega)-\mathrm{P})\\
&\quad\leq\frac{\epsilon}{2M}\sum_{i\in I}\exp\big(\sup(f|_{[i]})-\mathrm{P}\big)+2B\sum_{i\in I\setminus I_\epsilon:A_{i\omega_1}=1}\exp\big(\sup(f|_{[i]})-\mathrm{P}\big)\\
&\quad\leq\frac{\epsilon}{2}+\frac{\epsilon}{2}=\epsilon.
\end{aligned}$$

□

From now on we assume in this section that the incidence matrix A is finitely primitive. Then $\psi \geq R$ and therefore the *weighted normalized Perron–Frobenius operator* $\boldsymbol{T} : \mathcal{H}_0 \to \mathcal{H}_0$ given by the formula

$$\boldsymbol{T}(g) = \frac{1}{\psi}\mathcal{L}_0(g\psi) \tag{2.27}$$

is well defined. It is straightforward to check that $\boldsymbol{T}(\mathcal{H}_\beta) \subset \mathcal{H}_\beta$ (i.e./ $1/\psi$ and the product of any two functions in $\mathcal{H}_\beta$ are again in $\mathcal{H}_\beta$). The basic properties of the operator $\boldsymbol{T}$ which follow from Lemma 2.4.1, Theorem 2.3.5 and Theorem 2.4.3 are listed below.

Theorem 2.4.4 *Suppose a summable Hölder continuous function $f : E^\infty \to I\!R$ with an exponent $\beta > 0$ has a Gibbs state and the matrix A is finitely primitive. Defining the balancing functions*

$$u_n(\omega) = \exp(S_n f(\omega) - \mathrm{P}n)\frac{\psi(\omega)}{\psi(\sigma^n(\omega))}, \ n \geq 1,$$

we have for all $g \in \mathcal{H}_\beta$ and all $n \geq 1$

(a) $\boldsymbol{T}^n(g)(\omega) = \frac{1}{\psi(\omega)}\mathcal{L}_0^n(g\psi)(\omega) = \sum_{\tau\in E^n : A_{\tau_n\omega_1}=1} u_n(\tau\omega)g(\tau\omega)$.

(b) $\boldsymbol{T}^n(1\!\!1) = 1\!\!1$ *and* $||\boldsymbol{T}^n||_0 = 1$.

(c) $M = \sup_{n\geq 1}\{||\boldsymbol{T}^n||_\beta\} < \infty$.

(d) $\boldsymbol{T}^*(\tilde{\mu}_f) = \tilde{\mu}_f$. *In particular the closed subspaces* $\mathcal{H}_0^0 = \{g \in \mathcal{H}_0 : \tilde{\mu}_f(g) = 0\}$ *and* $\mathcal{H}_\beta^0 = \{g \in \mathcal{H}_\beta : \tilde{\mu}_f(g) = 0\}$ *are* $\boldsymbol{T}$*-invariant.*

(e) $\mathcal{H}_\beta = I\!R1\!\!1 \oplus \mathcal{H}_\beta^0$ $(g = \tilde{\mu}_f(g)1\!\!1 + (g - \tilde{\mu}_f(g)1\!\!1))$.

Denote

$$\mathcal{H}_\beta^{0,1} = \{g \in \mathcal{H}_\beta^0 : ||g||_\beta \leq 1\}.$$

We shall prove the following.

Lemma 2.4.5 *For each $n \geq 0$ let*

$$b_n = \sup\{||\boldsymbol{T}^n(g)||_\beta : g \in \mathcal{H}_\beta^{0,1}\}.$$

Then $\lim_{n\to\infty} b_n = 0$.

Proof. Define for every $n \geq 0$

$$a_n = \sup\{||\boldsymbol{T}^n(g)||_0 : g \in \mathcal{H}_\beta^{0,1}\}.$$

It immediately follows from Theorem 2.4.4(b) that the sequence $\{a_n\}_{n\geq 1}$ is nonincreasing. We shall show first that

$$\lim_{n\to\infty} a_n = 0.$$

Let Λ be a finite set of words of length q witnessing the primitivity of the incidence matrix A. For every $\alpha \in \Lambda$ fix a word $\overline{\alpha} \in E^\infty$ such that $\overline{\alpha}|_q = \alpha$. Suppose now on the contrary that $a = \lim_{n\to\infty} a_n > 0$. By Theorem 2.4.4(c), $\sup_{n\geq 0} b_n \leq M < \infty$. Therefore, there exists $k \geq 0$ such that if $\omega|_k = \tau|_k$, then

$$|\boldsymbol{T}^n(g)(\tau) - \boldsymbol{T}^n(g)(\omega)| \leq a/2 \tag{2.28}$$

for all $g \in \mathcal{H}_\beta^{0,1}$ and all $n \geq 0$. Let $\overline{a} = \min\{a, 1\}$. Since the measure $\tilde{\mu}_f$ is regular, there exists a compact set $Y \subset E^\infty$ such that $\tilde{\mu}_f(Y) \geq 1 - \frac{\overline{a}}{8}$. Fix now $g \in \mathcal{H}_\beta^{0,1}$ and $n \geq 0$. Suppose that $\boldsymbol{T}^n(g)(\tau) > a/4$ for all $\tau \in Y$. Since $\int \boldsymbol{T}^n g\, d\tilde{\mu}_f = 0$ and since $||\boldsymbol{T}^n(g)||_0 \leq ||g||_0 \leq 1$, we find

$$\begin{aligned} 0 &= \int_Y \boldsymbol{T}^n g\, d\tilde{\mu}_f + \int_{Y^c} \boldsymbol{T}^n g\, d\tilde{\mu}_f \geq \frac{a}{4}\left(1 - \frac{\overline{a}}{8}\right) - \frac{\overline{a}}{8} \geq \frac{\overline{a}}{4}\left(1 - \frac{\overline{a}}{4}\right) - \frac{\overline{a}}{8} \\ &= \frac{\overline{a}}{8} - \frac{\overline{a}^2}{16} > 0. \end{aligned}$$

This contradiction shows that $\boldsymbol{T}^n(g)(\tau) \leq a/4$ for some $\tau \in Y$ (depending on g and n). Fix this τ. Notice now that for every $\omega \in E^\infty$ there exists $\alpha(\omega) \in \Lambda$ such that $\rho(\omega) = \tau|_k \alpha(\omega)\omega$ is in E^∞. It follows from (2.28) that

$$\boldsymbol{T}^n(g)(\rho(\omega)) \leq \boldsymbol{T}^n(g)(\tau) + \frac{a}{2} \leq \frac{3}{4}a \leq a_n - \frac{a}{4}. \tag{2.29}$$

Put $p = q + k$. Since $\rho(\omega)|_k = \tau|_k$, applying Lemma 2.3.1 we get

$$\begin{aligned} \frac{u_p(\rho(\omega))}{u_p(\tau)} &= \exp\bigl(S_{q+k}f(\rho(\omega)) - S_{q+k}f(\tau)\bigr)\frac{\psi(\rho(\omega))\psi(\sigma^p(\tau))}{\psi(\tau)\psi(\sigma^p(\rho(\omega)))} \\ &\geq \frac{R^2}{Q^2}\exp\bigl(S_k f(\rho(\omega)) - S_k f(\tau)\bigr) \\ &\times \exp\bigl(S_q f(\rho(\sigma^k(\omega))) - S_q f(\sigma^k(\tau))\bigr) \\ &\geq R^2 Q^{-2} T(f)^{-1} \exp\bigl(S_q f(\alpha(\omega)\omega)\bigr)\exp(-q\sup(f)) \\ &\geq R^2 Q^{-2} T(f)^{-2} \exp\bigl(S_q f(\overline{\alpha(\omega)})\bigr)\exp(-q\sup(f)) \\ &\geq R^2 Q^{-2} T(f)^{-2} \exp\bigl(\max\{S_q f(\overline{\alpha}) : \alpha \in \Lambda\}\bigr)\exp(-q\sup(f)), \end{aligned}$$

where $\sup(f) < +\infty$ due to (2.16). Denoting the last (constant) expression appearing in this formula by U, we get $u_p(\rho(\omega)) \geq U u_p(\tau) \geq U \inf(u_p|_Y) > 0$ since u_p is positive continuous and Y is compact. Thus, using (2.29), we get

$$
\begin{aligned}
\boldsymbol{T}^p(\boldsymbol{T}^n g)(\omega) &= \boldsymbol{T}^n g(\rho) u_p(\rho(\omega)) + \sum_{\eta\in\sigma^{-p}(\omega)\setminus\{\rho(\omega)\}} \boldsymbol{T}^n g(\eta) u_p(\eta) \\
&\le \left(a_n - \frac{a}{8}\right) u_p(\rho(\omega)) + a_n \sum_{\eta\in\sigma^{-p}(\omega)\setminus\{\rho(\omega)\}} u_p(\eta) \\
&= a_n - \frac{a}{8} u_p(\rho(\omega)) \le a_n - \frac{a}{8} U \inf(u_p|_Y).
\end{aligned}
$$

Similarly we get $\boldsymbol{T}^p(\boldsymbol{T}^n g)(\omega) \ge -a_n + \frac{a}{4} U \inf(u_p|_Y)$ and in consequence, $||\boldsymbol{T}^{p+n} g||_0 \le a_n - \frac{a}{2} U \inf(u_p|_Y)$ or

$$
||\boldsymbol{T}^n g||_0 \le a_{n-p} - \frac{a}{2} U \inf(u_p|_Y)
$$

for every $n \ge p$. Taking the supremum over all $g \in \mathcal{H}_\beta^{0,1}$, we find $a_n \le a_{n-p} - \frac{a}{4} U \inf(u_p|_Y)$. So, $a = \lim_{n\to\infty} a_n \le \lim_{n\to\infty}\big(a_{n-p} - \frac{a}{4} U \inf(u_p|_Y)\big) = a - \frac{a}{4} U \inf(u_p|_Y) < a$. This contradiction shows that $\lim_{n\to\infty} a_n = 0$.

Fix now $\epsilon > 0$ and then an integer $v \ge 1$ so large that $a_v \le \frac{\epsilon}{2C}$ and $Q\mathrm{e}^{-\beta n} M \le \epsilon/2$ for all $n \ge v$. Then, in view of Lemma 2.4.1 for every $n \ge 2v$ and every $g \in \mathcal{H}_\beta^{0,1}$ we get

$$
\begin{aligned}
||\boldsymbol{T}^n g||_\beta \le ||\boldsymbol{T}^{n-v}(\boldsymbol{T}^v g)||_\beta &\le Q\mathrm{e}^{-\beta(n-v)} ||\boldsymbol{T}^v g||_\beta + C||\boldsymbol{T}^v g||_0 \\
&\le \frac{\epsilon}{2M} + C a_v \le \epsilon.
\end{aligned}
$$

So, $b_n \le \epsilon$. □

Theorem 2.4.6 *Suppose that $f : E^\infty \to I\!R$ is a summable Hölder continuous function, say with an exponent $\beta > 0$. Suppose also that the normalized conjugate Perron–Frobenius operator $\mathcal{L}_0^*$ has a Borel probability fixed point $\tilde{m}_f$. Assume that the incidence matrix A is finitely primitive. Then there exist constants $\overline{M} > 0$ and $0 < \gamma < 1$ such that for every $g \in \mathcal{H}_\beta$ and every $n \ge 0$*

(a)

$$
\|\boldsymbol{T}^n(g) - \int g \, d\tilde{\mu}_f\|_\beta \le \overline{M}\gamma^n \|g\|_\beta
$$

and

(b)

$$
\|\mathcal{L}_0^n(g) - (\int g \, d\tilde{m}_f)\psi\|_\beta \le \overline{M}\gamma^n \|g\|_\beta
$$

where $\boldsymbol{T}$ is the operator defined by (2.27), ψ is the fixed point for $\mathcal{L}_0$ and $\tilde{\mu}_f$ is the unique invariant Gibbs state of the potential f whose existence and uniqueness follow from Theorem 2.2.4 and Theorem 2.3.3.

Proof. Lemma 2.4.5 says that $\lim_{n\to\infty} \|\boldsymbol{T}|^n_{\mathcal{H}^0_\beta}\|_\beta = 0$. Thus, there exists $q \geq 1$ such that $\|\boldsymbol{T}|^q_{\mathcal{H}^0_\beta}\|_\beta \leq (1/2)$. By induction we have $\|\boldsymbol{T}|^{qn}_{\mathcal{H}^0_\beta}\|_\beta \leq (1/2)^n$. Consider now an arbitrary $n \geq 0$ and write $n = pq + r$, $0 \leq r \leq q-1$. Then, using in addition Theorem 2.4.4(c), we get for every $\zeta \in \mathcal{H}^0_\beta$

$$\begin{aligned}\|\boldsymbol{T}^n\zeta\|_\beta = \|\boldsymbol{T}^{pq}(\boldsymbol{T}^r\zeta)\|_\beta &\leq (1/2)^p\|\boldsymbol{T}^r\zeta\|_\beta \leq M(1/2)^p = M(1/2)^{\frac{n-r}{q}} \\ &\leq M(1/2)^{\frac{n-q+1}{q}} = M(1/2)^{\frac{1-q}{q}}(1/2)^{\frac{n}{q}}\end{aligned}$$

and therefore for every $n \geq 0$, $\|\boldsymbol{T}|^n_{\mathcal{H}^0_\beta}\|_\beta \leq M(1/2)^{\frac{1-q}{q}}\gamma^n$, where $\gamma = (1/2)^{1/q} < 1$. If now $g \in \mathcal{H}_\beta$, then $g - \tilde{\mu}_f(g) \in \mathcal{H}^0_\beta$ and $\|g - \tilde{\mu}_f(g)\|_\beta \leq \|g\|_\beta + \|\tilde{\mu}_f(g)\|_\beta \leq 2\|g\|_\beta$. Thus, for every $n \geq 0$

$$\|\boldsymbol{T}^n(g - \tilde{\mu}_f(g))\|_\beta \leq 2^{2q-1/q}M\gamma^n\|g\|_\beta$$

and the proof of Theorem 2.4.6(a) is complete. Part (b) is an immediate consequence of part (a). □

The next proposition, the last result of this section, explains the real dynamical meaning of the fixed points of the normalized Perron–Frobenius operator $\mathcal{L}_0$.

Proposition 2.4.7 *Assume that the operator conjugate to the normalized Perron–Frobenius operator $\mathcal{L}_0 = e^{-P(f)}\mathcal{L}_f$ has a Borel probability fixed point $\tilde{m}$. Let*

$$Fix(\mathcal{L}_0) = \{g \in L_1(\tilde{m}) : \mathcal{L}_0(g) = g, \int g d\tilde{m} = 1, \text{ and } g \geq 0\}$$

and let

$$AI(\tilde{m}) = \{g \in L_1(\tilde{m}) : g\tilde{m} \circ \sigma^{-1} = g\tilde{m}, \int g d\tilde{m} = 1, \text{ and } g \geq 0\}$$

denote the invariant absolutely continuous probability density functions. Then $Fix(\mathcal{L}_0) = AI(\tilde{m})$.

Proof. It follows from (2.21) that for every $i \in I$ and every $\omega \in E^\infty$ with $A_{i\omega_1} = 1$, we have

$$\frac{d\tilde{m} \circ i}{d\tilde{m}}(\omega) = \exp\big(f(i\omega) - P(f)\big),$$

where we treat $i : \{\omega \in E^\infty : A_{i\omega_1} = 1\} \to E^\infty$ as the map defined by the formula $i(\omega) = i\omega$. Therefore, the Perron–Frobenius operator $\mathcal{L}_0$ sends the density of a measure $\tilde{\mu}$ absolutely continuous with respect to $\tilde{m}$ to the density of the measure $\tilde{\mu} \circ \sigma^{-1}$. The proposition follows. □

2.5 Stochastic laws

In this section, in a fashion somewhat similar to section 3 of [DU1], we reap some of the fruits coming from the work done in the two preceding sections. Let Γ be a finite or countable measurable partition of a probability space $(Y, \mathcal{F}, \nu)$ and let $S : Y \to Y$ be a measure preserving transformation. For $0 \le a \le b \le \infty$, set $\Gamma_a^b = \bigvee_{a \le l \le b} S^{-l}\Gamma$. The measure ν is said to be absolutely regular with respect to the filtration defined by Γ, if there exists a sequence $\beta(n) \searrow 0$ such that

$$\int_Y \sup_a \sup_{A \in \Gamma_{a+n}^\infty} |\nu(A|\Gamma_0^a) - \nu(A)| d\nu \le \beta(n).$$

The numbers $\beta(n)$, $n \ge 1$, are called coefficients of absolute regularity. Let α be the partition of E^∞ into initial cylinders of length 1. Using Theorem 2.4.6, and proceeding exactly as in the proof of [Ry, §3 of Theorem 2.5] we derive the following (with the notation of previous sections).

Theorem 2.5.1 *The measure $\tilde{\mu}_f$ is absolutely regular with respect to the filtration defined by the partition α. The coefficients of absolute regularity decrease to 0 at an exponential rate.*

Theorem 2.5.1 says in particular that the dynamical system $(\sigma, \tilde{\mu}_f)$ is *weak-Bernoulli* (see [Or]). As an immediate consequence of this theorem and the results proved in [Or] we get the following.

Theorem 2.5.2 *The natural extension of the dynamical system $(\sigma, \tilde{\mu}_f)$ is isomorphic with some Bernoulli shift.*

It follows from this theorem that the theory of absolutely regular processes applies ([IL], [PS]). We sketch this application briefly. We say that a measurable function $g : E^\infty \to \mathbb{R}$ belongs to the space $L^*(\sigma)$ if there exist constants $\delta, \gamma, M > 0$ such that $\int \|g\|_0^{2+\delta} \, d\tilde{\mu}_f < \infty$ and

$$\int \|g - E_{\tilde{\mu}_f}(g|(\alpha)^n)\|_0^{2+\delta} \, d\tilde{\mu}_f \le M n^{-2-\gamma}$$

for all $n \geq 1$, where $E_{\tilde{\mu}_f}(g|(\alpha)^n)$ denotes the conditional expectation of g with respect to the partition $(\alpha)^n$ and the measure $\tilde{\mu}_f$. $L^*(\sigma)$ is a linear space. It follows from Theorem 2.5.1, [IL] and [PS] that with $\tilde{\mu}_f(g) = \int g\, d\tilde{\mu}_f$, the series

$$\sigma^2 = \sigma^2(g) = \int_{E^\infty} (g - \tilde{\mu}_f(g))^2\, d\tilde{\mu}_f$$
$$+ 2\sum_{n=1}^{\infty} \int_{E^\infty} (g - \tilde{\mu}_f(g))(g \circ \sigma^n - \tilde{\mu}_f(g))\, d\tilde{\mu}_f$$

is absolutely convergent and non-negative. The reader should not be confused by two different meanings of the symbol σ: the number defined above and the shift map. Then the process $(g \circ \sigma^n : n \geq 1)$ exhibits an *exponential decay of correlations*, and if $\sigma^2 > 0$ it satisfies the *central limit theorem.*

Theorem 2.5.3 *If $u, v \in L^*(\sigma)$ then there are constants $C, \theta > 0$ such that for every $n \geq 1$ we have*

$$\int (u - Eu)\big((v - Ev) \circ \sigma^n\big)\, d\mu_f \leq Ce^{-\theta n},$$

where $Eu = \int u\, d\tilde{\mu}_f$ and $Ev = \int v\, d\tilde{\mu}_f$.

Theorem 2.5.4 *If $g \in L^*(\sigma)$ and $\sigma^2(g) > 0$, then for all r*

$$\tilde{\mu}_f\left(\left\{\omega \in E^\infty : \frac{\sum_{j=0}^{n-1} g \circ \sigma^j(\omega) - nEg}{\sqrt{n}} < r\right\}\right)$$
$$\to \frac{1}{\sigma(g)\sqrt{2\pi}} \int_{-\infty}^{r} e^{-t^2/2\sigma(g)^2}\, dt.$$

A most fruitful geometric application is the *almost sure invariance principle* and therefore we devote more time to it. This principle means that one can redefine the process $(g \circ \sigma^n : n \geq 1)$ on some probability space on which there is defined a standard Brownian motion $(B(t) : t \geq 0)$ such that for some $\lambda > 0$ we have $\tilde{\mu}_f$-a. e.

$$\sum_{0 \leq j \leq t} [g \circ \sigma^j - \tilde{\mu}_f(g)] - B(\sigma^2 t) = O(t^{\frac{1}{2} - \lambda}).$$

Let $h : [1, \infty) \longrightarrow \mathbb{R}$ be a positive non-decreasing function. The function

h is said to belong to the *lower class* if

$$\left|\int_1^\infty \frac{h(t)}{t}\exp(-\frac{1}{2}h(t)^2)dt\right| < \infty$$

and to the *upper class* if

$$\int_1^\infty \frac{h(t)}{t}\exp(-\frac{1}{2}h(t)^2)dt = \infty.$$

Well-known results for Brownian motion imply (see Theorem A in [PS]) the following generalization of the *law of the interated logarithm.*

Theorem 2.5.5 *If $g \in L^*(\sigma)$ and $\sigma^2(g) > 0$ then*

$$\tilde{\mu}_f\Bigg(\{\omega \in E^\infty : \sum_{j=0}^{n-1}\big(g(\sigma^j(\omega)) - \tilde{\mu}_f(g)\big)$$
$$> \sigma(g)h(n)\sqrt{n} \text{ for infinitely many } n \geq 1\}\Bigg)$$
$$= \begin{cases} 0 & \text{if } h \text{ belongs to the lower class} \\ 1 & \text{if } h \text{ belongs to the upper class.} \end{cases}$$

Our last goal in this section is to provide a sufficient condition for a function ψ to belong to the space $L^*(\sigma)$.

Lemma 2.5.6 *Each Hölder continuous function which has some finite moment greater than 2 belongs to $L^*(\sigma)$.*

Proof. It suffices to show that any Hölder continuous function $\psi : E^\infty \to I\!R$ satisfies the requirement $\int \|\psi - E_{\tilde{\mu}_f}(\psi|(\alpha)^n)\|_0^3\, d\tilde{\mu}_f \leq Mn^{-2-\gamma}$. So, given $n \geq 1$ suppose that $\omega, \tau \in A$ for some $A \in \alpha^n$. In particular $\omega|_n = \tau|_n$. Hence $|\psi(\omega) - \psi(\tau)| \leq V_\beta(\psi)\mathrm{e}^{-\beta n}$ which means that $\psi(\tau) - V_\beta(\psi)\mathrm{e}^{-\beta n} \leq \psi(\omega) \leq \psi(\tau) + V_\beta(\psi)\mathrm{e}^{-\beta n}$. Integrating these inequalities against the measure $\tilde{\mu}_f$ and keeping ω fixed, we obtain

$$\int_A \psi d\tilde{\mu}_f - V_\beta(\psi)\mathrm{e}^{-\beta n}\tilde{\mu}_f(A) \leq \psi(\omega)\tilde{\mu}_f(A) \leq \int_A \psi d\tilde{\mu}_f + V_\beta(\psi)\mathrm{e}^{-\beta n}\tilde{\mu}_f(A).$$

Dividing these inequalities by $\tilde{\mu}_f(A)$ we deduce that

$$\left|\psi(\omega) - \frac{1}{\tilde{\mu}_f(A)}\int_A \psi d\tilde{\mu}_f\right| \leq V_\beta(\psi)\mathrm{e}^{-\beta n}.$$

Thus, $\int \|\psi(\omega) - E_{\tilde{\mu}_f}(\psi|(\alpha)^n)\|_0^3\, d\tilde{\mu}_f \leq V_\beta(\psi)^3\mathrm{e}^{-3\beta n}$ and we are done. □

2.6 Analytic properties of pressure and the Perron–Frobenius operator

In this section $\mathcal{K}_\beta$ is the space of all complex-valued Hölder continuous functions of order β on E^∞ and

$$\mathcal{K}_\beta^s = \{f \in \mathcal{K}_\beta : \sum_{i\in I} \exp\big(\sup(\mathrm{Re}(f|_{[i]}))\big) < \infty\}.$$

$L(\mathcal{K}_\beta)$ denotes the space of all bounded (continuous) operators on $\mathcal{K}_\beta$. We give a complete detailed proof of the real analyticity of the Perron–Frobenius operator, cf. [EM], [HMU] and [UZi]. We start with the following.

Lemma 2.6.1 *If for every $\omega \in E^\infty$, the function $t \mapsto f_t(\omega) \in \mathbb{C}$ is holomorphic on a domain $G \subset \mathbb{C}$ and the map $t \mapsto \mathcal{L}_{f_t} \in L(\mathcal{K}_\beta)$ is continuous on G, then the map $t \mapsto \mathcal{L}_{f_t} \in L(\mathcal{K}_\beta)$ is holomorphic on G.*

Proof. Let $\gamma \subset G$ be a simple closed, contractible in G, rectifiable curve. Fix $g \in \mathcal{K}_\beta$ and $\omega \in E^\infty$. Let $W \subset G$ be a bounded open set such that $\gamma \subset W \subset \overline{W} \subset G$. Since for each $e \in I$ such that $A_{e\omega_1} = 1$, the function $t \mapsto g(e\omega)\exp\big(f_t(\omega)\big) \in \mathbb{C}$, $t \in G$, is holomorphic and since for every $t \in W$

$$\begin{aligned}\|\sum_{e:A_{e\omega_1}=1} g(e\omega)\exp\big(f_t(\omega)\big)\|_\infty &\le \|\sum_{e:A_{e\omega_1}=1} g(e\omega)\exp\big(f_t(\omega)\big)\|_\beta \\ &\le \|g\|_\beta \sup\{\|\mathcal{L}_{f_z}\|_\beta : z \in \overline{W}\} < \infty\end{aligned}$$

by compactness of $\overline{W}$ and continuity of $t \mapsto \mathcal{L}_{f_t}$, we conclude that the function

$$t \mapsto \mathcal{L}_{f_t}(g)(\omega) = \sum_{e:A_{e\omega_1}=1} g(e\omega)\exp\big(f_t(\omega)\big) \in \mathbb{C},\ t \in W,$$

is holomorphic. Hence by Cauchy's theorem $\int_\gamma \mathcal{L}_{f_t} g(\omega)dt = 0$. Since the function $t \mapsto \mathcal{L}_{f_t} g \in \mathcal{K}_\beta$ is continuous, the integral $\int_\gamma \mathcal{L}_{f_t} g dt$ exists, and for every $\omega \in E^\infty$, we have $\int_\gamma \mathcal{L}_{f_t} g dt(\omega) = \int_\gamma \mathcal{L}_{f_t} g(\omega) dt = 0$. Hence $\int_\gamma \mathcal{L}_{f_t} g dt = 0$. Now, since $t \mapsto \mathcal{L}_{f_t} \in L(\mathcal{K}_\beta)$ is continuous, the integral $\int_\gamma \mathcal{L}_{f_t} dt$ exists, and for every $g \in \mathcal{K}_\beta$, $\int_\gamma \mathcal{L}_{f_t} dt(g) = \int_\gamma \mathcal{L}_{f_t} g dt = 0$. Thus $\int_\gamma \mathcal{L}_{f_t} dt = 0$ and in view of Morera's theorem the map $t \mapsto \mathcal{L}_{f_t} \in L(\mathcal{K}_\beta)$ is holomorphic on G. □

In order to prove the main result of this section we need several elementary lemmas. In order to formulate them we need to define a certain class of mappings. Namely, given $i \in I$, we define the mapping

$$i : \{\omega \in E^\infty : A_{i\omega_1} = 1\} \to E^\infty$$

by setting

$$i(\omega) = i\omega.$$

If $g : E^\infty \to \mathbb{C}$, then by $g \circ i : E^\infty \to \mathbb{C}$ we mean the function defined by the following formula.

$$g \circ i(\omega) = \begin{cases} g(i\omega) & \text{if } A_{i\omega_1} = 1 \\ 0 & \text{if } A_{i\omega_1} = 0. \end{cases}$$

Lemma 2.6.2 *If $i \in I$ and $\rho \in \mathcal{K}_\beta$ then the operator $A_{i,\rho}$ given by the formula*

$$A_{i,\rho}(g)(\omega) = \rho \circ i(\omega) \cdot g \circ i(\omega)$$

acts on the space $\mathcal{K}_\beta$, is continuous and $\|A_{i,\rho}\|_\beta \le 3\|\rho \circ i\|_\beta$.

Proof. Fix $g \in \mathcal{K}_\beta$, $\omega \in E^\infty$ and suppose that $A_{i\omega_1} = 1$. Then

$$|A_{i,\rho}(g)(\omega)| = |\rho(i\omega)||g(i\omega)| \le \|\rho \circ i\|_0 \|g\|_0 \le \|\rho \circ i\|_\beta \|g\|_\beta. \tag{2.30}$$

Fix now in addition $\tau \in E^\infty \setminus \{\omega\}$ such that $|\omega \wedge \tau| \ge 1$. Then

$$\begin{aligned} |A_{i,\rho}(g)(\omega) - A_{i,\rho}(g)(\tau)| &= |\rho(i\omega)g(i\omega) - \rho(i\tau)g(i\tau)| \\ &= |\rho(i\omega)(g(i\omega) - g(i\tau)) + g(i\tau)(\rho(i\omega) - \rho(i\tau))| \\ &\le \|\rho \circ i\|_0 |g(i\omega) - g(i\tau)| + \|g\|_0 |\rho(i\omega) - \rho(i\tau)| \\ &\le \|\rho \circ i\|_\beta \|g\|_\beta e^{-\beta|\omega\wedge\tau|} + \|g\|_\beta \|\rho \circ i\|_\beta e^{-\beta|\omega\wedge\tau|}. \end{aligned}$$

Hence $V_\beta(A_{i,\rho}(g)) \le 2\|\rho \circ i\|_\beta \|g\|_\beta$ and, combining this with (2.30), we conclude that $\|A_{i,\rho}(g)\| \le 3\|\rho \circ i\|_\beta \|g\|_\beta$. consequently $A_{i,\rho}$ acts on the space $\mathcal{K}_\beta$, is continuous and $\|A_{i,\rho}\|_\beta \le 3\|\rho \circ i\|_\beta$. □

Similarly (but more easily) one proves the following.

Lemma 2.6.3 *If $\rho, g \in \mathcal{K}_\beta$, then $\rho g \in \mathcal{K}_\beta$ and $\|\rho g\|_\beta \le 3\|\rho\|_\beta \|g\|_\beta$.*

Lemma 2.6.4 *If $f \in \mathcal{K}_\beta$ then $e^f \in \mathcal{K}_\beta$, and if $\rho : Y \to \mathcal{K}_\beta$ is a continuous mapping defined on a compact set Y then $e^\rho : Y \to \mathcal{K}_\beta$ is also continuous.*

Proof. By Lemma 2.6.3, $\|f^n\|_\beta \le 3^n \|f\|_\beta^n$. Hence, the series $e^f = \sum_{n=0}^\infty \frac{f^n}{n!}$ converges in $\mathcal{K}_\beta$ and the first part of our lemma is proved.

The second part follows now from the remark that for every $y \in Y$, $\|\rho(y)\|_\beta \leq \sup\{\|\rho(x)\|_\beta : x \in Y\} < \infty$ and the series $\sum_{n=0}^{\infty} \frac{\rho(y)^n}{n!}$ converges uniformly on Y. □

Lemma 2.6.5 *For every $R > 0$ there exists $M = M_R \geq 1$ such that if $|z - \xi| \leq R$, then $|\mathrm{e}^\xi - \mathrm{e}^z| \leq M\mathrm{e}^{\mathrm{Re} z}|z - \xi|$.*

Proof. Looking at the Taylor's series expansion of the exponential function about 0, we see that there exists a constant $M \geq 1$ such $|\mathrm{e}^w - 1| \leq M|w|$, if $|w| \leq R$. Hence $|\mathrm{e}^\xi - \mathrm{e}^z| = |\mathrm{e}^z||\mathrm{e}^{z-\xi} - 1| \leq \mathrm{e}^{\mathrm{Re} z}$ $M|z - \xi|$. □

Lemma 2.6.6 *If $f \in \mathcal{K}_\beta$, then for every $i \in I$*

$$\|e^{f\circ i}\|_\beta \leq 2M_{\|f\|_\beta} \exp\big(\sup \mathrm{Re}(f|_{[i]})\big)\|f\|_\beta.$$

Proof. Fix $\omega \in E^\infty$ such that $A_{i\omega_1} = 1$. Then $|\mathrm{e}^{f(i\omega)}| = \mathrm{e}^{\mathrm{Re} f(i\omega)} \leq \exp\big(\sup \mathrm{Re}(f|_{[i]})\big)$, whence

$$\|\mathrm{e}^{f\circ i}\|_\beta \leq \exp\big(\sup \mathrm{Re}(f|_{[i]})\big). \tag{2.31}$$

Fix now in addition $\tau \in E^\infty \setminus \{\omega\}$ with $|\tau \wedge \omega| \geq 1$. Using the Lemma 2.6.5, we get

$$|\mathrm{e}^{f(i\omega)} - \mathrm{e}^{f(i\tau)}| \leq M_{\|f\|_\beta}\mathrm{e}^{\mathrm{Re} f(\tau)}|f(i\omega) - f(i\tau)| \leq M_{\|f\|_\beta} \times \exp\big(\sup \mathrm{Re}(f|_{[i]})\big)\|f\|_\beta \mathrm{e}^{-\beta|\tau\wedge\omega|}.$$

Thus, $V_\beta(\mathrm{e}^{f\circ i}) \leq \exp\big(\sup \mathrm{Re}(f|_{[i]})\big)\|f\|_\beta$. Combining this and (2.31) completes the proof. □

Lemma 2.6.7 *If $\rho : Y \to \mathcal{K}_\beta$ is a continuous mapping defined on a metric space Y, then for every $i \in I$, the function $y \mapsto A_{i,\rho(y)} \in L(\mathcal{K}_\beta)$, $y \in Y$, is continuous.*

Proof. Fix $y_0 \in Y$ and take $\delta > 0$ so small that for every $y \in B(y_0, \delta)$, $\|\rho(y) - \rho(y_0)\|_\beta \leq \epsilon/3$. Then for $y \in B(y_0, \delta)$, we have in view of Lemma 2.6.2 the following.

$$\|A_{i,\rho(y)} - A_{i,\rho(y_0)}\|_\beta = \|A_{i,\rho(y)-\rho(y_0)}\|_\beta \leq 3\|\rho(y) - \rho(y_0)\|_\beta \leq 3.$$

□

Our main theorem in this section is the following.

Theorem 2.6.8 *If G is an open connected subset of $\mathbb{C}$, the function $t \mapsto f_t \in \mathcal{K}_\beta^s$, $t \in G$, is continuous and the function $t \mapsto f_t(\omega) \in \mathbb{C}$, $t \in G$,* *is holomorphic for every* $\omega \in E^\infty$, *then the function* $t \mapsto \mathcal{L}_{f_t} \in L(\mathcal{K}_\beta)$, $t \in G$, *is holomorphic.*

Proof. In view of Lemma 2.6.2 it suffices to demonstrate that the function $t \mapsto \mathcal{L}_{f_t} \in L(\mathcal{K}_\beta)$, $t \in G$, is continuous. So, fix $t_0 \in G$ and $\delta > 0$ so small that $B(t_0, 2\delta) \subset G$ and $\|f_t - f_{t_0}\|_\infty \le \|f_t - f_{t_0}\|_\beta \le 1$ for all $t \in B(t_0, 2\delta)$. By Lemmas 2.6.7 and 2.6.4, for all $i \in I$, the function $t \mapsto A_{i,e^{f_t}} \in L(\mathcal{K}_\beta)$, $t \in \overline{B(t_0,\delta)}$, is continuous. Since

$$\mathcal{L}_{f_t} = \sum_{i \in I} A_{i,e^{f_t}},$$

it therefore suffices to demonstrate that the series $\sum_{i\in I} A_{i,e^{f_t}}$ converges uniformly on $\overline{B(t_0,\delta)}$. And indeed, in view of Lemma 2.6.2 and Lemma 2.6.6, for every $i \in I$ and every $t \in \overline{B(t_0,\delta)}$ we have

$$\|A_{i,e^{f_t}}\|_\beta \le 3\|\exp(f_t \circ i)\|_\beta \le 6M \exp\big(\sup \mathrm{Re}\big(f_t|_{[i]}\big)\big) M_1,$$

where $M_1 = \sup\{\|f_t\|_\beta : t \in \overline{B(t_0,\delta)}\}$ is finite by continuity of the function $t \mapsto f_t \in \mathcal{K}_\beta^s$ on the compact set $\overline{B(t_0,\delta)}$, and $M = M_{M_1}$ in the sense of Lemma 2.6.5. Now, in view of our choice of δ, we can continue the above estimates as follows.

$$\begin{aligned}
\|A_{i,e^{f_t}}\|_\beta &\le 6MM_1 \exp\big(\sup \mathrm{Re}\big(f_{t_0}|_{[i]}\big) + \|f_t - f_{t_0}\|_0\big) \\
&\le 6MM_1 \exp\big(\sup \mathrm{Re}\big(f_{t_0}|_{[i]}\big) + 1\big) \\
&\le 6MM_1 \exp\big(\sup \mathrm{Re}\big(f_{t_0}|_{[i]}\big)\big).
\end{aligned}$$

Since by summability of the function f_{t_0}, the series $\sum_{i\in I} \exp\big(\sup \mathrm{Re}\big(f_{t_0}|_{[i]}\big)\big)$ converges. □

As an immediate consequence of Theorem 2.6.8, we get the following.

Corollary 2.6.9 *If* G *is an open connected subset of* $\mathbb{C}$, *and the function* $t \mapsto f_t \in \mathcal{K}_\beta^s$, $t \in G$, *is holomorphic, then the function* $t \mapsto \mathcal{L}_{f_t} \in L(\mathcal{K}_\beta)$, $t \in G$, *is also holomorphic.*

In the sequel we will need the following much more special result following immediately from Corollary 2.6.9.

Corollary 2.6.10 *If* G *is an open connected subset of* $\mathbb{C}$, *and for each* $t \in G$, *the potential* $f_t = f + t\psi \in \mathcal{K}_\beta^s$ *for some* $f, \psi \in \mathcal{K}_\beta$, *then the function* $t \mapsto \mathcal{L}_{f_t} \in L(\mathcal{K}_\beta)$, $t \in G$, *is holomorphic.*

Remark 2.6.11 *Since each real analytic function $t \mapsto f_t \in \mathcal{K}_\beta^s$ defined on an open interval $\Delta \subset I\!R$ extends uniquely to a holomorphic function defined on an open connected neighborhood of Δ in $\mathbb{C}$ and since $\mathcal{K}_\beta^s$ is an open subset of $\mathcal{K}_\beta$, Corollary 2.6.9 and 2.6.10 remain true if G is replaced by Δ and holomorphicity by real analyticity.*

As an immediate consequence of Remark 2.6.11, Corollary 2.6.9, Theorem 2.3.3 and Theorem 2.4.6 (implying that $e^{P(f)}$ is an isolated simple eigenvalue of the Perron–Frobenius operator $\mathcal{L}_f$) and the perturbation theory of analytic dependence of an isolated simple eigenvalue (see [Ka]), we get the following remarkable result.

Theorem 2.6.12 *If $\Delta \subset I\!R$ is an open interval and $t \mapsto f_t \in \mathcal{K}_\beta^s$, $t \in \Delta$, is a real-analytic family of real-valued functions, then the function $t \mapsto P(f_t) \in I\!R$, $t \in \Delta$, is also real analytic.*

Proposition 2.6.13 *If $q_0 \in I\!R$ and $f, \psi \in \mathcal{K}_\beta$ are such that $q_0 f + \psi \in \mathcal{K}_\beta^s$ and $\int -(qf + \psi) d\tilde{\mu}_{q_0} < \infty$ for all q in an open neighborhood of q_0, then*

$$\frac{dP}{dq}(q_0) = \int f d\tilde{\mu},$$

where $\tilde{\mu} = \tilde{\mu}_{q_0 f + \psi}$.

Proof. Since $\mathcal{K}_\beta^s$ is an open subset of $\mathcal{K}_\beta$ by Theorem 2.6.12 we know that the derivative $\frac{dP}{dq}(q_0)$ exists. Since the function $q \mapsto P(q)$ is convex, in order to complete the proof it is therefore enough to demonstrate that

$$P(q) \geq P(q_0) + \int f d\tilde{\mu}(q - q_0)$$

on an open neighborhood of q_0. And indeed, in view of our assumptions, Theorem 2.1.8 and Theorem 2.2.9, there exists an open neighborhood U of q_0 such that for every $q \in U$ we have

$$\begin{aligned} P(q) \geq h_{\tilde{\mu}} + \int (qf + \psi) d\tilde{\mu} &= h_{\tilde{\mu}} + \int q_0 f d\tilde{\mu} + \int \psi d\tilde{\mu} + \int f d\tilde{\mu}(q - q_0) \\ &= P(q_0) + \int f d\tilde{\mu}(q - q_0). \end{aligned}$$

□

Fix now $f, \psi \in \mathcal{K}_\beta$ and $(q, t) \in I\!R^2$ such that the function $qf + t\psi$ is summable. Using the notation, $\tilde{\mu}_{q,t} = \tilde{\mu}_{qf+t\psi}$ and $\int g d\tilde{\mu}_{q,t} = \tilde{\mu}_{q,t}(g)$, set

$$\sigma_{q,t}^2(f,\psi) = \sum_{k=0}^{\infty} \left(\tilde{\mu}_{q,t}(f\cdot\psi\circ\sigma^k) - \tilde{\mu}_{q,t}(f)\tilde{\mu}_{q,t}(\psi)\right)$$
$$= \sum_{k=0}^{\infty} \left(\tilde{\mu}_q(\psi\cdot f\circ\sigma^k) - \tilde{\mu}_{q,t}(f)\tilde{\mu}_q(\psi)\right).$$

If $f = \psi$ we simply write $\sigma_{q,t}^2(f)$ for $\sigma_{q,t}^2(f,f)$. The last result in this section is the following, which can be proved by proceeding as in the case of a subshift of finite type over a finite alphabet (see [Ru], [PU]).

Proposition 2.6.14 *If $f,\psi \in \mathcal{K}_\beta$, $q_0 f + t_0\psi$ is summable and $\int -(qf + t\psi)d\tilde{\mu}_{q_0,t_0} < \infty$, for all pairs (q,t) in an open neighborhood of (q_0,t_0) in $\mathbb{R}^2$, then*

$$\frac{\partial^2 \mathrm{P}}{\partial q\partial t}|_{(q_0,t_0)} = \lim_{n\to\infty}\frac{1}{n}\int S_n(f - \int f d\tilde{\mu}_{q_0,t_0})S_n(\psi - \int \psi d\tilde{\mu}_{q_0,t_0})$$
$$= \sigma_{q_0,t_0}^2(f,\psi).$$

2.7 The existence of eigenmeasures of the conjugate Perron–Frobenius operator and of Gibbs states

So far the character of this chapter has been slightly conditional. To remedy this we prove here the main result concerning the existence of eigenmeasures of the conjugate Perron–Frobenius operator. Throughout the section we assume that $f : E^\infty \to \mathbb{R}$ is a summable Hölder continuous bounded function with some exponent $\beta > 0$ which allows us to define the Perron–Frobenius operator $\mathcal{L}_f$ and its conjugate $\mathcal{L}_f^*$. In order to simplify notation we will drop the subscript f. We begin with the following result, whose first proof can be found in [B1].

Lemma 2.7.1 *If the alphabet I is finite and the incidence matrix is irreducible, then there exists an eigenmeasure $\tilde{m}$ of the conjugate operator $\mathcal{L}_f^*$.*

Proof. By our assumption $\mathcal{L}_f$ is a strictly positive operator (in the sense that it maps strictly positive functions into strictly positive functions). In particular the formula

$$\nu \mapsto \frac{\mathcal{L}_f^*(\nu)}{\mathcal{L}_f^*(\nu)(\mathbb{1})}$$

defines a continuous map of the space of Borel probability measures on E^∞ into itself. Since E^∞ is a compact metric space, the Schauder–Tikhonov theorem applies, and as its consequence, we conclude that the map defined above has a fixed point, say $\tilde{m}$. Then $\mathcal{L}_f^*(\tilde{m}) = \lambda\tilde{m}$, where $\lambda = \mathcal{L}_f^*(\tilde{m})(1\!\!1)$. □

In Theorem 2.7.3, the main result of this section, we will need a simple fact about irreducible matrices. We will provide a short proof for the sake of completeness. It is more natural and convenient to formulate it in the language of directed graphs. Let us recall that a directed graph is said to be *strongly connected* if and only if its incidence matrix is irreducible. In other words, it means that any two vertices can be joined by a path of admissible edges.

Lemma 2.7.2 *If* $\Gamma = \langle E, V\rangle$ *is a strongly connected directed graph, then there exists a sequence of strongly connected subgraphs* $\langle E_n, V_n\rangle$ *of* Γ *such that all the vertices* $V_n \subset V$ *and all the edges* E_n *are finite,* $\{V_n\}_{n=1}^\infty$ *is an increasing sequence of vertices,* $\{E_n\}_{n=1}^\infty$ *is an increasing sequence of edges,* $\bigcup_{n=1}^\infty V_n = V$ *and* $\bigcup_{n=1}^\infty E_n = E$.

Proof. Indeed, let $V = \{v_n : n \geq 1\}$ be a sequence of all vertices of Γ. and let $E = \{e_n : n \geq 1\}$ be a sequence of edges of Γ. We will proceed inductively to construct the sequences $\{V_n\}_{n=1}^\infty$ and $\{E_n\}_{n=1}^\infty$. In order to construct $\langle E_1, V_1\rangle$ let α be a path joining v_1 and v_2 ($i(\alpha) = v_1$, $t(\alpha) = v_2$) and let β be a path joining v_2 and v_1 ($i(\beta) = v_2$, $t(\beta) = v_1$). These paths exist since Γ is strongly connected. We define $V_1 \subset V$ to be the set of all vertices of paths α and β and $E_1 \subset E$ to be the set of all edges from α and β enlarged by e_1 if this edge is among all the edges joining the vertices of V_1. Obviously $\langle E_1, V_1\rangle$ is strongly connected and the first step of the inductive procedure is complete. Suppose now that a strongly connected graph $\langle E_n, V_n\rangle$ has been constructed. If $v_{n+1} \in V_n$, we set $V_{n+1} = V_n$ and E_{n+1} is then defined to be the union of E_n and all the edges from $\{e_1, e_2, \ldots, e_n, e_{n+1}\}$ that are among all the edges joining the vertices of V_n. If $v_{n+1} \notin V_n$, let α_n be a path joining v_n and v_{n+1} and let β_n be a path joining v_{n+1} and v_n. We define V_{n+1} to be the union of V_n and the set of all vertices of α_n and β_n. E_{n+1} is then defined to be the union of E_n, all the edges building the paths α_n and β_n and all the edges from $\{e_1, e_2, \ldots, e_n, e_{n+1}\}$ that are among all the edges joining the vertices of V_{n+1}. Since $< E_n, V_n >$ was strongly connected, so is $\langle E_{n+1}, V_{n+1}\rangle$. The inductive procedure is complete. It immediately follows from the construction that $V_n \subset V_{n+1}$, $E_n \subset E_{n+1}$, $\bigcup_{n=1}^{\infty} V_n = V$ and $\bigcup_{n=1}^{\infty} E_n = E$. $\square$

Our main result is the following.

Theorem 2.7.3 *Suppose that $f : E^\infty \to I\!R$ is a Hölder continuous bounded function such that*

$$\sum_{e \in I} \exp(\sup(f|_{[e]})) < \infty,$$

and the incidence matrix is irreducible. Then there exists a Borel probability eigenmeasure $\tilde{m}$ of the conjugate operator $\mathcal{L}_f^*$.

Proof. Without loss of generality we may assume that $I = I\!N$. Since the incidence matrix is irreducible, it follows from Lemma 2.7.2 that we can reorder the set $I\!N$ so that there exists an increasing unbounded sequence $\{l_n\}_{n\geq 1}$ such that for every $n \geq 1$ the matrix $A|_{\{1,\ldots,l_n\}\times\{1,\ldots,l_n\}}$ is irreducible. Then, in view of Lemma 2.7.1, there exists an eigenmeasure $\tilde{m}_n$ of the operator $\mathcal{L}_n^*$, conjugate to Perron–Frobenius operator

$$\mathcal{L}_n : C(E_{l_n}^\infty) \to C(E_{l_n}^\infty)$$

associated to the function $f|_{E_{l_n}^\infty}$, where, for any $q \geq 1$,

$$\begin{aligned} E_q^\infty &= E^\infty \cap \{1, \ldots, q\}^\infty \\ &= \{(e_k)_{k\geq 1} : 1 \leq e_k \leq q \text{ and } A_{e_k e_{k+1}} = 1 \text{ for all } k \geq 1\}. \end{aligned}$$

Occasionally we will also treat $\mathcal{L}_n$ as acting on $C(E^\infty)$ and $\mathcal{L}_n^*$ as acting on $C^*(E^\infty)$. Our first aim is to show that the sequence $\{\tilde{m}_n\}_{n\geq 1}$ is tight, where $\tilde{m}_n$, $n \geq 1$, are treated here as Borel probability measures on E^∞. Let $\mathrm{P}_n = \mathrm{P}(\sigma|_{E_{l_n}^\infty}, f|_{E_{l_n}^\infty})$. Obviously $\mathrm{P}_n \geq \mathrm{P}_1$ for all $n \geq 1$. For every $k \geq 1$ let $\pi_k : E^\infty \to I\!N$ be the projection onto the k-th coordinate, i.e. $\pi_k(\{(e_u)_{u\geq 1}\}) = e_k$. By Theorem 2.3.3, $\mathrm{e}^{\mathrm{P}_n}$ is the eigenvalue of $\mathcal{L}_n^*$ corresponding to the eigenmeasure $\tilde{m}_n$. Therefore, applying (2.18), we obtain for every $n \geq 1$, every $k \geq 1$, and every $e \in I\!N$ that

$$\begin{aligned} \tilde{m}_n(\pi_k^{-1}(e)) &= \sum_{\omega \in E_{l_n}^k : \omega_k = e} \tilde{m}_n([\omega]) \leq \sum_{\omega \in E_{l_n}^k : \omega_k = e} \exp\big(\sup(S_k f|_{[\omega]}) - \mathrm{P}_n k\big) \\ &\leq \mathrm{e}^{-\mathrm{P}_n k} \sum_{\omega \in E_{l_n}^k : \omega_k = e} \exp\big(\sup(S_{k-1} f|_{[\omega]}) + \sup(f|_{[e]})\big) \\ &\leq \mathrm{e}^{-\mathrm{P}_1 k} \left(\sum_{i \in I\!N} \mathrm{e}^{\sup(f|_{[i]})} \right)^{k-1} \mathrm{e}^{\sup(f|_{[e]})}. \end{aligned}$$

Therefore,

$$\tilde{m}_n(\pi_k^{-1}([e+1,\infty))) \le \mathrm{e}^{-\mathrm{P}_1 k}\left(\sum_{i\in I\!N}\mathrm{e}^{\sup(f|_{[i]})}\right)^{k-1}\sum_{j>e}\mathrm{e}^{\sup(f|_{[j]})}.$$

Fix now $\epsilon>0$ and for every $k\ge 1$ choose an integer $n_k\ge 1$ such that

$$\mathrm{e}^{-\mathrm{P}_1 k}\left(\sum_{i\in I\!N}\mathrm{e}^{\sup(f|_{[i]})}\right)^{k-1}\sum_{j>n_k}\mathrm{e}^{\sup(f|_{[j]})}\le\frac{\epsilon}{2^k}.$$

Then, for every $n\ge 1$ and every $k\ge 1$, $\tilde{m}_n(\pi_k^{-1}([n_k+1,\infty)))\le\frac{\epsilon}{2^k}$. Hence

$$\begin{aligned}\tilde{m}_n\left(E^\infty\cap\prod_{k\ge 1}[1,n_k]\right) &\ge 1-\sum_{k\ge 1}\tilde{m}_n(\pi_k^{-1}([n_k+1,\infty)))\\ &\ge 1-\sum_{k\ge 1}\frac{\epsilon}{2^k}=1-\epsilon.\end{aligned}$$

Since $E^\infty\cap\prod_{k\ge 1}[1,n_k]$ is a compact subset of E^∞, the tightness of the sequence $\{\tilde{m}_n\}_{n\ge 1}$ is therefore proved. Thus, in view of Prohorov's theorem there exists $\tilde{m}$, a weak-limit point of the sequence $\{\tilde{m}_n\}_{n\ge 1}$. Let now $\mathcal{L}_{0,n}=\mathrm{e}^{-\mathrm{P}_n}\mathcal{L}_n$ and $\mathcal{L}_0=\mathrm{e}^{-\mathrm{P}(f)}\mathcal{L}$ be the corresponding normalized operators. Fix $g\in C_b(E^\infty)$ and $\epsilon>0$. Let us now consider an integer $n\ge 1$ so large that the following requirements are satisfied.

$$\sum_{i>n}\|g\|_0\exp\big(\sup(f|_{[i]})-\mathrm{P}(f)\big)\le\frac{\epsilon}{6}, \tag{2.32}$$

$$\sum_{i\ge 1}\|g\|_0\exp\big(\sup(f|_{[i]})\big)\mathrm{e}^{-\mathrm{P}_1}(\mathrm{P}(f)-\mathrm{P}_n)\le\frac{e}{6}, \tag{2.33}$$

$$|\tilde{m}_n(g)-\tilde{m}(g)|\le\frac{\epsilon}{3}, \tag{2.34}$$

and

$$\left|\int\mathcal{L}_0(g)d\tilde{m}-\int\mathcal{L}_0(g)d\tilde{m}_n\right|\le\frac{\epsilon}{3}. \tag{2.35}$$

It is possible to make condition (2.33) satisfied since, due to Theorem 2.1.5, $\lim_{n\to\infty}\mathrm{P}_n=\mathrm{P}(f)$. Let $g_n=g|_{E^\infty_{I_n}}$. The first two observations are the following.

$$\begin{aligned}\mathcal{L}^*_{0,n}\tilde{m}_n(g) &= \int_{E^\infty} \sum_{i\le n: A_{i\omega_n}=1} g(i\omega)\exp(f(i\omega)-\mathrm{P}_n)d\tilde{m}_n(\omega)\\ &= \int_{E^\infty_{l_n}} \sum_{i\le n: A_{i\omega_n}=1} g(i\omega)\exp(f(i\omega)-\mathrm{P}_n)d\tilde{m}_n(\omega)\\ &= \int_{E^\infty_{l_n}} \sum_{i\le n: A_{i\omega_n}=1} g_n(i\omega)\exp(f(i\omega)-\mathrm{P}_n)d\tilde{m}_n(\omega)\\ &= \mathcal{L}^*_{0,n}\tilde{m}_n(g_n) = \tilde{m}_n(g_n)\end{aligned} \tag{2.36}$$

and

$$\tilde{m}_n(g_n) - \tilde{m}_n(g) = \int_{E^\infty_{l_n}} (g_n - g)d\tilde{m}_n = \int_{E^\infty_{l_n}} 0 d\tilde{m}_n = 0. \tag{2.37}$$

Using the triangle inequality we get the following.

$$\begin{aligned}|\mathcal{L}^*_0\tilde{m}(g) - \tilde{m}(g)| \le{}& |\mathcal{L}^*_0\tilde{m}(g) - \mathcal{L}^*_0\tilde{m}_n(g)|\\ &+ |\mathcal{L}^*_0\tilde{m}_n(g) - \mathcal{L}^*_{0,n}\tilde{m}_n(g)|\\ &+ |\mathcal{L}^*_{0,n}\tilde{m}_n(g) - \tilde{m}_n(g_n)|\\ &+ |\tilde{m}_n(g_n) - \tilde{m}_n(g)| + |\tilde{m}_n(g) - \tilde{m}(g)|\end{aligned} \tag{2.38}$$

Let us now look at the second summand. Applying (2.33) and (2.32) we get

$$\begin{aligned}&|\mathcal{L}^*_0\tilde{m}_n(g) - \mathcal{L}^*_{0,n}\tilde{m}_n(g)|\\ &\quad= \Bigg| \int_{E^\infty} \sum_{i\le n: A_{i\omega_n}=1} g(i\omega)\big(\exp(f(i\omega)-\mathrm{P}(f))\\ &\quad - \exp(f(i\omega)-\mathrm{P}_n)\big)\big)d\tilde{m}_n(\omega)\\ &\quad + \int_{E^\infty} \sum_{i> n: A_{i\omega_n}=1} g(i\omega)\exp(f(i\omega)-\mathrm{P}(f))d\tilde{m}_n(\omega)\Bigg|\\ &\quad \le \sum_{i\le n} \|g\|_0 \mathrm{e}^{f(i\omega)}\mathrm{e}^{-\mathrm{P}_n}(\mathrm{P}(f)-\mathrm{P}_n)\\ &\quad + \sum_{i>n} \|g\|_0 \exp\big(\sup(f|_{[i]} - \mathrm{P}(f)\big)\\ &\quad \le \sum_{i\ge 1} \|g\|_0 \exp\big(\sup(f|_{[i]})\big)\mathrm{e}^{-\mathrm{P}_1}(\mathrm{P}(f)-\mathrm{P}_n) + \frac{\epsilon}{6}\\ &\quad \le \frac{\epsilon}{6} + \frac{\epsilon}{6} = \frac{\epsilon}{3}.\end{aligned} \tag{2.39}$$

Combining now in turn (2.35), (2.39), (2.36), (2.37) and (2.34) we get from (2.38) that

$$|\mathcal{L}_0^*\tilde{m}(g) - \tilde{m}(g)| \le \frac{\epsilon}{3} + \frac{\epsilon}{3} + \frac{\epsilon}{3} = \epsilon.$$

Letting $\epsilon \searrow 0$ we therefore get $\mathcal{L}_0^*\tilde{m}(g) = \tilde{m}(g)$ or $\mathcal{L}_f^*\tilde{m}(g) = e^{P(f)}\tilde{m}(g)$. Thus, $\mathcal{L}_f^*\tilde{m} = e^{P(f)}\tilde{m}$. □

As an immediate consequence of this theorem and Theorem 2.3.3, we get the following important result.

Corollary 2.7.4 *Suppose that $f : E^\infty \to I\!R$ is a Hölder continuous bounded function such that*

$$\sum_{e\in I} \exp(\sup(f|_{[e]})) < \infty$$

and the incidence matrix is finitely irreducible. Then there exists a Gibbs state for f.

As an immediate consequence of Theorem 2.7.3, Theorem 2.3.6, Theorem 2.3.3, and Theorem 2.2.4, we get the following.

Corollary 2.7.5 *Suppose that $f : E^\infty \to I\!R$ is a Hölder continuous bounded function such that*

$$\sum_{e\in I} \exp(\sup(f|_{[e]})) < \infty$$

and the incidence matrix is finitely irreducible. Then

(a) There exists a unique eigenmeasure $\tilde{m}_f$ of the conjugate Perron–Frobenius operator $\mathcal{L}_f^*$ and the corresponding eigenvalue is equal to $e^{P(f)}$.

(b) The eigenmeasure $\tilde{m}_f$ is a Gibbs state for f.

(c) The function $f : E^\infty \to I\!R$ has a unique σ-invariant Gibbs state $\tilde{\mu}_f$. In case the matrix is finitely primitive, this Gibbs state is completely ergodic and the stochastic laws presented in Section 2.5 are satisfied.

3

Hölder Families of Functions and F-Conformal Measures

In this chapter we come back to the setting from Chapter 1. Our aim here is to define summable Hölder families of functions and with the help of the machinery developed in Chapter 2 to construct F-conformal measures, which in turn will be used in Chapter 4 to construct geometrically significant conformal measures.

3.1 Summable Hölder families

Let $F = \{f^{(e)} : X_{t(e)} \to I\!R : e \in I\}$ be a family of real-valued functions. For every $n \geq 1$ and $\beta > 0$ let

$$V_n(F) = \sup_{\omega \in E^n} \sup_{x,y \in X_{t(\omega)}} \{|f^{(\omega_1)}(\phi_{\sigma(\omega)}(x)) - f^{(\omega_1)}(\phi_{\sigma(\omega)}(y))|\} e^{\beta(n-1)}.$$

We have made the conventions that the empty word $\emptyset$ is the only word of length 0 and $\phi_\emptyset = \mathrm{Id}_X$. Thus, $V_1(F) < \infty$ simply means the diameters of the sets $f^i(X)$ are uniformly bounded. The collection F is called a *Hölder family of functions* (of order β) if

$$V_\beta(F) = \sup_{n \geq 1}\{V_n(F)\} < \infty. \tag{3.1}$$

The Hölder family F is called summable (of order β) if (3.1) is satisfied and

$$\sum_{e \in E} \|e^{f^{(e)}}\|_0 < \infty. \tag{3.2}$$

We define the topological pressure $\mathrm{P}(F)$ of F by setting

$$\mathrm{P}(F) = \lim_{n \to \infty} \frac{1}{n} \log \sum_{\omega \in n} \left\| \exp\left(\sum_{j=1}^{n} f^{\omega_j} \circ \phi_{\sigma^j \omega} \right) \right\|_0$$

$$= \lim_{n\to\infty} \frac{1}{n} \log \sum_{\omega\in E^n} \exp\left(\sup_X \sum_{j=1}^n f^{\omega_j} \circ \phi_{\sigma^j \omega}\right),$$

where $\|\cdot\|_0$ denotes the supremum norm. Given $n \geq 1$ and $\omega \in E^n$ we define the function $S_\omega(F) : X_{t(\omega)} \to I\!R$ by declaring

$$S_\omega(F) := \sum_{j=1}^n f^{(\omega_j)} \circ \phi_{\sigma^j \omega}.$$

Repeating with obvious modifications the proof of Theorem 2.1.3 we find the following.

Theorem 3.1.1 *We have*

$$\mathrm{P}(F) = \inf\left\{ t \in I\!R : \sum_{\omega\in E^*} \exp\bigl(\sup(S_\omega(F))\bigr)\, e^{-t|\omega|} < \infty \right\}.$$

Let us now prove the following version of the bounded distortion property.

Lemma 3.1.2 *If $\omega \in E^*$ and $x, y \in X_{t(\omega)} \cap \phi_\tau(X_{t(\tau)})$ for some $\tau \in E^*$, then*

$$|S_\omega(F)(x) - S_\omega(F)(y)| \leq \frac{V(F)\mathrm{e}^\beta}{1-\mathrm{e}^{-\beta}} \mathrm{e}^{-\beta|\tau|}$$

Proof. Let $n = |\omega|$. Write $x = \phi_\tau(u)$, $y = \phi_\tau(w)$, where $u, w \in X_{t(\tau)}$. By (3.1) we get

$$\begin{aligned}
&\left| \sum_{j=1}^n f^{(\omega_j)}(\phi_{\sigma^j\omega}(x)) - \sum_{j=1}^n f^{(\omega_j)}(\phi_{\sigma^j\omega}(y)) \right| \\
&= \left| \sum_{j=1}^n f^{(\omega\tau)_j} \circ \phi_{\sigma^j\omega\tau}(u) - \sum_{j=1}^n f^{(\omega\tau)_j} \circ \phi_{\sigma^j\omega\tau}(w) \right| \\
&\leq \sum_{j=1}^n \left| f^{(\omega\tau)_j} \circ \phi_{\sigma^j\omega\tau}(u) - f^{(\omega\tau)_j} \circ \phi_{\sigma^j\omega\tau}(w) \right| \\
&\leq \sum_{j=1}^n V(F)\mathrm{e}^{-\beta(n+|\tau|-j-1)} \\
&\leq \frac{V(F)\mathrm{e}^\beta}{1-\mathrm{e}^{-\beta}} \mathrm{e}^{-\beta|\tau|}.
\end{aligned}$$

□

Set

$$T(F) = \exp\left(\frac{V(F)e^{\beta}}{1-e^{-\beta}}\right).$$

In order to connect with the previous chapter we now introduce a potential function or *amalgamated function*, $f : E^\infty \to I\!R$, induced by the family of functions F as follows.

$$f(\omega) = f^{(\omega_1)}(\pi(\sigma(\omega))).$$

Our convention will be to use lower case letters for the potential function corresponding to a given Hölder system of functions. The following lemma is a straightforward consequence of Lemma 3.1.2.

Lemma 3.1.3 *If F is a Hölder family (of order β) then the amalgamated function f is Hölder continuous (of order β).*

The following result demonstrates that the concept of topological pressure $\mathrm{P}(F)$ can, in fact, be reduced to the topological pressure introduced in the previous chapter.

Proposition 3.1.4 $\mathrm{P}(F) = \mathrm{P}(f)$.

Proof. First notice that for every $n \geq 0$, every $\omega \in E^n$ and every $\tau \in E^\infty$ with $A_{\omega_n\tau_1} = 1$

$$\begin{aligned} S_\omega(F)(\tau) &= \sum_{j=1}^{n} f^{(\omega_j)} \circ \phi_{\sigma^j\omega}(\tau) = \sum_{j=1}^{n} f(\sigma^{j-1}\omega\tau) \\ &= \sum_{j=0}^{n-1} f(\sigma^j\omega\tau) = S_n f(\omega\tau). \end{aligned}$$

Hence,

$$\begin{aligned} \|\exp S_\omega(F)\|_0 &\geq \exp\big(\sup\{S_n f(\omega\tau) : \tau \in E^\infty, A_{\omega_n\tau_1} = 1\}\big) \\ &= \exp\big(\sup(S_n f|_{[\omega]})\big). \end{aligned}$$

And therefore $\mathrm{P}(F) \geq \mathrm{P}(f)$. Fix now an arbitrary $\tau \in E^\infty$ with $A_{\omega_n\tau_1} = 1$ and consider an arbitrary $x \in X$. Then, by Lemma 3.1.2 we have

$$\begin{aligned} \exp S_\omega(F)(x) &\leq e^{T(F)} \exp\big(S_\omega(F)(\pi(\tau))\big) \\ &\leq e^{T(F)} S_n f(\omega\tau) \leq e^{T(F)} \exp\big(\sup(S_n f|_{[\omega]})\big). \end{aligned}$$

Therefore $\exp\big(\sup S_\omega(F)\big) \le \mathrm{e}^{T(F)} \exp\big(\sup(S_n f|_{[\omega]})\big)$ and consequently, $\mathrm{P}(F) \le \mathrm{P}(f)$. □

3.2 F-conformal measures

A Borel probability measure m is said to be F-*conformal* provided it is supported on the limit set J and the following two conditions are satisfied. For every $\omega \in E^*$ and for every Borel set $A \subset X_{t(\omega)}$

$$m(\phi_\omega(A)) = \int_A \exp\big(S_\omega(F) - \mathrm{P}(F)|\omega|\big)\, dm \tag{3.3}$$

and for all incomparable words $\omega, \tau \in E^*$

$$m\big(\phi_\omega(X_{t(\omega)}) \cap \phi_\tau(X_{t(\omega)})\big) = 0. \tag{3.4}$$

A simple inductive argument shows that instead of (3.3) and (3.4) it is enough to require that for every $i \in I$ and for every Borel set $A \subset X_{t(i)}$

$$m(\phi_i(A)) = \int_A \exp\big(f^{(i)} - \mathrm{P}(F)\big)\, dm \tag{3.5}$$

and for all $i, j \in I$, $i \ne j$

$$m\big(\phi_i(X_{t(i)}) \cap \phi_j(X_{t(j)})\big) = 0. \tag{3.6}$$

Given $a \in I$ set

$$X_a = \bigcup_{b \in I: A_{ab}=1} \phi_b(X_{t(b)}).$$

We shall prove the following auxiliary result.

Lemma 3.2.1 *If m is a Borel probability measure on J such that (3.4) holds and (3.3) is satisfied for all Borel sets $A \subset X_i$, $i \in I$, then m is F-conformal.*

Proof. Since for every $i \in I$, $(X_{t(i)} \setminus X_i) \cap J = \emptyset$, we have

$$m(X_{t(i)} \setminus X_i) = 0. \tag{3.7}$$

Take now an arbitrary element $i \in I$ and consider an arbitrary element $x \in \phi_i\big(X_{t(i)} \setminus X_i\big) \cap J$. Then $x = \pi(\omega) = \phi_{\omega_1}(\pi(\sigma(\tau)))$ for some $\omega \in E^\infty$ with $\omega_1 \ne i$. Hence $J \cap \phi_i\big(X_{t(i)} \setminus X_i\big) \subset \bigcup_{j \ne i} \phi_j\big(X_{t(j)}\big) \cap \phi_i\big(X_{t(i)}\big)$ and consequently by (3.6)

$$\begin{aligned} m\big(\phi_i\big(X_{t(i)} \setminus X_i\big)\big) &= m\big(J \cap \phi_i\big(X_{t(i)} \setminus X_i\big)\big) \\ &\le \sum_{j \ne i} m\big(\phi_j\big(X_{t(j)}\big) \cap \phi_i\big(X_{t(i)}\big)\big) = 0. \end{aligned}$$

Therefore using (3.7) we obtain by induction, for every $\omega \in E^*$ and every $A \subset X_{t(\omega)}$,

$$m(\phi_\omega(A)) = m(\phi_\omega(A \cap X_{\omega_{|\omega|}})) = \int_{A \cap X_{\omega_{|\omega|}}} \exp\big(S_\omega(F) - \mathrm{P}(F)|\omega|\big)\, dm$$
$$= \int_A \exp\big(S_\omega(F) - \mathrm{P}(F)|\omega|\big)\, dm. \tag{3.8}$$

□

We shall now provide some sufficient conditions for the existence (and uniqueness) of F-conformal measures. Our first condition comes from the following definition. Namely, we say that an iterated function system $\{\phi_i : i \in I\}$ satisfies the *strong separation condition* if

$$\phi_i(X) \cap \phi_j(X) = \emptyset$$

for all $i, j \in I$, $i \neq j$. Our second condition is somewhat technical and may seem strange at first sight. Its real geometrical meaning will become clear in the next chapter, which is devoted to conformal systems.

Definition 3.2.2 *Given $q \geq 1$, we say that two different words $\rho, \tau \in E^*$ of the same length, say $n > q$, form a pair of q-codes of a point $x \in X$ if $x \in \phi_\rho(X_{t(\rho)}) \cap \phi_\tau(X_{t(\tau)})$ and $\rho|_{n-q} = \tau|_{n-q}$. The* GDMS S *is said to be conformal-like if for every $q \geq 1$ there is no point in X (or equivalently in J) with arbitrarily long pairs of q-codes.*

Of course each system satisfying the strong open set condition is conformal-like. It will turn out in the next chapter that each conformal system is conformal-like.

Theorem 3.2.3 *Suppose that the GDMS $\{\phi_i : i \in I\}$ is conformal-like and that the incidence matrix is finitely primitive. If F is a summable Hölder family of functions, then there exists a unique F-conformal measure m_F. Moreover $m_F = \tilde{m}_F \circ \pi^{-1}$, where $\tilde{m}_F$ is the eigenmeasure of the conjugate Perron–Frobenius operator $\mathcal{L}_f^*$.*

Proof. By Corollary 2.7.5(a) there exists $\tilde{m}$, an eigenmeasure of the conjugate operator $\mathcal{L}_f^*$ associated with the amalgamated function $f : E^\infty \to I\!R$, and the corresponding eigenvalue is equal to $e^{\mathrm{P}(f)}$. Let $m_F = \tilde{m}_F \circ \pi^{-1}$. We shall show that m_F is an F-conformal measure. And indeed, suppose on the contrary that $m_F(\phi_\rho(X_{t(\rho)}) \cap \phi_\tau(X_{t(\tau)})) > 0$

for some two incomparable words $\rho, \tau \in I^*$ with $i(\rho) = i(\tau) = v$ for some $v \in V$. We may assume without loss of generality that ρ and τ are of the same length, say $q \geq 1$. Let $E = \phi_\rho(X_{t(\rho)}) \cap \phi_\tau(X_{t(\tau)})$ and for every $n \geq 1$, let $E_n = \bigcup_{\omega \in E^n : t(\omega) = v} \phi_\omega(E)$. Since each element of E_n admits at least two q-codes of length $n + q$, we conclude that

$$\bigcap_{k=1}^{\infty} \bigcup_{n=k}^{\infty} E_n = \emptyset.$$

On the other hand, $E_n = \bigcup_{\omega \in E^n : t(\omega) = v} \phi_\omega(E) \supset \pi(\sigma^{-n}(\pi^{-1}(E)))$, which implies that $\pi^{-1}(E_n) \supset \sigma^{-n}(\pi^{-1}(E))$. In view of Corollary 2.7.5(c) there exists a σ-invariant Gibbs state $\tilde{\mu}$ for f which, by Proposition 2.2.2, is equivalent with $\tilde{m} = \tilde{m}_F$. Hence $\tilde{\mu}(\pi^{-1}(E_n)) \geq \tilde{\mu}(\sigma^{-n}(\pi^{-1}(E)) = \tilde{\mu} \circ \pi^{-1}(E) > 0$. Thus,

$$\tilde{\mu} \circ \pi^{-1} \left(\bigcap_{k=1}^{\infty} \bigcup_{n=k}^{\infty} E_n \right) \geq \tilde{\mu} \circ \pi^{-1}(E) > 0.$$

In particular, $\bigcap_{k=1}^{\infty} \bigcup_{n=k}^{\infty} E_n \neq \emptyset$. This contradiction shows that property (3.4) holds. In order to prove (3.3) fix $\omega \in E^*$, say $\omega \in E^n$, and a Borel set $B \subset X_{t(\omega)}$. Put

$$\pi_\omega^{-1}(B) = \{\tau \in \pi^{-1}(B) : A_{\omega_n \tau_1} = 1\}$$

and consider an element $\tau \in \pi^{-1}(B) \setminus \pi_\omega^{-1}(B)$. Then $A_{\omega_n \tau_1} = 0$ and, on the other hand, $\pi(\tau) \in B \subset X_{t(\omega)}$ which implies that $\pi(\tau) = \phi_i(X_{t(i)})$ for some $i \in I$ such that $A_{\omega_n i} = 1$. Hence $i \neq \tau_1$ and $\pi(\tau) \in \phi_i(X_{t(i)}) \cap \phi_{\tau_1}(X_{t(\tau_1)})$. Therefore

$$\pi\big(\pi^{-1}(B) \setminus \pi_\omega^{-1}(B)\big) \subset \bigcup_{i \neq j} \phi_i(X_{t(i)}) \cap \phi_j(X_{t(j)})$$

and, using (3.4), we consequently get

$$\begin{aligned} \tilde{m}\big(\pi^{-1}(B) \setminus \pi_\omega^{-1}(B)\big) &\leq \tilde{m} \circ \pi^{-1} \left(\bigcup_{i \neq j} \phi_i(X_{t(i)}) \cap \phi_j(X_{t(j)}) \right) \\ &= m_F \left(\bigcup_{i \neq j} \phi_i(X_{t(i)}) \cap \phi_j(X_{t(j)}) \right) = 0 \end{aligned} \tag{3.9}$$

It is straightforward to check that

$$\pi^{-1}(\phi_\omega(B)) \supset [\omega \pi_\omega^{-1}(B)].$$

Consider then an element $\rho \in \pi^{-1}(\phi_\omega(B)) \setminus [\omega\pi_\omega^{-1}(B)]$. Then $\pi(\rho) = \phi_{\omega i}(x)$, where $A_{\omega_n i} = 1$, $\phi_i(x) \in B$ and $\rho \notin [\omega\pi_\omega^{-1}(B)]$. If $\rho|_{n+1} = \omega i$, then $x = \pi(\sigma^{n+1}\rho)$ and we deduce that $\pi(\rho^n\rho) = \phi_i(\pi(\sigma^{n+1}\rho)) = \phi_i(x)$, and consequently $\rho = \omega i\sigma^{n+1}(\omega) \in [\omega\pi_\omega^{-1}(B)]$. Therefore $\rho|_{n+1} \neq \omega i$ and we get that

$$\pi^{-1}(\phi_\omega(B)) \setminus [\omega\pi_\omega^{-1}(B)] \subset \pi^{-1}\left(\bigcup_{\tau,\eta\in E^{n+1}:\tau\neq\eta} \phi_\tau(X_{t(\tau)}) \cap \phi_\eta(X_{t(\eta)})\right).$$

Using (3.4) and the definition of the measure m_F we therefore deduce that

$$\tilde{m}\left(\pi^{-1}(\phi_\omega(B)) \setminus [\omega\pi_\omega^{-1}(B)]\right) = 0.$$

Combining this equality, (3.9) and (2.18), we can write

$$\begin{aligned}
m_F(\phi_\omega(B)) &= \tilde{m}\circ\pi^{-1}(\phi_\omega(B)) = \tilde{m}\big([\omega\pi_\omega^{-1}(B)]\big) \\
&= \int_{\pi_\omega^{-1}(B)} \exp\big(S_n f(\omega\rho) - \mathrm{P}(f)n\big)\, d\tilde{m}(\rho) \\
&= \int_{\pi_\omega^{-1}(B)} \exp\big(S_\omega F(\pi(\rho)) - \mathrm{P}(F)n\big)\, d\tilde{m}(\rho) \\
&= \int_{\pi^{-1}(B)} \exp\big(S_\omega F(\pi(\rho)) - \mathrm{P}(F)n\big)\, d\tilde{m}(\rho) \\
&= \int_B \exp\big(S_\omega F(x) - \mathrm{P}(F)n\big)\, d\tilde{m}\circ\pi^{-1}(x) \\
&= \int_B \exp\big(S_\omega F - \mathrm{P}(F)n\big)\, dm_F.
\end{aligned}$$

So, by Lemma 3.2.1, the proof of the existence part is complete.

In order to prove uniqueness suppose that ν is another F-conformal measure. In view of (3.4) there exists a unique probability measure $\tilde{\nu}$ on E^∞ such that $\nu = \tilde{\nu}\circ\pi^{-1}$. Fix $n \geq 1$, $\omega \in E^n$ and a Borel set $A \subset E^\infty$ such that $A_{\omega_n\tau_{-1}} = 1$ for every $\tau \in A$. Then $\pi(A) \subset X_\omega$, and using (3.3), we get

$$\begin{aligned}
\tilde{\nu}([\omega A]) = \nu(\pi([\omega A])) = \nu(\phi_\omega(\pi(A))) &= \int_{\pi(A)} \exp(S_\omega F - \mathrm{P}(F)n)\, d\nu \\
&= \int_A \exp(S_n f(\omega\rho) - \mathrm{P}(f)n)\, d\tilde{\nu}.
\end{aligned}$$

Therefore, in view of Remark 2.3.2, $\tilde{\nu}$ is an eigenmeasure of the operator $\mathcal{L}_f^*$ corresponding to the eigenvalue $e^{\mathrm{P}(f)}$. Hence, by Theorem 2.3.6, $\tilde{\nu} = \tilde{m}$. Thus $\nu = \tilde{\nu}\circ\pi^{-1} = \tilde{m}\circ\pi^{-1} = m_F$. □

We shall now prove a result which, though simple, is rich in geometric consequences.

Lemma 3.2.4 *With the same assumptions as in Theorem 3.1.7 we have*

$$M = \min\{m_F(X_v) : v \in V\} > 0.$$

Proof. Since the graph Γ is strongly connected, there exists $e \in I$ such that $i(e) = v$. Using (2.20) we then get

$$m_F(X_v) \geq \tilde{m}_F([e]) \geq CT(f)^{-1}(\#\Lambda)^{-1}\exp\big(\sup(f|_{[e]}) - \mathrm{P}(F)\big) > 0.$$

□

Keeping the assumption that the system S is conformal-like and F is a summable Hölder family of functions we put

$$\mu_F = \tilde{\mu}_f \circ \pi^{-1}. \tag{3.10}$$

Since μ_F is equivalent with m_F with Radon-Nikodym derivatives bounded away from zero and infinity, we call μ_F *the S-invariant version of* m_F.

Remark 3.2.5 *We would like to draw the reader's attention to the fact that in proving uniqueness in the previous theorem we have, in fact, demonstrated more. Namely, if we assume that a measure m supported on J satisfies (3.4) and (3.3) with an arbitrary constant L replacing the constant $\mathrm{P}(F)$, then $m \circ \pi$ is an eigenmeasure of the conjugate Perron–Frobenius operator $\mathcal{L}_f^*$ corresponding to the eigenvalue e^L. It then in turn follows from Theorem 2.3.3 that $L = \mathrm{P}(f) = \mathrm{P}(F)$, and consequently, m is F-conformal.*

4
Conformal Graph Directed Markov Systems

This is the central chapter of our book. We present here the basic and more refined geometric properties of limit sets of conformal Graph directed Markov systems.

4.1 Some properties of conformal maps in $I\!R^d$ with $d \geq 2$

In this section we study in detail analytic, geometric and especially distortion properties of conformal maps in any dimension $d \geq 3$. We say that a C^1 diffeomorphism (homeomorphism) $\phi : U \to I\!R^d$, $d \geq 1$, from an open connected set $U \subset I\!R^n$ to $I\!R^d$ is *conformal* if its derivative at every point of U is a similarity map. By $\phi'(z) : I\!R^d \to I\!R^d$ we denote the derivative of ϕ evaluated at the point z and by $|\phi'(z)|$ its norm, which in the conformal case coincides with the coefficient of similarity. Note that for $d = 1$ C^1-conformality means that all the maps ϕ_e, $e \in I$, are C^1-diffeomorphisms and monotone, for $d \geq 2$ the words C^1-conformal mean holomorphic or antiholomorphic, and for $d \geq 3$ the maps ϕ_e, $e \in I$ are Möbius transformations. More precisely each C^1-conformal homeomorphism ϕ defined on an open connected subset of $I\!R^d$, $d \geq 3$, extends to the entire space $I\!R^d$ and takes the form

$$\phi = \lambda A \circ i_{a,r} + b, \tag{4.1}$$

where $0 < \lambda \in R$ is a positive scalar, A is a linear isometry in $I\!R^d$, $i_{a,r}$ is either *inversion* with respect to the sphere centered at a point a and with radius r, or the identity map, and $b \in I\!R^d$. In the sequel λ will be called the scalar factor and $a = \phi^{-1}(\infty)$ will be called the center of inversion. If A is the identity map, ϕ will be called a *conformal affine homeomorphism.* The proof of this characterization of C^1-conformal homeomorphisms in the dimension ≥ 3 can be found in [BP] for example, where it is called *Liouville's theorem.* In particular it is easy to compute that

$$|\phi'(z)| = \frac{\lambda r^2}{\|z-a\|^2} \text{ and } \frac{\phi'(z)}{|\phi'(z)|} = \mathrm{Id} - 2Q(z-a), \tag{4.2}$$

where $Q(z)$ is the matrix given by the formula

$$Q(z)_{ij} = \frac{z_i z_j}{\|z\|^2}.$$

In dimension $d = 2$ we will extensively use the following celebrated *Koebe's distortion theorem* (see e.g. [CG]).

Theorem 4.1.1 *If $w \in \mathbb{C}$ and $g : B(w,R) \to \mathbb{C}$ is a univalent holomorphic function, then for all $z \in B(w,R)$,*

$$\frac{1-\frac{|z-w|}{R}}{\left(1+\frac{|z-w|}{R}\right)^3} \leq \frac{|g'(z)|}{|g'(w)|} \leq \frac{1+|z-w|}{\left(1-\frac{|z-w|}{R}\right)^3}.$$

Fix $0 < \gamma \leq 1/2$ so small that

$$K_\gamma = \max\left\{\frac{1+\gamma}{(1-\gamma)^3}, \left(\frac{1-\gamma}{(1+\gamma)^3}\right)^{-1}\right\} \leq \sqrt{2}. \tag{4.3}$$

We will need the following fact, which is in a sense an improvement of Koebe's distortion theorem (Theorem 4.1.1).

Theorem 4.1.2 *There exists a constant $K_3 \geq 1$ such that if $x \in \mathbb{C}$, $R > 0$ and $\phi : B(x,R) \to \mathbb{C}$ is a univalent conformal map, then*

$$|\phi'(y) - \phi'(x)| \leq \frac{K_3}{R}|\phi'(x)||y-x|$$

for all $y \in B\left(x, (\min\{\gamma, \frac{\pi}{24}\})R\right)$.

Proof. Since each conformal map in the plane is either holomorphic or antiholomorphic and since complex conjugation is an isometry, we may assume without loss of generality that ϕ is holomorphic. A holomorphic function $g : \mathbb{D} \to \mathbb{C}$ ($\mathbb{D} = \{z \in \mathbb{C} : |z| < 1\}$) is called a *Bloch function* if

$$\|g\|_{\mathcal{B}} = \sup_{z \in \mathbb{D}}\{(1-|z|^2)|g'(z)|\} < \infty.$$

If $f : \mathbb{D} \to \mathbb{C}$ is a univalent holomorphic univalent map, then (see [Po2, p.73]) $\log f'$ is a Bloch function and $\|\log f'\|_{\mathcal{B}} \leq 6$ for every branch of logarithm of f'. Since the function $z \mapsto \phi(x + Rz)$ is univalent and holomorphic on $\mathbb{D}$, using the chain rule twice, we conclude that

$$|(\log \phi')'(w)| \leq \frac{8}{R} \tag{4.4}$$

for all $w \in \overline{B}(x, R/2)$. Denote by $\log\left(\frac{\phi'(w)}{\phi'(x)}\right)$ the branch of the logarithm of the function $w \mapsto \frac{\phi'(w)}{\phi'(x)}$ defined by the formula $w \mapsto \log \phi'(w) - \log \phi'(x)$. Since $\log \phi'(y) - \log \phi'(x) = \int_x^y (\log \phi')'(z)dz$ for all $y \in B(x, R)$, using (4.4) we get

$$\left|\log\left(\frac{\phi'(y)}{\phi'(x)}\right)\right| = |\log \phi'(y) - \log \phi'(x)| \leq \frac{8}{R}|y - x| \tag{4.5}$$

for all $y \in \overline{B}(x, R/2)$. If now $y \in B(x, \gamma R)$, then in view of Theorem 4.1.1 and the choice of γ, we find

$$\frac{1}{\sqrt{2}} \leq \left|\frac{\phi'(y)}{\phi'(x)}\right| \leq \sqrt{2}. \tag{4.6}$$

If now $y \in B\left(x, \frac{\pi R}{24}\right)$, then using (4.5) we obtain

$$\left|\mathrm{Arg}\left(\frac{\phi'(y)}{\phi'(x)}\right)\right| \leq \frac{8}{R} \cdot \frac{\pi R}{24} = \frac{\pi}{3} \tag{4.7}$$

If $\log w$, for w in the right half plane is the branch of the logarithm sending 1 to 0, then the function $\frac{w-1}{\log w}$ is analytic on the right plane. Therefore putting

$$F = \{z \in \mathbb{C} \colon \frac{1}{\sqrt{2}} \leq |z| \leq \sqrt{2} \text{ and } |\mathrm{Arg}(z)| \leq \pi/3\}$$

we get that

$$C := \sup\left\{\left|\frac{w-1}{\log w}\right| : w \in F\right\} < \infty. \tag{4.8}$$

Since all branches of the logarithm differ by $2\pi i n$, $n \in \mathbb{Z}$, we conclude that (4.8) remains true (perhaps with a larger constant C) if by $\log w$ we mean any value of the logarithm. Combining (4.5)–(4.8), we get

$$\left|\frac{\phi'(y)}{\phi'(x)} - 1\right| \leq C_1 \left|\log\left(\frac{\phi'(y)}{\phi'(x)}\right)\right| \leq \frac{12C_1}{R}|y - x|$$

for all $y \in B\left(x, (\min\{\gamma, \frac{\pi}{24}\})R\right)$ Therefore, putting $K_3 = 8C_1$ completes the proof. □

In the dimension $d \geq 3$ we have the following.

Theorem 4.1.3 *Suppose that Y is a bounded subset of $\mathbb{R}^d$, $d \geq 3$, and W is an open subset of $\mathbb{R}^d$ containing $\overline{Y}$. Then there exists a constant $K_4 \geq 1$ such that if $\phi : \overline{\mathbb{R}^d} \to \overline{\mathbb{R}^d}$ is a conformal diffeomorphism such that $\phi(W) \subset W$, then*

$$|\phi'(y)| - |\phi'(x)| \leq K_4 |\phi'(x)\| \cdot |y - x\|$$

for all $x, y \in Y$. In particular

$$\frac{|\phi'(y)|}{|\phi'(x)|} \leq 1 + K_4 \text{diam}(Y).$$

Proof. In view of (4.1) there then exist $\lambda > 0$, a linear isometry $A : \mathbb{R}^d \to \mathbb{R}^d$, an inversion (or the identity map) $i = i_{a,r}$ and a vector $b \in \mathbb{R}^d$ such that $\phi = \lambda A \circ i + b$. When i is the identity map the statement of our theorem is obvious. So, we may assume that i is an inversion. Since $\phi(W) \subset W \subset \mathbb{R}^d$, $a \notin W$. Therefore for all $x, y \in Y$

$$\begin{aligned} \frac{\|x-a\|}{\|y-a\|} &\leq \frac{\|x-y\| + \|y-a\|}{\|y-a\|} = 1 + \frac{\|x-y\|}{\|y-a\|} \\ &\leq 1 + \frac{\text{diam}(Y)}{\text{dist}(Y, \partial W)}. \end{aligned} \tag{4.9}$$

Thus, using (4.2), we get

$$\frac{|\phi'(y)|}{|\phi'(x)|} \leq \left(1 + \frac{\text{diam}(Y)}{\text{dist}(Y, \partial W)}\right)^2.$$

The proof of the second part of our theorem is complete. In order to prove the first part we may assume without loosing generality that $|\phi'(x)| \leq |\phi'(y)|$. Using (4.2) and (4.9) we then get

$$\begin{aligned} |\phi'(y)| - |\phi'(x)| &\leq |\phi'(x)| \left(\frac{|\phi'(y)|}{|\phi'(x)|} - 1 \right) = |\phi'(x)| \left(\frac{\|x-a\|^2}{\|y-a\|^2} - 1 \right) \\ &= |\phi'(x)| \left(\frac{\|x-a\|}{\|y-a\|} - 1 \right) \left(\frac{\|x-a\|}{\|y-a\|} + 1 \right) \\ &\leq |\phi'(x)| \left(2 + \frac{\text{diam}(Y)}{\text{dist}(Y, \partial W)} \right) \frac{\|x-y\|}{\|y-a\|} \\ &\leq |\phi'(x)| \left(2 + \frac{\text{diam}(Y)}{\text{dist}(Y, \partial W)} \right) \frac{\|x-y\|}{\text{dist}(Y, \partial W)} \\ &\leq \left(2 + \frac{\text{diam}(Y)}{\text{dist}(Y, \partial W)} \right) \left\{ \frac{1}{\text{dist}(Y, \partial W)} \right\} |\phi'(x)| \cdot \|y - x\| \end{aligned}$$

The proof is thus complete. □

As an immediate consequence of Theorem 4.1.2 and Theorem 4.1.3, we obtain the following bounded distortion result.

Corollary 4.1.4 *Suppose that Y is a bounded subset of $\mathbb{R}^d$, $d \geq 2$, and W is an open subset of $\mathbb{R}^d$ containing $\overline{Y}$. Then for every $\epsilon > 0$ there exists $\delta > 0$ such that if $\phi : \overline{\mathbb{R}^d} \to \overline{\mathbb{R}^d}$ is a conformal diffeomorphism such that $\phi(W) \subset W$, then*

$$(1+\epsilon)^{-1} \leq \frac{|\phi'(y)|}{|\phi'(x)|} \leq 1+\epsilon$$

for all $x \in Y$ and all $y \in B(x, \delta)$.

As an immediate consequence of Theorem 4.1.2 alone we get the following.

Theorem 4.1.5 *If $x \in \mathbb{C}$, $R > 0$ and $\phi : B(x, R) \to \mathbb{C}$ is a univalent holomorphic map, then*

$$|\phi(y) - \phi(x) - \phi'(x)(y-x)| \leq \frac{K_3}{R}|\phi'(x) \cdot \|y-x\|^2$$

for all $y \in B\left(x, (\min\left\{\gamma, \frac{\pi}{24})\right\} R\right)$, where γ is defined by (4.3) and K_3 comes from Theorem 4.1.2.

Proof. In view of Theorem 4.1.2 we get

$$\begin{aligned}
&|\phi(y) - \phi(x) - \phi'(x)(y-x)| \\
&= |\int_x^y \phi'(z)dz - \int_x^y \phi'(x)dz| = \| \int_x^y (\phi'(z) - \phi'(x))dz| \\
&\leq \frac{K_3}{R}|\phi'(x)| \cdot |y-x| \cdot |y-x| = \frac{K_3}{R}|\phi'(x)||y-x|^2.
\end{aligned}$$

The proof is finished. □

In the case $d \geq 3$ we have a similar result.

Theorem 4.1.6 *Suppose that Y is a bounded subset of $\mathbb{R}^d$, $d \geq 3$, and W is an open subset of $\mathbb{R}^d$ containing $\overline{Y}$. Then there exists a constant $K_5 \geq 1$ such that if $\phi : \overline{\mathbb{R}^d} \to \overline{\mathbb{R}^d}$ is a conformal diffeomorphism such that $\phi(W) \subset W$, then for all $x, y \in Y$ we have*

$$\|\phi(y) - \phi(x) - \phi'(x)(y-x)\| \leq K_5|\phi'(x)| \cdot \|y-x\|^2.$$

Proof. Write $\phi = \lambda A \circ i_{a,r} + b$. If $i_{a,r}$ is the identity map, then the left-hand side of the claimed inequality is equal to 0 and we are done. Otherwise, putting $i = i_{a,r}$ and using (4.2), we get

$$\begin{aligned}
&\phi(y) - \phi(x) - \phi'(x)(y-x) \\
&\quad = \lambda A \circ i(y) - \lambda A \circ i(x) - \lambda A \circ i'(x)(y-x) \\
&\quad = \lambda A(i(y) - i(x) - i'(y-x)) \\
&\quad = \lambda A \left(r^2 \frac{y-a}{\|y-a\|^2} - r^2 \frac{x-a}{\|x-a\|^2} \right. \\
&\quad \left. - \frac{r^2}{\|x-a\|^2} (\mathrm{Id} - 2Q(x-a))(y-x) \right) \\
&\quad = \lambda r^2 A \left(\frac{y-a}{\|y-a\|^2} - \frac{x-a}{\|x-a\|^2} - \frac{(\mathrm{Id} - 2Q(x-a))(y-x)}{\|x-a\|^2} \right).
\end{aligned} \tag{4.10}$$

Since it is straightforward to check that

$$Q(x-a)u = \frac{\langle x-a, u \rangle}{\|x-a\|^2}(x-a),$$

where $\langle \cdot, \cdot \rangle$ is the standard inner product in $\mathbb{R}^d$, we can write

$$\begin{aligned}
\Delta &:= \frac{y-a}{\|y-a\|^2} - \frac{x-a}{\|x-a\|^2} - \frac{(\mathrm{Id} - 2Q(x-a))(y-x)}{\|x-a\|^2} \\
&= \frac{y-x}{\|y-a\|^2} + \frac{x-a}{\|y-a\|^2} - \frac{x-a}{\|x-a\|^2} - \frac{y-x}{\|x-a\|^2} \\
&\quad + 2\frac{\langle x-a, y-x \rangle}{\|x-a\|^4}(x-a) \\
&= \left(\frac{1}{\|y-a\|^2} - \frac{1}{\|x-a\|^2} \right)(y-x) \\
&\quad + \left(\frac{1}{\|y-a\|^2} - \frac{1}{\|x-a\|^2} + 2\frac{\langle x-a, y-x \rangle}{\|x-a\|^4} \right)(x-a).
\end{aligned}$$

Write now

$$x - a = \alpha(y-x) + \beta(y-x)^{\perp},$$

where $(y-x)^{\perp}$ is a vector perpendicular to $y-x$ with $\|(y-x)^{\perp}\| = \|y-x\|$. Then

$$\begin{gathered}
\|x-a\|^2 = (\alpha^2 + \beta^2)\|y-x\|^2, \; \|y-a\|^2 = ((\alpha+1)^2 + \beta^2)\|y-x\|^2 \\
\text{and} \\
\langle x-a, y-x \rangle = \alpha\|y-x\|^2.
\end{gathered} \tag{4.11}$$

Thus

$$\Delta = \left(\frac{1}{\|y-a\|^2} - \frac{1}{\|x-a\|^2} + \alpha \left(\frac{1}{\|y-a\|^2} - \frac{1}{\|x-a\|^2} \right.\right.$$
$$\left.\left. + \frac{2\alpha\|y-x\|^2}{\|x-a\|^4} \right)\right) \times (y-x)$$
$$+ \beta \left(\frac{1}{\|y-a\|^2} - \frac{1}{\|x-a\|^2} + \frac{2\alpha\|y-x\|^2}{\|x-a\|^4} \right) (y-x)^\perp$$
$$= \frac{1}{\|x-a\|^2} \left[\left(\frac{\alpha^2+\beta^2}{(\alpha+1)^2+\beta^2} - 1 + \alpha \left(\frac{\alpha^2+\beta^2}{(\alpha+1)^2+\beta^2} - 1 \right) \right.\right.$$
$$\left.\left. + \frac{2\alpha^2}{\alpha^2+\beta^2} \right) (y-x) + \beta \left(\frac{\alpha^2+\beta^2}{(\alpha+1)^2+\beta^2} - 1 + \frac{2\alpha}{\alpha^2+\beta^2} \right) (y-x)^\perp \right]$$
$$= \frac{1}{\|x-a\|^2} \left[\left(\frac{(1+\alpha)(\alpha^2+\beta^2)}{(\alpha+1)^2+\beta^2} - (1+\alpha) + \frac{2\alpha^2}{\alpha^2+\beta^2} \right) (y-x) \right.$$
$$\left. + \beta \left(\frac{\alpha^2+\beta^2}{(\alpha+1)^2+\beta^2} - 1 + \frac{2\alpha}{\alpha^2+\beta^2} \right) (y-x)^\perp \right]. \tag{4.12}$$

Now, a direct calculation using (4.11) shows that

$$\left| \frac{(1+\alpha)(\alpha^2+\beta^2)}{(\alpha+1)^2+\beta^2} - (1+\alpha) + \frac{2\alpha^2}{\alpha^2+\beta^2} \right| = \frac{|\alpha^3+\alpha^2-3\alpha\beta^2-\beta^2|}{(\alpha^2+\beta^2)((\alpha+1)^2+\beta^2)}$$
$$= \frac{\|y-x\|^4}{\|x-a\|^2\|y-a\|^2} |\alpha^3+\alpha^2-3\alpha\beta^2-\beta^2|$$
$$\le \frac{\|y-x\|^4}{\|x-a\|^2\|y-a\|^2} (|\alpha|^3+\alpha^2+3|\alpha|\beta^2+\beta^2)$$
$$\le \frac{\|y-x\|^4}{\|x-a\|^2\|y-a\|^2} \left(4 \left(\frac{\|x-a\|}{\|y-x\|} \right)^3 + \frac{\|x-a\|^2}{\|y-x\|^2} \right)$$
$$\le \frac{\|y-x\|}{\|y-a\|} \left(\frac{4\|x-a\|}{\|y-a\|} + \frac{\|y-x\|}{\|y-a\|} \right)$$
$$\le \frac{\|y-x\|}{\|y-a\|} \left(\frac{4(\|y-x\|+\|y-a\|)}{\|y-a\|} + \frac{\|y-x\|}{\|y-a\|} \right)$$
$$\le \frac{\|y-x\|}{\|y-a\|} \left(5\frac{\|y-x\|}{\|y-a\|} + 4 \right)$$
$$\le \frac{1}{R} \left(\frac{5\mathrm{diam}(Y)}{R} + 4 \right) \|y-x\|,$$

where $R = \text{dist}(Y, \partial W)$ and

$$\begin{aligned}
&|\beta(\frac{\alpha^2+\beta^2}{(\alpha+1)^2+\beta^2} - 1 + \frac{2\alpha}{\alpha^2+\beta^2})| \\
&= \frac{|\beta||3\alpha^2 - \beta^2 + 2\alpha|}{(\alpha^2+\beta^2)((\alpha+1)^2+\beta^2)} \leq \frac{|\beta|(3\alpha^2+\beta^2+2|\alpha|)}{\|x-a\|^2\|y-a\|^2}\|y-x\|^4 \\
&\leq \frac{1}{\|x-a\|^2\|y-a\|^2}\left(4\left(\frac{\|x-a\|}{\|y-x\|}\right)^3 + 2\left(\frac{\|x-a\|}{\|y-x\|}\right)^2\right)\|y-x\|^4 \\
&\leq \frac{\|y-x\|}{\|y-a\|^2}(4\|x-a\| + 2\|y-x\|) \leq \frac{\|y-x\|}{\|y-a\|^2}(4\|y-a\| + 6\|y-x\|) \\
&\leq \frac{1}{\|y-a\|}\left(4 + 6\frac{\|y-x\|}{\|y-a\|}\right)\|y-x\| \leq \frac{1}{R}\left(4 + \frac{6\text{diam}(Y)}{R}\right)\|y-x\|.
\end{aligned}$$

Therefore, combining these two estimates with (4.12), we find there is a constant K_5 such that

$$\|\Delta\| \leq \frac{1}{\|x-a\|^2}K_5\|y-x\|^2.$$

It now follows from this along with (4.10) and (4.2) that

$$\|\phi(y)-\phi(x)-\phi'(x)(y-x)\| \leq \frac{\lambda r^2}{\|x-a\|^2}K_5\|y-x\|^2 = K_5|\phi'(x)|\cdot\|y-x\|^2.$$

□

Given a point $x \in \mathbb{R}^d$, a vector $u \in \mathbb{R}^d$ and an angle $\alpha \in (0, \pi/2)$ we put

$$\begin{aligned}
&\text{Con}(x, \alpha, u) \\
&\quad = \{y \in \mathbb{R}^d : \angle(y-x, u) \leq \alpha \text{ and } \|y-x\| \leq \|u\|\} \\
&\quad = \{y \in \mathbb{R}^d : \cos\angle(y-x, u) > \cos\alpha \text{ and } \|y-x\| \leq \|u\|\} \\
&\quad = \{y \in \mathbb{R}^d : \langle y-x, u\rangle > \cos\alpha\|u\| \cdot \|y-x\| \text{ and } \|y-x\| \leq \|u\|\}
\end{aligned}$$

and we call $\text{Con}(x, \alpha, u)$ the *cone generated by* x, α *and* u. We prove the following geometrical fact about conformal maps.

Theorem 4.1.7 *Suppose that Y is a bounded subset of $\mathbb{R}^d$, $d \geq 2$, and W is an open subset of $\mathbb{R}^d$ containing $\overline{Y}$. Then for every $\epsilon \in (0, \pi/2)$ there exists $\delta > 0$ such that if $y \in Y$, $\|u\| < \delta$ and $\alpha \in (0, \pi/2 - \epsilon)$, and $\phi : W \to W$ is a C^1 conformal diffeomorphism, then*

$$\phi(\text{Con}(x, \alpha, u)) \subset \text{Con}(\phi(x), \alpha + \epsilon, 2\phi'(x)u).$$

Proof. Since in the case when $d = 2$, the conformal diffeomorphism is either holomorphic or antiholomorphic and since complex conjugacy is an isometry, we may assume that if $d = 2$, then ϕ is holomorphic. In view of Theorem 4.1.6 and Theorem 4.1.5, putting $K_6 = \max\left\{K_5, \frac{K_3}{\mathrm{dist}(Y,\partial W)}\right\}$ we see that if $x \in Y$ and $\|y - x\| < \min\{\gamma, \frac{\pi}{24}\mathrm{dist}$ $(Y, \partial W)\}$, then

$$\|\phi(y) - \phi(x) - \phi'(x)(y - x)\| \le K_6|\phi'(x)| \cdot \|y - x\|^2. \tag{4.13}$$

Fix $\eta > 0$. In view of Theorem 4.1.3 and Theorem 4.1.2 there exists $9 < \delta < \min\{\gamma, \frac{\pi}{24}\mathrm{dist}(Y, \partial W)\}$ (independent of ϕ) such that

$$(1 + \eta)^{-1} \le \frac{|\phi'(y)|}{|\phi'(x)|} \le 1 + \eta$$

for all $x \in Y$ and all $y \in B(x, \delta)$. Fix in turn $u \in I\!R^d$ with $\|u\| \le \delta$. If $\|y - x\| \le u$, then

$$\begin{aligned} \|\phi(y) - \phi(x)\| &\le (1 + \eta)|\phi'(x)\| \cdot \|y - x\| \le (1 + \eta)|\phi'(x)\| \cdot \|u\| \\ &= (1 + \eta)\|\phi'(x)u\|. \end{aligned} \tag{4.14}$$

Now,

$$\begin{aligned} &\langle \phi(y) - \phi(x), (1 + \eta)\phi'(x)u\rangle \\ &= \langle \phi'(x)(y - x), (1 + \eta)\phi'(x)u\rangle \\ &+ \langle \phi(y) - \phi(x) - \phi'(x)(y - x), (1 + \eta)|\phi'(x)u > . \end{aligned} \tag{4.15}$$

If $y \in \mathrm{Con}(x, \alpha, u)$, then using (4.14), we get

$$\begin{aligned} &\langle \phi'(x)(y - x), (1 + \eta)\phi'(x)u\rangle \\ &= (1 + \eta)|\phi'(x)|^2\langle y - x, u\rangle \\ &\ge (1 + \eta)|\phi'(x)|^2 \cos\alpha\|u\|\|y - x\| \\ &\ge \cos\alpha\|\phi'(x)u\|\|\phi(y) - \phi(x)\|. \end{aligned} \tag{4.16}$$

On the other hand, it follows from (4.13) that

$$\begin{aligned} &|\langle \phi(y) - \phi(x) - \phi'(x)(y - x), \phi'(x)u\rangle| \\ &\le \|\phi(y) - \phi(x) - \phi'(x)(y - x)\| \cdot \|\phi'(x)u\| \\ &\le K_6|\phi'(x) \cdot \|y - x\|^2 \cdot \|\phi'(x)u\|. \end{aligned} \tag{4.17}$$

Assuming $\delta > 0$ to be smaller than $(2K_6)^{-1}$, we obtain from (4.13) that

$$\begin{aligned}\|\phi(y)-\phi(x)\| &\geq |\phi'(x)\cdot\|y-x\| - K_6|\phi'(x)|\cdot\|y-x\|^2\\ &\geq |\phi'(x)|\cdot\|y-x\| - K_6|\phi'(x)|\cdot\|y-x\|(2K_6)^{-1}\\ &\geq \frac{1}{2}|\phi'(x)|\cdot\|y-x\|.\end{aligned}$$

Therefore, we can continue (4.17) as follows

$$\begin{aligned}&|\langle\phi(y)-\phi(x)-\phi'(x)(y-x),\phi'(x)u\rangle|\\ &\leq 2K_6\|\phi'(x)u\|\cdot\|\phi(y)-\phi(x)\|\cdot\|y-x\|\\ &\leq 2K_6\delta(1+\eta)^{-1}(\|(1+\eta)\phi'(x)u\|)\|\phi(y)-\phi(x)\|.\end{aligned}$$

Combining this and (4.15) we get

$$\begin{aligned}&\langle\phi(y)-\phi(x),(1+\eta)\phi'(x)u\rangle\\ &\geq (1+\eta)^{-1}(\cos\alpha-2K_6\delta)\big(\|(1+\eta)\phi'(x)u\|)\|\phi(y)-\phi(x)\|\big).\end{aligned}$$

Taking now $0<\eta<1/2$ and $\delta>0$ so small that $\arccos\big(1+\eta)^{-1}(\cos\alpha-2K_6\delta)\big)<\alpha+\epsilon$, this inequality along with (4.14) shows that

$$\begin{aligned}\phi(\mathrm{Con}(x,\alpha,u)) &\subset \mathrm{Con}(\phi(x),\alpha+\epsilon,(1+\eta)\phi'(x)u)\\ &\subset \mathrm{Con}(\phi(x),\alpha+\epsilon,2\phi'(x)u).\end{aligned}$$

□

4.2 Conformal measures; Hausdorff and box dimensions

We are now in position to introduce and explore the central object of this book. Namely, we call a GDMS conformal *(CGDMS)* if the following conditions are satisfied.

(4a) For every vertex $v\in V$, X_v is a compact connected subset of a Euclidean space $I\!R^d$ (the dimension d common for all $v\in V$) and $X_v=\overline{\mathrm{Int}(X_v)}$.

(4b) *(Open set condition)(OSC)* For all $a,b\in E$, $a\neq b$,

$$\phi_a(\mathrm{Int}(X_{t(a)})\cap\phi_b(\mathrm{Int}(X_{t(b)})=\emptyset.$$

(4c) For every vertex $v\in V$ there exists an open connected set $W_v\supset X_v$ such that for every $e\in I$ with $t(e)=v$, the map ϕ_e extends to a C^1 conformal diffeomorphism of W_v into $W_{i(e)}$.

(4d) *(Cone property)* There exist $\gamma, l > 0$, $\gamma < \pi/2$, such that for every $x \in X \subset I\!R^d$ there exists an open cone $\mathrm{Con}(x, \gamma, l) \subset \mathrm{Int}(X)$ with vertex x, central angle of measure γ, and altitude l.

(4e) There are two constants $L \geq 1$ and $\alpha > 0$ such that

$$\big||\phi_e'(y)| - |\phi_e'(x)|\big| \leq L\|(\phi_e')^{-1}\|^{-1}\|y - x\|^\alpha$$

for every $e \in I$ and every pair of points $x, y \in X_{t(e)}$, where $|\phi_\omega'(x)|$ means the norm of the derivative.

Since for every $0 < r < \max\{\mathrm{dist}(X_v, \partial W_v)\}$ and every $e \in I$, we have $\phi_e(B(X_{t(e)}, r)) \subset B(X_{t(i)}, r)$ and since the set of vertices V is finite, as an immediate consequence of Theorem 4.1.2 and Theorem 4.1.3, we get the following remarkable result.

Proposition 4.2.1 *If $d \geq 2$ and a family $S = \{\phi_e\}_{e \in I}$ satisfies conditions (4a) and (4c), then it also satisfies condition (4e) with $\alpha = 1$.*

As a rather straightforward consequence of (4e) we get the following.

Lemma 4.2.2 *If $S = \{\phi_e\}_{e \in I}$ is a CGDMS, then for all $\omega \in E^*$ and all $x, y \in W_{t(\omega)}$, we have*

$$\big|\log|\phi_\omega'(y)| - \log|\phi_\omega'(x)|\big| \leq \frac{L}{1-s}\|y - x\|^\alpha.$$

Proof. For every $\omega \in E^*$, say $\omega \in E^n$, and every $z \in W_{t(\omega)}$ put $z_k = \phi_{\omega_{n-k+1}} \circ \phi_{\omega_{n-k+2}} \circ \cdots \circ \phi_{\omega_n}(z)$; put also $z_0 = z$. In view of (4e) for any two points $x, y \in W_{t(\omega)}$ we have

$$\begin{aligned}
&\big|\log(|\phi_\omega'(y)|) - \log(|\phi_\omega'(x)|)\big| \\
&= \left|\sum_{j=1}^n \log\left(1 + \frac{|\phi_{\omega_j}'(y_{n-j})| - |\phi_{\omega_j}'(x_{n-j})|}{|\phi_{\omega_j}'(x_{n-j})|}\right)\right| \\
&\leq \sum_{j=1}^n \|(\phi_{\omega_j}')^{-1}\|\big||\phi_{\omega_j}'(y_{n-j})| - |\phi_{\omega_j}'(x_{n-j})|\big| \qquad (4.18) \\
&\leq \sum_{j=1}^n L|y_{n-j} - x_{n-j}|^\alpha \\
&\leq L\sum_{j=1}^n s^{\alpha(n-j)}\|y - x\|^\alpha \leq \frac{L}{1-s}\|y - x\|^\alpha.
\end{aligned}$$

□

As a straightforward consequence of (4e) we get the following.

(4f) (Bounded distortion property). There exists $K \geq 1$ such that for all $\omega \in E^*$ and all $x, y \in X_{t(\omega)}$

$$|\phi'_\omega(y)| \leq K|\phi'_\omega(x)|.$$

We shall now prove some basic geometric consequences of the properties (4a)–(4f) and the results proved in the previous section.

We would like to point out that in the case $d \geq 2$ property (4f) was already established in Corollary 4.1.4. As an immediate of the Mean value inequality and property (4f) we get the following. For all finite words $\omega \in E^*$, all convex subsets C of $W_{t(\omega)}$, all $x \in X_{t(\omega)}$ and all radii $r \leq \text{dist}(X_{t(\omega)}, \partial W_{t(\omega)})$

$$\text{diam}(\phi_\omega(C)) \leq \|\phi'_\omega\|\text{diam}(C), \ \ \phi_\omega(B(x,r)) \subset B(\phi_\omega(x), \|\phi'_\omega\|r). \quad (4.19)$$

We shall prove that there exists a constant $D \geq 1$ such that

$$\text{diam}\big(\phi_\omega(W_{t(\omega)})\big) \leq D\|\phi'_\omega\|, \quad (4.20)$$

for all finite words $\omega \in E^*$. And indeed, take $0 < r < \frac{1}{2}\min\{\text{dist}(X_v, \partial W_v) : v \in V\}$. Then $\phi_e\big(B(X_{t(e)}, r)\big) \subset B(X_{t(e)}, r)$ for all $e \in I$, and as W_v we may take the ball $B(X_v, r)$ for every $v \in V$. Since each set $\overline{B(X_v, r)}$ is compact and connected, we may cover it by finitely many balls $B(x_1, 2r), \ldots, B(x_{q(v)}, 2r)$ for some $q(v) \geq 1$ with the centers $x_1, \ldots, x_{q(v)}$ in X_v. Using then (4.19) we conclude that

$$\text{diam}\big(\phi_\omega(W_{t(\omega)})\big) \leq q(t(\omega))\|\phi'_\omega\|4r \leq D\|\phi'_\omega\|,$$

where $D = 4r\max\{q(v) : v \in V\}$.

We shall now prove the following formula.

$$\phi_\omega(B(x,r)) \supset B(\phi_\omega(x), K^{-1}\|\phi'_\omega\|r), \quad (4.21)$$

for all finite words $\omega \in E^*$, all $x \in X_{t(\omega)}$ and all $0 < \text{dist}(X_{t(\omega)}, \partial W_{t(\omega)})$. And indeed, $B(x,r) \subset W_{t(\omega)})$. Take also any $\omega \in E^*$ and let $R \geq 0$ be the maximal radius such that

$$B(\phi_\omega(x), R) \subset \phi_\omega(B(x,r)). \quad (4.22)$$

Then $\partial\big(B(\phi_\omega(x), R)\big) \cap \partial\big(\phi_\omega(B(x,r))\big) \neq \emptyset$, and in view of (4f) we have

$$\phi_\omega^{-1}\big(B(\phi_\omega(x), R)\big) \subset B(x, \|(\phi_\omega^{-1})'\|R) \subset B(x, K\|\phi'_\omega\|^{-1}R)$$

which implies that $B(\phi_\omega(x), R) \subset \phi_\omega\big(B(x, K\|\phi'_\omega\|^{-1}R)\big)$. Therefore $K\|\phi'_\omega\|^{-1}R \geq r$ and using (4.22) we obtain (4.21). We shall now prove the following inequality, perhaps with a larger constant D.

$$\operatorname{diam}(\phi_\omega(X_{t(\omega)})) \ge D^{-1}\|\phi'_\omega\| \tag{4.23}$$

for all finite words $\omega \in E^*$. And indeed, for every $v \in V$ put

$$r = \min\{\operatorname{dist}(X_v, \partial W_v) : v \in V\}.$$

Fix $x_v \in X_v$ and $y_v \in (X_v \setminus \{v\} \cap B(x_v, K^{-1}r)$. For every $\omega \in E^*$ we have by (4.21), $\phi_\omega(B(x_{t(\omega)}, r)) \supset B(\phi_\omega(x_{t(\omega)}), K^{-1}\|\phi'_\omega\|r)$ and by (4.19), $\phi_\omega(y_{t(\omega)}) \in B(\phi_\omega(x_{t(\omega)}), K^{-1}||\phi'_\omega||r)$. Therefore, applying the mean value inequality to the map ϕ_ω^{-1} restricted to the convex set $B(\phi_\omega(x_{t(\omega)}), K^{-1}||\phi'_\omega||r)$ along with (4f), we obtain the following.

$$\begin{aligned} ||y_{t(\omega)} - x_{t(\omega)}|| &= ||\phi_\omega^{-1}(\phi_\omega(y_{t(\omega)})) - (\phi_\omega^{-1}(\phi_\omega(x_{t(\omega)}))|| \\ &\le K||(\phi_\omega^{-1})'|| \cdot ||\phi_\omega(y_{t(\omega)}) - (\phi_\omega(x_{t(\omega)})|| \\ &\le K||\phi'_\omega||^{-1}||\phi_\omega(y_{t(\omega)}) - (\phi_\omega(x_{t(\omega)})||. \end{aligned}$$

Thus

$$\begin{aligned} \operatorname{diam}(\phi_\omega(X_{t(\omega)})) &\ge ||\phi_\omega(y_{t(\omega)}) - \phi_\omega(x_{t(\omega)})|| \\ &\ge K^{-1}||\phi'_\omega|| \cdot ||y_{t(\omega)} - x_{t(\omega)}|| \\ &\ge K^{-1}\min\{||y_v - x_v|| : v \in V\}||\phi'_\omega|| \end{aligned}$$

and we are done.

In the case when $d \ge 2$ the following proposition follows immediately from Theorem 4.1.7 and if $d = 1$ it is an immediate consequence of (4f).

Proposition 4.2.3 *For every $\epsilon > 0$ there exists $\delta > 0$ such that if $\omega \in E^*$, $x \in X_{t(\omega)}$, $||u|| \le \delta$ and $\alpha \in (0, \pi/2 - \epsilon)$, then*

$$\phi_\omega(\operatorname{Con}(x, \alpha, u)) \subset \operatorname{Con}(\phi_\omega(x), \alpha + \epsilon, 2\phi'(x)u).$$

We shall prove the following. There exists $\beta > 0$ such that if $D \ge 1$ is sufficiently large, then for all $\omega \in E^*$ and $x \in X_{t(\omega)}$, we have

$$\begin{aligned} \phi_\omega(X_{t(\omega)}) &\supset \operatorname{Con}(\phi_\omega(x), \beta, D^{-1}\phi'(x)u_x) \\ &\supset \operatorname{Con}(\phi_\omega(x), \beta, D^{-2}\operatorname{diam}(\phi_\omega(W_{t(\omega)}))u_x). \end{aligned} \tag{4.24}$$

And indeed, if $d = 1$, the first inclusion is an immediate consequence of (4.21). So, in proving this inclusion we may assume that $d \ge 2$. Notice that then there exists an integer $q \ge 1$ such that for every $x \in X$ there exist unit vectors $u_1(x), \dots, u_q(x)$ such that all tuples

$(u_x, u_1(x), \ldots, u_q(x))$ are mutually isometric,

$$\mathrm{Con}(x,\gamma,\mathbb{R}_+u_x) \cup \bigcup_{j=1}^{q} \mathrm{Con}(x,\gamma,\mathbb{R}_+u_j(x)) \subset \mathbb{R}^d \tag{4.25}$$

and

$$\mathrm{Con}(x,\gamma/2,\mathbb{R}_+u_x) \cap \bigcup_{j=1}^{q} \mathrm{Con}(x,\gamma,\mathbb{R}_+u_j(x)) = \emptyset.$$

Then, there exists an $\epsilon > 0$ such that

$$\mathrm{Con}(x,\gamma/4,\mathbb{R}_+\tilde{u}_x) \cap \bigcup_{j=1}^{q} \mathrm{Con}(x,\gamma+\epsilon,\mathbb{R}_+\tilde{u}_j(x)) = \emptyset \tag{4.26}$$

for all $x \in X$ and all tuples $(\tilde{u}_x, \tilde{u}_1(x), \ldots, \tilde{u}_q(x))$ equivalent with $(u_x, u_1(x), \ldots, u_q(x))$ by a similarity map. Let $0 < \delta < \min\{\mathrm{dist}(X_v, \partial W_v) : v \in V\}$ be ascribed to this ϵ according to Proposition 4.2.3. In view of this proposition, (4.26) and the fact that $\phi'_\omega(x) : \mathbb{R}^d \to \mathbb{R}^d$ is a similarity map, we get

$$\mathrm{Con}(\phi_\omega(x),\gamma/4,\mathbb{R}_+\phi'_\omega\tilde{u}_x) \cap \bigcup_{j=1}^{q} \phi_\omega(\mathrm{Con}(x,\gamma,\delta u_j(x))) = \emptyset \tag{4.27}$$

for all $\omega \in E^*$. In view of (4.21) and (4.25) we find

$$\phi_\omega\left(\mathrm{Con}(x,\gamma,\frac{\delta}{l}u_x)\right) \cup \bigcup_{j=1}^{q} \phi_\omega(\mathrm{Con}(x,\gamma,\delta u_j(x))) \supset B(\phi_\omega(x)K^{-1}||\phi'_\omega||\delta).$$

This and (4.27) imply that

$$\begin{aligned}\phi_\omega\left(\mathrm{Con}(x,\gamma,\frac{\delta}{l}u_x)\right) &\supset \mathrm{Con}(\phi_\omega(x),\gamma/4,K^{-1}||\phi'_\omega||\delta|\phi'_\omega(x)|^{-1}\phi'_\omega(x)u_x) \\ &\supset \mathrm{Con}(\phi_\omega(x),\gamma/4,(Kl)^{-1}\delta\phi'_\omega(x)u_x)\end{aligned}$$

and we can take any $D \geq KL\delta^{-1}$ and $\beta = \gamma/4$ for the first inclusion in formula (4.24) to hold. The second inclusion obviously holds if D is large enough. We are done.

As an immediate consequence of (4.24) we get the following.

Lemma 4.2.4 *If S is a* CGDMS, *then*

$$\sup_{n\geq 1}\sup_{x\in X} \#\{\omega \in E^n : x \in \phi_\omega(X_{t(\omega)})\} \leq \frac{\lambda_{d-1}(S^{d-1})}{\beta} < \infty,$$

where S^{d-1} is the $(d-1)$-dimensional unit sphere and λ_{d-1} is the $(d-1)$-dimensional Lebesgue measure on S^{d-1}.

In fact we have more.

Proposition 4.2.5 *Every conformal* GDMS S *is conformal-like.*

Proof. Suppose on the contrary that S is not conformal-like. It means that there exist a point $x \in X$, an integer $q \geq 1$ and an unbounded increasing sequence $\{n_k\}_{k\geq 1}$ along with pairwise different words $\rho^{(k)}, \tau^{(k)} \in E^{n_k}$ such that

$$x \in \phi_{\tau^{(k)}}(X_{t(\tau^{(k)})}) \cap \phi_{\rho^{(k)}}(X_{t(\rho^{(k)})})$$

and $\rho^{(k)}|_{n_k-q} = \tau^{(k)}|_{n_k-q}$. Passing to a subsequence we may assume that $n_{k+1} - n_k > q$ for every $k \geq 1$. We shall construct by induction with respect to $k \geq 1$ a sequence $\{C_k\}_{k\geq 1}$ such that for every k, C_k consists of at least $k+1$ incomparable words from $\{\rho^{(j)}, \tau^{(j)} : j \leq k\}$.

Indeed, set $C_1 = \{\rho^{(1)}, \tau^{(1)}\}$. Suppose now that C_k has been defined. If $\rho^{(k+1)}$ does not extend any word in C_k, then we form C_{k+1} by adding $\rho^{(k+1)}$ to C_k. We can do a similar thing in case $\tau^{(k+1)}$ does not extend any word in C_k. If, on the other hand, $\rho^{(k+1)}$ extends some word κ in C_k and $\tau^{(k+1)}$ extends a word η in C_k, then κ and η are both extended by ρ^{k+1} since for $j \leq k$, $|\rho^{(j)}|, |\tau^{(j)}| \leq n_j \leq n_k$, $n_{k+1} > n_k + q$, and $\rho^{(k+1)}|_{n_{k+1}-q} = \tau^{(k+1)}|_{n_{k+1}-q}$. Since the words in C_k are incomparable, $\kappa = \eta$ and this is the only word in C_k which is extended by both $\rho^{(k+1)}$ and $\tau^{(k+1)}$. In this case we form C_{k+1} by taking away κ and adding both $\rho^{(k+1)}$ and $\tau^{(k+1)}$. Now the sets $\{\phi_\kappa(X_{t(\kappa)}) : \kappa \in C_k\}$ are non-overlapping since the words are incomparable. By (4.24) we get $k+1$ pairwise disjoint open cones each with vertex x and opening angle β. Since this is impossible if k is large enough, the proof is finished. □

We will frequently make use of the following simple estimate. This sort of estimate was used by Moran [Mo] in the study of geometric constructions using similarity maps. This includes the iteration of finitely many similarity maps and was used by Hutchinson [Hu]. The estimate was extended to the random case in [GMW],to directed graph constructions using similarity maps in [MW2] and to the iteration of infinitely many conformal maps in [MU1]. This estimate remains valid when the cone property is replaced in these arguments by the more general “neighobhood boundedness property” as defined and used in [GMW].

Lemma 4.2.6 *If S is a* CGDMS, *then for every $x \in X$, every $\kappa > 0$ and every $r > 0$, the cardinality of any collection of mutually incomparable words $\omega \in E^*$ satisfying $B(x,r) \cap \phi_\omega(X_{t(\omega)}) \neq \emptyset$ and* $\mathrm{diam}(\phi_\omega(X_{t(\omega)})) \geq$ *κr is bounded from above by the number $V_d \kappa^{-d} D^{2d} \beta^{-1}(1+\kappa D^{-2})$.*

Proof. Let F be such a collection. It follows from (4.24) that

$$\phi_\omega(X_{t(\omega)}) \supset \mathrm{Con}(\phi_\omega(x_\omega), \beta, D^{-2}\kappa r) \subset B(x, (1+\kappa D^{-2})r)$$

for all $\omega \in F$, where $\phi_\omega(x_\omega)$ is on the boundary of $\phi_\omega(X_{t(\omega)}) \cap B(x,r)$. Hence the cones $\mathrm{Con}(\phi_\omega(x_\omega), \beta, \kappa D^{-2} r)$, $\omega \in F$, are mutually disjoint and therefore

$$\begin{aligned} V_d(1+\kappa D^{-2})^d r^d &= \lambda_d(B(x,(1+\kappa D^{-2})r)) \\ &\geq \sum_{\omega \in F} \lambda_d\big(\mathrm{Con}(\phi_\omega(x_\omega), \beta, \kappa D^{-2} r)\big) \\ &= \#F \beta (\kappa D^{-2} r)^d. \end{aligned}$$

Now, the required estimate follows. □

We shall now introduce the most important technical tool in our explorations of geometry of limit sets. It was originated in the context of Fuchsian groups by S. Patterson in [Patt], adopted by D. Sullivan in [Su1] to the case of rational functions and by the authors to the case of conformal iterated function systems (see [MU1]). So, given $t \geq 0$ a Borel probability measure m is said to be *t-conformal* provided it is supported on the limit set J and the following two conditions are satisfied. For every $\omega \in E^*$ and for every Borel set $A \subset X_{t(\omega)}$

$$m(\phi_\omega(A)) = \int_A |\phi_\omega'|^t \, dm \tag{4.28}$$

and for all incomparable words $\omega, \tau \in E^*$

$$m\big(\phi_\omega(X_{t(\omega)}) \cap \phi_\tau(X_{t(\tau)})\big) = 0. \tag{4.29}$$

A simple inductive argument shows that instead of (4.28) and (4.29) it is enough to require that for every $e \in I$ and for every Borel set $A \subset X_{t(e)}$

$$m(\phi_e(A)) = \int_A |\phi_e'|^t \, dm \tag{4.30}$$

and for all $a, b \in I$, $a \neq b$

$$m\big(\phi_a(X_{t(a)}) \cap \phi_b(X_{t(b)})\big) = 0. \tag{4.31}$$

Given $t \geq 0$ and $n \geq 1$, we denote

$$Z_n(t) = \sum_{|\omega|=n} ||\phi'_\omega||^t \text{ and } \mathrm{P}(t) = \mathrm{P}(t\mathrm{Log}),$$

where

$$\mathrm{Log} = \{\log |\phi'_e|\}_{e \in I}.$$

We then simply call $\mathrm{P}(t)$ the pressure function. The following simple but crucial fact about the family Log is an immediate consequence of Lemma 4.2.2.

Proposition 4.2.7 *Log is a Hölder family of order* $\alpha \log s$.

Let now

$$F(S) = \{t \geq 0 : \mathrm{P}(t) < \infty\} \text{ and } \theta(S) = \inf(F(S)).$$

The basic straightforward properties of the pressure function are contained in the following (the first and the last one implied by Proposition 2.1.9).

Proposition 4.2.8 *Suppose that the incidence matrix A is finitely primitive. Then*

(a) $\inf\{t : Z_1(t) < \infty\} = \theta(S)$.
(b) *The topological pressure function* $\mathrm{P}(t)$ *is non-increasing on* $[0, \infty)$, *strictly decreasing on* $[\theta, \infty)$ *to negative infinity, convex and continuous on* $F(S)$.
(c) $\mathrm{P}(0) = \infty$ *if and only if I is infinite.*
(d) $\mathrm{P}(t) = \inf\left\{u \geq 0 : \sum_{\omega \in E^*} ||\phi'_\omega||^t e^{-u|\omega|} < \infty\right\}$.

Following the terminology introduced in [MU1] in the context of iterated function systems we call the system S *regular* if there exists $t \geq 0$ such that $\mathrm{P}(t) = 0$. It follows from Proposition 4.2.8 that there exists at most one such t. We shall prove the following useful characterization of regularity.

Theorem 4.2.9 *A finitely primitive* CGDMS *is regular if and only if there exists a t-conformal measure. If such a measure exists, then this measure is unique,* $\mathrm{P}(t) = 0$ *and the t-conformal measure is tLog-conformal.*

Proof. Suppose that a t-conformal measure exists. Then in view of Remark 3.2.5, $\mathrm{P}(t) = 0$ and m is $t\mathrm{Log}$-conformal. In view of Theorem 3.2.3(uniqueness) it is therefore left to show that if $\mathrm{P}(t) = 0$, then a t-conformal measure exists. And indeed, if $\mathrm{P}(t) = 0$, then in particular, $\mathrm{P}(t) < \infty$, and in view of Proposition 4.2.7 along with Proposition 2.1.9, $t\mathrm{Log}$ is a summable Hölder family of functions. Hence, we conclude from Theorem 3.2.3 and Proposition 4.2.5 that there exists a unique $t\mathrm{Log}$-conformal measure. Since $\mathrm{P}(t) = 0$, this measure is t-conformal. □

The unique t-conformal measure will be denoted in the sequel by m, its invariant version by μ and their lifts to the coding space respectively by $\tilde{m}$ and $\tilde{\mu}$. As an immediate consequence of this theorem and Lemma 3.2.4, we get the following.

Lemma 4.2.10 *If a finitely primitive* CGDMS *is regular,* $\mathrm{P}(h) = 0$, *and* m *is the unique* h*-conformal measure, then*

$$M = \min\{m(X_v) : v \in V\} > 0.$$

The next result establishes a strong geometric meaning of conformal measure in the finite case. Its first version can be found in [B2] and in the context of conformal iterated function systems it essentially appeared in [Be].

Theorem 4.2.11 *If* S *is a finite* CGDMS *whose incidence matrix is primitive, then* S *is regular and there exists* $C \geq 1$ *such that*

$$C^{-1} \leq \frac{m(B(x,r))}{r^h} \leq C$$

for all $x \in J$ *and* $0 < r < \frac{1}{2}\min\{\mathrm{diam}(X_v) : v \in V\}$, *where* h *is the unique zero of the pressure function. Then* $h = \mathrm{HD}(J)$ *and in particular,* $0 < \mathrm{H_h}(J), \Pi_{\mathrm{h}}(J) < \infty$.

Proof. Since I is finite and S is primitive, $0 < \mathrm{P}(t) < \infty$. Therefore, it follows from Proposition 4.2.8 that there exists a unique $h > 0$ such that $\mathrm{P}(h) = 0$. It then in turn follows from Theorem 4.2.9 that there exists an h-conformal measure for S. Since S is finite,

$$\xi = \inf\{||\phi_i'|| : i \in I\} > 0.$$

Consider $x = \pi(\omega)$, $\omega \in E^\infty$, $0 < r < \frac{1}{2}\min\{\mathrm{diam}(X_v) : v \in V\}$, and let $n \geq 0$ be the least integer such that $\phi_{\omega|_n}(X_{t(\omega_n)}) \subset B(x,r)$. Then

$n \geq 1$ and by conformality of m, (4f) and Lemma 4.2.10

$$m(B(x,r)) \geq m\big(\phi_{\omega|_n}(X_{t(\omega_n)})\big) \geq K^{-h}||\phi'_{\omega|_n}||^h m\big(X_{t(\omega_n)})\big) \\ \geq MK^{-h}||\phi'_{\omega|_n}||^h, \tag{4.32}$$

where M is the constant coming from Lemma 4.2.10. From the choice of n we conclude that $\phi_{\omega|_{n-1}}(X_{t(\omega_{n-1})})$ is not contained in $B(x,r)$. Thus, by (4.20) and (4f), we get

$$r \leq \operatorname{diam}\big(\phi_{\omega|_{n-1}}(X_{t(\omega_{n-1})})\big) \leq D||\phi'_{\omega|_{n-1}}|| \\ \leq DK||\phi'_{\omega_n}||^{-1}||\phi'_{\omega|_n}|| \leq DK\xi^{-1}||\phi'_{\omega|_n}||.$$

Combining this and (4.32) we obtain

$$m(B(x,r)) \geq MK^{-2h}D^h\xi r^h.$$

So, the first part of our theorem is proved. Let now Z be the family of all minimal (in the sense of length) words $\omega \in E^*$ such that

$$\phi_\omega(X_{t(\omega)}) \cap B(x,r) \neq \emptyset \quad \text{and} \quad \phi_\omega(X_{t(\omega)}) \subset B(x,2r). \tag{4.33}$$

Then $\operatorname{diam}(\phi_{\omega|_{|\omega|-1}}(X_{t(\omega_{|\omega|-1})})) \geq r$ and let

$$R = \{\omega|_{|\omega|-1} : \omega \in Z\}.$$

Note that R is finite and therefore we can find a finite subfamily R^* of R consisting of mutually incomparable words such that each element of R is an extension of an element from R^*. Then by Lemma 4.2.6, $\#R^* \leq V_d D^{2d}\beta^{-1}(1+D^{-2})^d$. Since $J \cap (B(x,r) \subset \bigcup\{\tau \in R^*\}\phi_\tau(X_\tau)$ and since for every $\tau \in R^*$,

$$||\phi'_\tau|| \leq K||\phi_{\tau e}|| \cdot ||\phi'_e||^{-1} \leq K\xi^{-1}||\phi_{\tau e}|| \leq 4DK\xi^{-1}r,$$

where $e \in I$ is such that $\tau e \in Z$, and the last inequality follows from (4.33) and (4.20), we conclude that

$$\begin{aligned} m(B(x,r)) = m(J \cap B(x,r)) &\leq m\left(\bigcup_{\tau \in R^*} \phi_\tau(X_{t(\tau)})\right) \\ &\leq \sum_{\tau \in R^*} m(\phi_\tau(X_{t(\tau)})) \leq \sum_{\tau \in R^*} ||\phi'_\tau||^h m(X_{t(\tau)}) \\ &\leq \sum_{\tau \in R^*} ||\phi'_\tau||^h \leq \sum_{\tau \in R^*} (4DK\xi^{-1}r)^h \\ &= \#R^*(4\xi^{-1}DK)^h r^h = V_d D^{2d}\beta^{-1}(1+D^{-2})^d(4DK\xi^{-1})^h r^h. \end{aligned}$$

□

In the proof of the next theorem, the main result of this section, we will need the following simple fact.

Lemma 4.2.12 *If S is a regular finitely primitive* CGDMS, *then for every $n \geq 1$*

$$1 \leq \sum_{\omega \in E^n} ||\phi_\omega||^h \leq K^h M^{-1},$$

where h is the unique zero of the pressure function $\mathrm{P}(t)$ *and M is the constant coming from Lemma 4.2.10.*

Proof. Let m be the unique h-conformal measure, which exists by Theorem 4.2.9. The h-conformality of m along with Lemma 4.2.10 imply now that for every $n \geq 1$

$$\begin{aligned}\sum_{\omega \in E^n} ||\phi_\omega||^h &\leq \sum_{\omega \in E^n} K^h \frac{m(\phi_\omega(X_{t(\omega)}))}{m(X_{t(\omega)})} \\ &\leq K^h M^{-1} \sum_{\omega \in E^n} m(\phi_\omega(X_{t(\omega)})) \leq K^h M^{-1}\end{aligned}$$

and

$$\sum_{\omega \in E^n} ||\phi_\omega||^h \geq \sum_{\omega \in E^n} \frac{m(\phi_\omega(X_{t(\omega)}))}{m(X_{t(\omega)})} \geq \sum_{\omega \in E^n} m(\phi_\omega(X_{t(\omega)})) = 1.$$

□

We would like to immediately add that (see Example 5.2.6, Example 5.2.7 and Example 5.2.8) this theorem fails in the infinite case already for iterated function systems. There are irregular systems.

Let $\mathcal{F}in(I)$ denote the family of all finite subsets of I. We shall now prove a dynamical characterization of the Hausdorff dimension of the limit set J, whose various versions have a long past and which goes along the line continued in [Be], [B2], [DU2], [MM], [MW1], [Mo], [MU1] and others.

Theorem 4.2.13 *If the conformal* GDMS S *is finitely primitive, then:*

$$\mathrm{HD}(J) = \inf\{t \geq 0 : \mathrm{P}(t) < 0\} = \sup\{h_F : F \in \mathcal{F}in(I)\} \geq \theta.$$

If $\mathrm{P}(t) = 0$, *then t is the only zero of the function* $\mathrm{P}(t)$ *and* $t = \mathrm{HD}(J)$.

Proof. Let $\xi = \inf\{t \geq 0 : \mathrm{P}(t) < 0\}$. Take $t > \xi$. Then, using (4.20), for every integer $n \geq 1$ sufficiently large we have

$$\sum_{\omega \in E^n} \mathrm{diam}(\phi_\omega(X_{t(\omega)}))^t \leq D^t \sum_{\omega \in E^n} ||\phi'_\omega||^t \leq D^t \exp(n\mathrm{P}(t)/2).$$

Since the family $\phi_\omega(X_{t(\omega)})$, $\omega \in E^n$, is a cover of J and since its diameters converge to 0 as $n \to \infty$, it follows from the estimate obtained that $\mathrm{H}_t(J) = 0$. Thus $\mathrm{HD}(J) \le \xi$. Let Λ and q be given by the finite primitivity of the system S. Set $\eta = \sup\{h_F : F \in \mathcal{F}in(I)\}$. In order to demonstrate that $\xi \le \eta$ we need an improvement of Lemma 4.2.12. For every $v \in V$ fix $e_v \in I$ such that $i(e_v) = v$. Consider then an arbitrary finite subset F of I such that $G^q \supset \Lambda$ and $G \supset \{e_v : v \in V\}$. Since G is finite and finitely primitive, the system S_G is regular and there exists an h_G-conformal measure m_G for S_G. Using (2.21) and proceeding as in the proof of Lemma 4.2.12, we find for every $n \ge 1$

$$
\begin{aligned}
\sum_{\omega \in E^n \cap G^n} ||\phi_\omega||^h &\le \sum_{\omega \in E^n \cap G^n} K^{h_G} \frac{m(\phi_\omega(X_{t(\omega)}))}{m(X_{t(\omega)})} \\
&\le K^{h_G} \sum_{\omega \in E^n \cap G^n} \min\{K^{-h_G}||\phi'_\alpha||^{h_G} : \alpha \in \Lambda\}^{-1} \\
&\times T(h_G \mathrm{Log}) \# \Lambda ||\phi_{e_{t(\omega)}}||^{-h_G} m(\phi_\omega(X_{t(\omega)})) \\
&\le K^{2h_G} K^{h_G}) \# \Lambda \max\{||\phi'_\alpha||^{-h_G} : \alpha \in \Lambda\} \\
&\times \max\{||\phi_{e_v}||^{-h_G} : v \in V\} \sum_{\omega \in E^n \cap G^n} m(\phi_\omega(X_{t(\omega)})) \\
&\le K^{3h} \# \Lambda \max\{||\phi'_\alpha||^{-h} : \alpha \in \Lambda\} \max\{||\phi_{e_v}||^{-h} : v \in V\}.
\end{aligned}
$$

Denote this last number by $\overline{M}$ and notice that it is independent of G. Denote by $H \subset I$ the minimal set such that $H^q \supset \Lambda$ and $H \supset \{e_v : v \in V\}$. Obviously $\eta = \sup\{h_F : F \in \mathcal{F}in(I)$ and $F \supset H\}$. Since $h_F \le h = h_I$ for every $F \subset I$ we have $M_F \le M_I$ for every $F \in \mathcal{F}in(I)$ such that $F^q \subset \Lambda$. Using the above estimate we can write as follows for any $t > \eta$ and any $n \ge 1$.

$$
\begin{aligned}
\sum_{\omega \in E^n} ||\phi'_\omega||^t &= \sup_{F \in \mathcal{F}in(I): F \supset H} \sum_{\omega \in F^n \cap E^n} ||\phi'_\omega||^t \\
&\le \sup_F \{ \sum_{\omega \in F^n \cap E^n} ||\phi'_\omega||^{h_F} s^{n(t-h_F)} \} \\
&\le s^{(t-\eta)n} \sup_F \{ \sum_{\omega \in F^n} ||\phi'_\omega||^{h_F} \} \le \overline{M} s^{(t-\eta)n}.
\end{aligned}
$$

Hence $\mathrm{P}(t) \le (t-\eta)\log s < 0$, which gives $t \ge \xi$ and consequently $\eta \ge \xi$. Obviously (cf. Lemma 4.2.10) $\eta \le \mathrm{HD}(J)$, and since we have proved that $\mathrm{HD}(J) \le \xi$, the proof of the "equality" part of the theorem is completed. The inequality $\theta \le \xi$ follows immediately from the definitions of both numbers. Finally, the last statement of the theorem is true since $\mathrm{P}(t)$ is continuous and strictly decreasing on (θ,∞). □

It easily follows from Theorem 4.2.11 that if the set of vertices is finite then the box-counting dimension and the Hausdorff dimension of the limit set coincide. As Example 5.2.3 shows, this equality fails in the infinite case. We shall however now provide a complete solution of the problem of determining the *Minkowski or box-counting* dimension and the packing dimension (see Appendix 2, [Ma], or [Fa2]) of the limit sets for all conformal primitive GDMS. Our first step is to demonstrate that these numbers are equal even if we assume only that the contractions forming the system S are bi-Lipschitz continuous.

Theorem 4.2.14 *If S is a* CGDMS *and all contractions ϕ_i are bi-Lipschitz continuous, then* $\mathrm{PD}(J) = \overline{\mathrm{BD}}(J) = \mathrm{PD}(\overline{J}) = \overline{\mathrm{BD}}(\overline{J})$.

Proof. The inequalities $\mathrm{PD}(J) \le \mathrm{PD}(\overline{J}) \le \overline{\mathrm{BD}}(\overline{J})$ and $\mathrm{PD}(J) \le \overline{\mathrm{BD}}(J) = \overline{\mathrm{BD}}(\overline{J})$ are obvious. Thus to complete the proof it suffices to show that $\mathrm{PD}(J) \ge \overline{\mathrm{BD}}(J)$. Indeed, fix $t < \overline{\mathrm{BD}}(J)$ and consider an arbitrary countable cover $\{Y_n : n \ge 1\}$ of J. Since the metric space E^∞ is complete, there exists $q \ge 1$ such that $\pi^{-1}(\overline{Y_q})$ has non-empty interior in E^∞. Therefore there exists an $\omega \in E^*$ such that $[\omega] \subset \pi^{-1}(\overline{Y_q})$, whence $\phi_\omega(J) = \pi([\omega]) \subset \overline{Y_q}$. Since $t < \overline{\mathrm{BD}}(J)$, we have $\Pi_t^*(J) = \infty$. Since ϕ_ω is bi-Lipschitz continuous, $\Pi_t^*(\phi_\omega(J)) = \infty$. We therefore get $\Pi_t^*(Y_q) = \Pi_t^*(\overline{Y_q}) \ge \Pi_t^*(\phi_\omega(J)) = \infty$. Thus $\sum_{n\ge1} \Pi_t^*(Y_n) = \infty$ and consequently $\Pi_t(J) = \infty$, which completes the proof. □

We now come back to conformal systems. We begin with the following definition. For every $R \subset E^*$ and every $Y \subset X$ we set

$$L_R(Y) = \bigcup_{w\in R} \phi_\omega(Y \cap X_{t(\omega)}).$$

If $R = E^n$ for some $n \ge 1$, we simply write $L_n(Y)$ for $L_{E^n}(Y)$. We recall that $N_r(E)$ is the minimum number of balls of radius $\le r$ needed to cover a set E.

In order to prove our characterization we need several lemmas.

Lemma 4.2.15 *Let $\{\phi_i : i \in I\}$ be a conformal finitely primitive GDMS. Let Y be an arbitrary subset of X such that $Y \cap X_v$ is a singleton of every*

$v \in V$. Then

$$\mathrm{PD}(J) = \overline{\mathrm{BD}}(J) = \max\{\mathrm{HD}(J), \sup_{n\geq 1} \overline{\mathrm{BD}}(L_n(Y))\}.$$

Proof. Theorem 4.2.14 says that $\mathrm{PD}(J) = \overline{\mathrm{BD}}(J)$. Let

$$M = \max\{\mathrm{HD}(J),\ \overline{\mathrm{BD}}(L_n(Y))\}.$$

Fix $t > M$. By Theorem 4.2.13 $\mathrm{P}(t) < 0$. Thus there is some Q such that if $q \geq Q$, then $Z_q(t) < 4^{-t}$ and if $|\omega| \geq Q$, then $||\phi'_\omega|| \leq 1/4$. Fix $q \geq Q$ and choose A such that for all $D \geq r > 0$, $N_r(L_q(Y)) \leq Ar^{-t}$. Now, choose B such that if $1 \leq r \leq D$, then $N_r(J) \leq Br^{-t}$ and such that $B \geq 4^t A/(1 - 4^t Z_q(t))$. We will show by induction that for each $n \in I\!N$, if $1/n \leq r \leq D$, then $N_r(J) \leq Br^{-t}$. This inequality holds for $n = 1$. Suppose it holds for n and $1/(n+1) \leq r < 1/n$. Let $C_{n+1} = \{\omega \in E^q : \mathrm{diam}(\phi_\omega(J_{t(\omega)})) \leq 1/(2(n+1))\}$. Since $J = \left(\bigcup_{\omega\in C_{n+1}} \phi_\omega(J_{t(\omega)})\right) \cup \left(\bigcup_{\omega\in E^q\setminus C_{n+1}} \phi_\omega(J(J_{t(\omega)})\right)$, we get

$$N_r(J) \leq N_{1/(n+1)}(J) \leq N_{1/(n+1)}\left(\bigcup_{\omega\in C_{n+1}} \phi_\omega(J)\right) + \sum_{\omega\in E^q\setminus C_{n+1}} N_{1/(n+1)}(\phi_\omega(J)).$$

For $\omega \in E^q \setminus C_{n+1}$, we have

$$N_{1/(n+1)}(\phi_\omega(J_{t(\omega)})) \leq N_{1/((n+1)||\phi'_\omega||)}(J) \leq N_{1/(2(n+1)||\phi'_\omega||)}(J).$$

Since $||\phi'_\omega|| \leq 1/4 \leq (1/2)(n/n+1)$, we have $1/n \leq 1/(2(n+1)||\phi'_\omega||)$. Since $1/(2(n+1)) < \mathrm{diam}(\phi_\omega(J_{t(\omega)})) \leq D||\phi'_\omega||$, we get $1/(2(n+1)\ ||\phi'_\omega||) \leq D$. So, by the inductive hypothesis, $N_{1/(n+1)}(\phi_\omega(J_{t(\omega)})) \leq B\left(2(n+1)||\phi'_\omega||\right)^t$. Next, we claim that

$$N_{1/(n+1)}(\bigcup_{\omega\in C_{n+1}} \phi_\omega(J_{t(\omega)})) \leq N_{1/(2(n+1))}(L_q(Y)).$$

In order to see this, let $B(y_j, 1/(2(n+1)))$ be a collection of balls of radius $1/(2(n+1))$ covering $L_q(Y)$. Suppose $z \in \phi_\omega({}_{t(\omega)}J)$, where $\omega \in C_{n+1}$. Then $|z - \phi_\omega(x_{t(\omega)})| \leq \mathrm{diam}(\phi_\omega(J_{t(\omega)})) \leq 1/(2(n+1))$, where $x_{t(\omega)}$ is the only element of the intersection $Y \cap X_{t(\omega)}$. For some j we have $|\phi_\omega(x_{t(\omega)}) - y_j| \leq 1/(2(n+1))$. So, the balls $B(y_j, 1/(n+1))$ cover $\bigcup_{\omega\in C_{n+1}} \phi_\omega({}_{t(\omega)}J)$. Our claim follows from this. Since $n+1 \leq 2/r$, we get

$$
\begin{aligned}
N_r(t) &\le A2^t(n+1)^t + \sum_{\omega\in E^q} B2^t(n+1)^t||\phi'_\omega||^t \\
&\le 4^t\left[A + BZ_q(t)\right]r^{-t} \le Br^{-t}.
\end{aligned}
$$

This completes the induction argument. It now follows that $\overline{\mathrm{BD}}(J) \le t$. The opposite inequality is obvious. □

Our goal is to show that we can replace the supremum in the previous lemma with a simple maximum. We use two lemmas to accomplish this.

Lemma 4.2.16 *If S is a* CGDMS, *then for all $v \in V$, all $x, y \in X_v$, and all $n \ge 1$*

$$\overline{\mathrm{BD}}(L_n(x)) = \overline{\mathrm{BD}}(L_n(y)).$$

Proof. First notice that it suffices to prove this equality for $n = 1$ since for every $n \ge 1$ the collection of maps $\{\phi_\omega : \omega \in E^n\}$ forms a conformal graph directed Markov system with the same set of vertices again. Without loss of generality it is enough to show that $\overline{\mathrm{BD}}(L_1(y)) \le \overline{\mathrm{BD}}((L_1(x))$. Towards this goal, take $0 < r \le \min\{\mathrm{diam}(X_v) : v \in V\}$ and let $I_r = \{e \in I : \mathrm{diam}(\phi_e(X_{t(e)})) \le r/2\}$. Then $N_r(L_{I_r}(y)) \le N_{r/2}(L_{I_r}(x))$. Clearly, $N_r(L_{I\setminus I_r}(z)) \le \#(I \setminus I_r)$, for all $z \in X$. On the other hand by Lemma 4.2.6 applied with $\kappa = 1/2$, $N_r(L_{I\setminus I_r}(z)) \ge \#(I \setminus I_r)/V_d 2^d D^{2d}\beta^{-1}(1 + 2^{-1}D^{-2})$. Hence,

$$
\begin{aligned}
N_r(L_1(y)) &\le N_{r/2}(L_{I_r}(x)) + N_r(L_{I\setminus I_r}(y)) \\
&\le N_{r/2}(L_{I_r}(x)) + V_d 2^d D^{2d}\beta^{-1}(1 + 2^{-1}D^{-2})N_r(L_{I\setminus I_r}(x)) \\
&\le \left(1 + V_d 2^d D^{2d}\beta^{-1}(1 + 2^{-1}D^{-2})\right)N_{r/2}(L_1(x)).
\end{aligned}
$$

Therefore,

$$\overline{\mathrm{BD}}(L_1(y)) = \overline{\lim_{r\to 0}}\ \frac{\log N_r((L_1(y))}{\log r} \le \overline{\lim_{r\to 0}}\ \frac{\log N_r((L_1(x))}{\log r} = \overline{\mathrm{BD}}(L_1(x)).$$

□

Lemma 4.2.16 enables us to speak about the numbers $\mathrm{BD}_n(v) := \overline{\mathrm{BD}}(L_n(x))$ for all $n \ge 1$, all $x \in X_v$ and all $v \in V$. Let

$$\overline{\mathrm{OD}}(S) = \max\{\overline{\mathrm{BD}}_1(v) : v \in V\}$$

and let w be one of the vertices where this maximum is assumed. We shall prove the following.

Lemma 4.2.17 *If S is a* CGDMS, *then for all $v \in V$ and for all $n \geq 1$, $\overline{\mathrm{BD}}_n(v) \leq \overline{\mathrm{OD}}(S)$.*

Proof. For every $v \in V$ fix $x_v \in \mathrm{Int}(X_v)$. Then there exists $0 < \rho < \min\{\mathrm{dist}(X_u, X_v) : u, v \in V \text{ and } u \neq v\}$ such that $B(x_v, \rho) \subset \mathrm{Int}(X_v)$ for every $v \in V$. First, we shall show that $\theta(S) \leq \overline{\mathrm{BD}}(L_1(x))$. To see this, fix $t > t - s > \overline{\mathrm{BD}}(L_1(x_w)) \geq \overline{\mathrm{BD}}(L_1(x_v)$ for every $v \in V$. Then fix $0 < \epsilon < \rho/2$ and for every $v \in V$ consider the set

$$I_v(\epsilon) = \{e \in I : t(e) = v \text{ and } K\epsilon\rho^{-1} \leq ||\phi_e'(x_v)|| \leq 2K\epsilon\rho^{-1}\}.$$

Since for given $v \in V$, the balls $B(\phi_i(x_v), \epsilon)$ with $e \in I_v(\epsilon)$ are disjoint, $N_\epsilon(L_1(x_v)) \geq \#I_v(\epsilon)$. Since the set V is finite, we have for all $v \in V$ and all $\epsilon > 0$ small enough, $\epsilon^s \geq \epsilon^t N_\epsilon((L_1(x_v))$. Therefore, for all k large enough, say $k \geq k_0$, we get obtain

$$\begin{aligned}
\sum_{k\geq k_0}\sum_{v\in V}\sum_{i\in I_v(2^{-k})} ||\phi_i'||^t &\leq \sum_{k\geq k_0}\sum_{v\in V} 2^t K^t \rho^{-t} 2^{-kt} \#I_v(2^{-k}) \\
&\leq 2^t K^t \rho^{-t} \sum_{k\geq k_0}\sum_{v\in V} 2^{-kt} N_{2^{-k}}(L_1(x)) \\
&\leq (2K\rho^{-1})^t \sum_{v\in V}\sum_{k\geq k_0} 2^{-ks} \\
&\leq (2K\rho^{-1})^t \frac{\#V}{1-2^{-s}} < \infty.
\end{aligned}$$

Since $\lim_{i\in I} ||\phi_i'|| = 0$, the set $I \backslash \sum_{v\in V} \bigcup_{k\geq k_0} I(2^{-k})$ is finite, and therefore $t \geq \theta(S)$. Letting $t \to \overline{\mathrm{BD}}(L_1(x_w))$, we get $\theta(S) \leq \overline{\mathrm{BD}}(L_1(x_w))$.

Now, fix $t > \overline{\mathrm{BD}}(L_1(x_w))$ again. We shall show by induction that for all $n \geq 1$ there exists $0 < A_n < \infty$ such that

$$N_r(L_n(x_v)) \leq A_n r^{-t}, \tag{4.34}$$

for all $v \in V$ and all $0 < r \leq 2D$. Indeed, the existence of A_1 is immediate as $t > \overline{\mathrm{BD}}(L_1(x_w)) \geq \overline{\mathrm{BD}}(L_1(x_v))$. Suppose that $0 < A_n < \infty$ exists. To prove the existence of A_{n+1}, set $I_1 = \{\omega \in E^n : \mathrm{diam}(\phi_\omega(X_{t(\omega)}) < r/2\}$. Since for every $v \in V$, $\omega \in I_1$ and every $e \in I$ such that $A_{\omega_n e} = 1$ and $t(e) = v$, we have $\phi_{\omega j}(x_v) = \phi_\omega(\phi_j(x_v)) \in B(\phi_\omega(x_{i(e)}, r/2)$, we conclude that for all $v \in V$, $N_r(L_{(I_1\times I)\cap E^{n+1}}, x_v) \leq \sum_{u\in V} N_{r/2}(L_{I_1}(x)) \leq \sum_{u\in V} N_{r/2}(L_n(x)) \leq 2^t \#V A_n r^{-t}$. If $\omega \in E^n \setminus I_1$, then for all $v \in V$, $N_r(L_{(\{\omega\}\times I)\cap E^{n+1}}, x_v) \leq N_{r/||\phi_\omega'||}(L_1(x_v)) \leq A_1 ||\phi_\omega'||^t r^{-t}$, where the second inequality sign holds since $r/||\phi_\omega'|| \leq$

$2\text{diam}(\phi_\omega(X_{t(\omega)}))/||\phi'_\omega|| \le 2D$. Thus, since $t > \theta(S)$,

$$\begin{aligned} N_r(L_{n+1}(x)) &\le 2^t \# V A_n r^{-t} + A_1 r^{-t} \sum_{\omega \in E^n \setminus I_1} ||\phi'_\omega||^t \\ &\le 2^t \# V A_n r^{-t} + A_1 Z_n(t) r^{-t} \\ &= (2^t \# V A_n + A_1 Z_n(t)) r^{-t}. \end{aligned}$$

The proof of (4.34) is completed by setting $A_{n+1} = 2^t \# V A_n + A_1 Z_n(t)$. Hence $\overline{\text{BD}_n}(v) = \overline{\text{BD}_n}(L_n(x_v)) \le t$ and therefore $\overline{\text{BD}_n}(v) \le \overline{\text{OD}}(S)$. □

As a consequence of Lemma 4.2.15 and Lemma 4.2.17, we have a simple means of obtaining the packing and upper box-counting dimensions of the limit set.

Theorem 4.2.18 *Let $\{\phi_i : i \in I\}$ be a conformal GDMS. Then $PD(J) = \overline{\text{BD}}(J) = \overline{\text{OD}}(S)$.*

4.3 Strongly regular, hereditarily regular and irregular systems

In this section following the terminology introduced in [MU1] in the context of iterated function systems we classify GDMS into regular, hereditarily regular and irregular ones. The latter differentiate themselves by odd geometric features.

Definition 4.3.1 A CGDMS *is said to be strongly regular if there exists* $t \ge 0$ *such that* $0 < \text{P}(t) < \infty$.

A family $\{\phi_i\}_{i \in F}$ is said to be a *cofinite subsystem* of a system $= \{\phi_i\}_{i \in I}$ if $F \subset I$ and the difference $I \setminus F$ is finite.

Definition 4.3.2 A CGDMS *is said to be hereditarily regular if each of its cofinite subsystem is regular.*

Using Proposition 2.1.9 we get the following obvious result.

Lemma 4.3.3 *The following conditions are equivalent.*

(a) $\psi_S(t) < \infty$.
(b) There exists a cofinite subsystem S' of S such that $\psi_{S'}(t) < \infty$.
(c) For every cofinite subsystem S' of S we have $\psi_{S'}(t) < \infty$.
(d) $\text{P}_S(t) < \infty$.
(e) There exists a cofinite subsystem S' of S such that $\text{P}_{S'}(t) < \infty$.
(f) For every cofinite subsystem S' of S we have $\text{P}_{S'}(t) < \infty$.

A strong geometric meaning of hereditarily regular systems is provided by the following.

Theorem 4.3.4 *An infinite system S is hereditarily regular if and only if* $\mathrm{P}(\theta) = \infty \Leftrightarrow \psi(\theta) = \infty \Leftrightarrow \{t : \mathrm{P}(t) < \infty\} = (\theta, \infty) \Leftrightarrow \{t : \psi(t) < \infty\} = (\theta, \infty)$. *If S is hereditarily regular, then $h > \theta$.*

Proof. If $\{t : \mathrm{P}(t) < \infty\} = (\theta, \infty)$, then S is hereditarily regular in view of Lemma 4.3.3, Theorem 4.2.9, and Proposition 4.2.8. If $\psi(\theta) < \infty$, then there exists a cofinite subsystem S' of S such that $\psi_{S'}(\theta) < 1$, whence $\mathrm{P}_{S'}(\theta) < 0$. Therefore S' is not regular in view of Theorem 4.2.9, Theorem 4.2.13 and Proposition 4.2.8. All other equivalences involved in this theorem now follow from Lemma 4.3.3. □

Theorem 4.3.5 *Each strongly regular system is regular and each hereditarily regular system is strongly regular. In addition, for each strongly regular system S, $h_S > \theta_S$.*

Proof. It immediately follows from Proposition 4.2.8 that each strongly regular system S is regular and that $h_S > \theta_S$ for this system S. If S is hereditarily regular, then $\theta_S = \infty$ by Theorem 4.3.4, and S is therefore strongly regular by Proposition 4.2.8. □

We shall prove the following interesting characterization of the θ_S number.

Theorem 4.3.6

$$\lim_{T \in \mathcal{F}in(I)} h_{I \setminus T} = \inf_{T \in \mathcal{F}in(I)} h_{I \setminus T} = \theta_S.$$

Proof. In view of Lemma 4.3.3, $\theta_{S'} = \theta_S$ for every cofinite subsystem S' of S. Therefore, using Theorem 4.2.13, we conclude that $\lim_{T \in \mathcal{F}in(I)} h_{I \setminus T} \geq \theta_S$. In order to prove the opposite inequality fix $t > \theta_S$. Then $\psi_S(t) < \infty$, and therefore there exists $F \in \mathcal{F}in(I)$ such that $\psi_{I \setminus T}(t) < 1$ for every finite subset T of I containing F. Hence $\mathrm{P}_{I \setminus T}(t) < 0$ for every finite subset T of I containing F which shows that $\lim_{T \in \mathcal{F}in(I)} h_{I \setminus T} \leq t$. □

Definition 4.3.7 *If a* CGDMS *S is not regular we call it irregular.*

From Theorem 4.2.9 and Proposition 4.2.8 we get the following characterization of irregular systems.

Theorem 4.3.8 *A CGDMS S is irregular if and only if* $\mathrm{P}(h) < 0 \Leftrightarrow \mathrm{P}(\theta) < 0$.

As the next theorem shows, irregular systems also exhibit some hereditary features.

Theorem 4.3.9 *If S is irregular, then every cofinite subsystem S' of S is irregular and $h_{S'} = \theta_S$.*

Proof. In view of Theorem 4.3.8 $h_S = \theta_S$. In view of Lemma 4.3.3 and Theorem 4.3.8, $\mathrm{P}_{S'}(\theta_{S'}) = \mathrm{P}_{S'}(\theta_S) \le \mathrm{P}_S(\theta_S) < 0$ and therefore it follows from Theorem 4.3.8 that S' is irregular. Thus, using Lemma 4.3.3, $h_{S'} = \theta_{S'} = \theta_S$. □

We end this section with the following interesting characterization of strongly regular systems.

Theorem 4.3.10 *Suppose that $S = \{\phi_i : i \in I\}$ is a* CGDMS. *Then the following conditions are equivalent.*

(a) S is strongly regular.
(b) $h_S > \theta_S$.
(c) There exists a proper cofinite subsystem S' of S such that $h_{S'} < h_S$.
(d) For every proper subsystem S' of S we have $h_{S'} < h_S$.

Proof. It is obvious that (d) $\Rightarrow$ (c). The implication (c) $\Rightarrow$ (b) follows from Theorem 4.3.9, Proposition 4.2.8 and Theorem 4.2.13. The implication (b) $\Rightarrow$ (a) is an immediate consequence of Theorem 4.3.9 and Theorem 4.3.8. Thus, we are only left to show that (a) $\Rightarrow$ (d). So, consider a proper subsystem $S' = \{\phi_i : i \in I'\}$ of S. Suppose first that S' is irregular and fix any number $\alpha \in (\theta_S, h_S)$. Then $\mathrm{P}_{S'}(\alpha) < \infty$ and therefore by Theorem 4.2.13, $h_{S'} \le \alpha < h_S$ and we are done in this case. So, suppose that S' is regular and additionally suppose to the contrary that $h = h_{S'} = h_S$. Recall that by $\tilde{\mu}$ and $\tilde{\mu}'$ we denote the invariant Gibbs states respectively corresponding to the families $\{h \log|\phi_i'|\}_{i\in I}$ and $\{h \log|\phi_i'|\}_{i\in I'}$. Then $\tilde{\mu}'([\omega]) \le QQ'\tilde{\mu}([\omega])$ for every $\omega \in E'^*$ and some constants Q and Q'. Since $E'^\infty \subset E^\infty$, $\tilde{\mu}'$ can be regarded as a probability measure on E^∞ ($\tilde{\mu}'(A) = \tilde{\mu}(A \cap E'^\infty)$ for every Borel set $A \subset E^\infty$. Then this last inequality implies that $\tilde{\mu}'$ is absolutely continuous with respect to $\tilde{\mu}$ (even $d\tilde{\mu}'/d\tilde{\mu} \le QQ'$). Since, in addition, both measures $\tilde{\mu}^*$ and $\tilde{\mu}'^*$ are ergodic and σ-invariant on E^∞, they must coincide. This however is a contradiction as $\tilde{\mu}'^*([j]) = 0$ and $\tilde{\mu}^*([j])Q^{-1}||\phi_j'|| > 0$ for every $j \in I \setminus I'$. □

4.4 Dimensions of measures

Let us observe first that the argument given at the beginning of the proof of Theorem 3.2.3 also yields the following remarkable fact, which can be called a measure theoretic open set condition.

Theorem 4.4.1 *If μ is a Borel shift-invariant ergodic probability measure on E^∞, then*

$$\mu \circ \pi^{-1}\big(\phi_\omega(X_{t(\omega)}) \cap \phi_\tau(X_{t(\tau)})\big) = 0 \tag{4.35}$$

for all incomparable words $\omega, \tau \in E^*$.

Recall that if ν is a finite Borel measure on X, then $\mathrm{HD}(\nu)$, the Hausdorff dimension of ν, is the minimum of Hausdorff dimensions of sets of full ν measure. As before, by $\alpha = \{[i] : i \in I\}$, we denote the partition of E^∞ into initial cylinders of length 1. If μ is a Borel shift-invariant ergodic probability measure on E^∞, by $\mathrm{h}_\mu(\sigma)$ we denote its entropy with respect to the shift map $\sigma : E^\infty \to E^\infty$ and by $\chi_\mu(\sigma) = \int \zeta d\mu > 0$ its *characteristic Lyapunov exponent*, where

$$\zeta(\omega) = -\log|\phi_{\omega_1}'(\pi(\sigma(\omega)))|.$$

We start with the following main result of this section versions of which were established in various conformal contexts.

Theorem 4.4.2 *(Volume lemma) Suppose that μ is a Borel shift-invariant ergodic probability measure on E^∞ such that at least one of the numbers $\mathrm{H}_\mu(\alpha)$ or $\chi_\mu(\sigma)$ is finite, where $\mathrm{H}_\mu(\alpha)$ is the entropy of the partition α with respect to the measure μ. Then*

$$\mathrm{HD}(\mu \circ \pi^{-1}) = \frac{\mathrm{h}_\mu(\sigma)}{\chi_\mu(\sigma)}.$$

Proof. Suppose first that $\mathrm{H}_\mu(\alpha) < \infty$. Then the series $\sum_{i \in I} -\mu([i]) \log(||\phi_i'||_0)$ converges and using (4e) and (4f) we conclude that the function ζ is integrable. Since $\mathrm{H}_\mu(\alpha) < \infty$ and since α is a generating partition, the entropy $\mathrm{h}_\mu(\sigma) = \mathrm{h}_\mu(\sigma, \alpha) \leq \mathrm{H}_\mu(\alpha)$ is finite. Thus, in view of the Birkhoff ergodic theorem and the Breimann–Shannon–McMillan theorem there exists a set $Z \subset E^\infty$ such that $\mu(Z) = 1$,

$$\lim_{n\to\infty} \frac{1}{n}\sum_{j=0}^{n-1} \zeta\circ\sigma^j(\omega) = \chi_\mu(\sigma) \ \text{ and } \ \lim_{n\to\infty} \frac{-\log(\mu([\omega|_n])}{n} = \mathrm{h}_\mu(\sigma) \quad (4.36)$$

for all $\omega \in Z$. Fix now $\omega \in Z$ and $\eta > 0$. For $r > 0$ let $n = n(\omega, r) \geq 0$ be the least integer such that $\phi_{\omega|_n}(X_{t(\omega_n)}) \subset B(\pi(\omega), r)$. Then $\log(\mu \circ \pi^{-1}(B(\pi(\omega), r))) \geq \log(\mu \circ \pi^{-1}(\phi_{\omega|_n}(X_{t(\omega_n)}))) \geq \log(\mu([\omega|_n]) \geq -(\mathrm{h}_\mu(\sigma) + \eta)n$ for every r small enough (which implies that $n = n(\omega, r)$ is large enough) and $\mathrm{diam}(\phi_{\omega|_{n-1}}(X_{t(\omega_{n-1})})) \geq r$. The last inequality implies that

$$\begin{aligned}\log r &\leq \log\bigl(\mathrm{diam}(\phi_{\omega|_{n-1}}(X_{t(\omega_{n-1})}))\bigr) \leq \log\bigl(D|\phi'_{\omega|_{n-1}}(\pi(\sigma^{n-1}(\omega)))|\bigr) \\ &\leq \log D + \sum_{j=1}^{n-1} \log|\phi'_{\omega_j}(\pi(\sigma^j(\omega)))| \leq \log D - (n-1)(\chi_\mu(\sigma) - \eta)\end{aligned}$$

for all r small enough. Therefore, for these r

$$\begin{aligned}\frac{\log\bigl(\mu\circ\pi^{-1}(B(\pi(\omega), r))\bigr)}{\log r} &\leq \frac{-(\mathrm{h}_\mu(\sigma) + \eta)n}{\log D - (n-1)(\chi_\mu(\sigma) - \eta)} \\ &= \frac{\mathrm{h}_\mu(\sigma) + \eta}{\frac{-\log D}{n} + \frac{n-1}{n}(\chi_\mu(\sigma) - \eta)}.\end{aligned}$$

Hence letting $r \to 0$, and consequently $n \to \infty$, we obtain

$$\limsup_{r\to 0} \frac{\log\bigl(\mu\circ\pi^{-1}(B(\pi(\omega), r))\bigr)}{\log r} \leq \frac{\mathrm{h}_\mu(\sigma) + \eta}{\chi_\mu(\sigma) - \eta}.$$

Since η was an arbitrary positive number we finally obtain

$$\limsup_{r\to 0} \frac{\log\bigl(\mu\circ\pi^{-1}(B(\pi(\omega), r))\bigr)}{\log r} \leq \frac{\mathrm{h}_\mu(\sigma)}{\chi_\mu(\sigma)}$$

for all $\omega \in Z$. Hence (see Appendix, Section A.2), as $\mu\circ\pi^{-1}(\pi(Z)) = 1$, $\mathrm{HD}(\mu\circ\pi^{-1}) \leq \mathrm{h}_\mu(\sigma)/\chi_\mu(\sigma)$. Let now $J_1 \subset J$ be an arbitrary Borel set such that $\mu\circ\pi^{-1}(J_1) > 0$. Fix $\eta > 0$. In view of (4.36) and Egorov's theorem there exist $n_0 \geq 1$ and a Borel set $\tilde{J}_2 \subset \pi^{-1}(J_1)$ such that $\mu(\tilde{J}_2) > \mu(\pi^{-1}(J_1))/2 > 0$,

$$\mu([\omega|_n]) \leq \exp\bigl((-\mathrm{h}_\mu(\sigma) + \eta)n\bigr) \quad (4.37)$$

and $|\phi'_{\omega|_n}(\pi(\sigma^n(\omega))| \geq \exp\bigl((-\chi_\mu(\sigma) - \eta)n\bigr)$ for all $n \geq n_0$ and all $\omega \in \tilde{J}_2$. By (4.23), the last inequality implies that there exists $n_1 \geq n_0$ such that

$$\mathrm{diam}\bigl(\phi_{\omega|_n}(X_{t(\omega_n)})\bigr) \geq D^{-1}\mathrm{e}^{(-\chi_\mu(\sigma)-\eta)n} \geq \mathrm{e}^{-(\chi_\mu(\sigma)+2\eta)n} \quad (4.38)$$

for all $n \geq n_1$ and all $\omega \in \tilde{J}_2$. Given now $0 < r < \exp(-(\chi_\mu(\sigma)+2\eta)n_1)$ and $\omega \in \tilde{J}_2$ let $n(\omega, r)$ be the least number n such that $\mathrm{diam}\big(\phi_{\omega|_{n+1}}(X_{t(\omega_{n+1})})\big) < r$. Using (4.38) we deduce that $n(\omega,r)+1 > n_1$, hence $n(\omega, r) \geq n_1$ and $\mathrm{diam}\big(\phi_{\omega|_n}(X_{t(\omega_n)})\big) \geq r$. In view of Lemma 4.2.6 there exists a universal constant $L \geq 1$ such that for every $\omega \in \tilde{J}_2$ and $0 < r < \exp(-(\chi_\mu(\sigma)+2\eta)n_1)$ there exist points $\omega^{(1)}, \dots, \omega^{(k)} \in \tilde{J}_2$ with $k \leq L$ such that $\pi(\tilde{J}_2) \cap B(\pi(\omega), r) \subset \bigcup_{j=1}^k \phi_{\omega^{(j)}|_{n(\omega^{(j)},r)}}\left(X_{t\left(\omega^{(j)}|_{n(\omega^{(j)},r)}\right)}\right)$. Let $\tilde{\mu} = \mu|_{\tilde{J}_2}$ be the restriction of the measure μ to the set $\tilde{J}_2$. Using (4.35), (4.37) and (4.38) we get

$$\begin{aligned}
&\tilde{\mu} \circ \pi^{-1}(B(\pi(\omega), r)) \\
&\leq \sum_{j=1}^k \mu \circ \pi^{-1}\big(\phi_{\omega^{(j)}|_{n(\omega^{(j)},r)}}(X_{t(\omega^{(j)}|_{n(\omega^{(j)},r)})})\big) \\
&= \sum_{j=1}^k \mu\big([\omega^{(j)}|_{n(\omega^{(j)},r)}]\big) \leq \sum_{j=1}^k \exp\big((-\mathrm{h}_\mu(\sigma)+\eta)n(\omega^{(j)}, r)\big) \\
&= \sum_{j=1}^k \Big(\exp(-(\chi_\mu(\sigma)+2\eta)(n(\omega^{(j)}, r)+1))\Big)^{\frac{n(\omega^{(j)},r)}{n(\omega^{(j)},r)+1}\cdot\frac{-\mathrm{h}_\mu(\sigma)+\eta}{-(\chi_\mu(\sigma)+2\eta)}} \\
&\leq \sum_{j=1}^k \mathrm{diam}\left(\phi_{\omega^{(j)}|_{n(\omega^{(j)},r)+1}}\left(X_{t\left(\omega^{(j)}_{n(\omega^{(j)},r)+1}\right)}\right)\right)^{\frac{n(\omega^{(j)},r)}{n(\omega^{(j)},r)+1}\cdot\frac{\mathrm{h}_\mu(\sigma)-\eta}{\chi_\mu(\sigma)+2\eta}} \\
&\leq \sum_{j=1}^k r^{\frac{n(\omega^{(j)},r)}{n(\omega^{(j)},r)+1}\cdot\frac{\mathrm{h}_\mu(\sigma)-\eta}{\chi_\mu(\sigma)+2\eta}} \leq L r^{\frac{\mathrm{h}_\mu(\sigma)-2\eta}{\chi_\mu(\sigma)+2\eta}},
\end{aligned}$$

where the last inequality holds assuming n_1 to be so large that $\frac{n_1}{n_1+1} \cdot \frac{\mathrm{h}_\mu(\sigma)-\eta}{\chi_\mu(\sigma)+2\eta} \geq \frac{\mathrm{h}_\mu(\sigma)-2\eta}{\chi_\mu(\sigma)+2\eta}$. Hence (see Appendix 2) $\mathrm{HD}(J_1) \geq \mathrm{HD}(\pi(\tilde{J}_2)) \geq \frac{\mathrm{h}_\mu(\sigma)-2\eta}{\chi_\mu(\sigma)+2\eta}$ and since η was an arbitrary number $\mathrm{HD}(J_1) \geq \frac{\mathrm{h}_\mu(\sigma)}{\chi_\mu(\sigma)}$. Thus $\mathrm{HD}(\mu \circ \pi^{-1}) \geq \frac{\mathrm{h}_\mu(\sigma)}{\chi_\mu(\sigma)}$ and the proof is complete in the case of finite entropy. If $\mathrm{H}_\mu(\alpha) = \infty$ but $\chi_\mu(\sigma) < \infty$, then the above considerations would imply that $\mathrm{HD}(\mu) = \infty$ which is impossible, and the proof is finished. □

Remark 4.4.3 *Note that in proving* $\mathrm{HD}(\mu \circ \pi^{-1}) \geq \frac{\mathrm{h}_\mu(\sigma)}{\chi_\mu(\sigma)}$ *we did not use the property* $\mu([\omega]) = \mu \circ \pi^{-1}(\phi_\omega(X_{t(\omega)}))$, $\omega \in E^*$, *which is equivalent with (4.35).*

Remark 4.4.4 *It is worth noting that* $\mathrm{H}_\mu(\alpha) < \infty$ *if and only if* $\mathrm{H}_\mu(\alpha^q) < \infty$ *for some* $q \geq 1$ *and therefore it is sufficient to assume in Theorem 4.4.2 that* $\mathrm{H}_\mu(\alpha^q) < \infty$ *for some* $q \geq 1$.

As an immediate consequence of Theorem 4.4.2 and Remark 4.4.4 we get the following.

Corollary 4.4.5 *If* $F = \{f^{(i)} : i : i \in I\}$ *is a summable Hölder family of functions and the series*

$$\sum_{\omega \in E^q} -\tilde{m}_F(\phi_\omega(X)) \log(\tilde{m}_F(\phi_\omega(X)))$$

converges for some $q \geq 1$, *then*

$$\mathrm{HD}(m_F) = \mathrm{HD}(\mu_F) = \frac{\mathrm{h}_{\tilde{\mu}_F}(\sigma)}{\chi_{\tilde{\mu}_F}(\sigma)}.$$

Corollary 4.4.6 *If the* CGDMS $S = \{\phi_i\}_{i \in I}$ *is strongly regular or, more generally, if it is regular and* $\mathrm{H}_{\tilde{\mu}}(\alpha) < \infty$, *where* μ *is the* S*-invariant version of the* h*-conformal measure* m, *then*

$$\mathrm{HD}(m) = \mathrm{HD}(\mu) = \mathrm{HD}(J_S).$$

Proof. We want to remark first that for each strongly regular system S, $\mathrm{H}_{\tilde{\mu}}(\alpha) < \infty$. And indeed, since S is strongly regular, there exists $\eta > 0$ such that $Z_1(h - \eta) < \infty$, which means that $\sum_{i \in I} ||\phi_i'||^{h-\eta} < \infty$. Since $||\phi_i'||^{-\eta} \geq -h \log ||\phi_i'||$ for all but perhaps finitely many $i \in I$, the series $\sum_{i \in I} -h \log(||\phi_i'||) ||\phi_i'||^h$ converges. Hence, by the Gibbs property, $\sum_{i \in I} -\log(\tilde{\mu}([i]) \tilde{\mu}([i]) < \infty$, which means that $\mathrm{H}_{\tilde{\mu}}(\alpha) < \infty$. So, suppose that S is regular and $\mathrm{H}_{\tilde{\mu}}(\alpha) < \infty$. Since μ and m are equivalent, $\mathrm{HD}(\mu) = \mathrm{HD}(m)$. Since for every $\omega \in E^\infty$,

$$\lim_{n\to\infty} \frac{-\log(\tilde{\mu}([\omega|_n]))}{\sum_{j=0}^{n-1} \zeta \circ \sigma^j(\omega)} = \lim_{n\to\infty} \frac{h \sum_{j=0}^{n-1} \zeta \circ \sigma^j(\omega)}{\sum_{j=0}^{n-1} \zeta \circ \sigma^j(\omega)} = h,$$

using Birkhoff's ergodic theorem and the Breimann–Shannon–McMillan theorem, we see that $\mathrm{h}_\mu / \chi_\mu = h$. The proof is now concluded by invoking Theorem 4.4.2 and Remark 4.4.3. □

We end this short section with the proof of the following two facts showing that essentially $\tilde{\mu}_{-h\zeta}$ is the only invariant measure on E^∞ whose projection onto J_S has the maximal dimension $\mathrm{HD}(J_S)$.

Theorem 4.4.7 *Suppose that $S = \{\phi_i\}_{i\in I}$ is a regular conformal system such that $\chi_{\tilde{\mu}_{-h\zeta}} < \infty$. Suppose also that μ is a shift-invariant ergodic Borel probability measure on E^∞ such that $\mathrm{H}_\mu(\alpha) < \infty$. If $\mathrm{HD}(\mu \circ \pi^{-1}) = h := \mathrm{HD}(J)$, then $\mu = \tilde{\mu}_{-h\zeta}$.*

Proof. If $\chi_\mu = \infty$, then it follows from Remark 4.4.3 that $h = \mathrm{HD}(\mu \circ \pi^{-1}) = 0$, which is a contradiction. So, $\chi_\mu < \infty$ and it follows from Remark 4.4.3 that $\mathrm{h}_\mu - h\chi_\mu \geq 0$. Since, in view of Theorem 4.2.9 and Theorem 4.2.13, $\mathrm{P}(-h\zeta) = \mathrm{P}(h) = 0$, we therefore deduce from Theorem 2.2.9 with $f = -h\zeta$, that $\mu = \tilde{m}u_{-h\zeta}$. □

Corollary 4.4.8 *Suppose that $S = \{\phi_i\}_{i\in I}$ is a regular conformal system such that $\chi_{\tilde{\mu}_{-h\zeta}} < \infty$. Suppose also that $F = \{f^{(i)} : i \in I\}$ is a summable Hölder family of functions satisfying the assumptions of Corollary 4.4.5 (or equivalently $\mathrm{H}_{\tilde{\mu}_F}(\alpha) < \infty$). If $\mathrm{HD}(\mu_F) = h := \mathrm{HD}(J)$, then $\tilde{\mu}_F = \tilde{\mu}_{-h\zeta}$ and the difference between the amalgamated function $f : E^\infty \to I\!R$ and the function $-h\zeta : E^\infty \to I\!R$ is cohomologous to a constant in the class of bounded Hölder continuous functions on E^∞.*

Proof. Since $\mu_F = \tilde{\mu}_F \circ \pi^{-1}$, all the assumptions of Theorem 4.4.7 are satisfied. It therefore follows from this theorem that $\tilde{\mu}_F = \tilde{\mu}_{-h\zeta}$. As an immediate application of Theorem 2.2.7 we now conclude that $f + h\zeta$ is cohomologous to a constant in the class of bounded Hölder continuous functions. □

4.5 Hausdorff, packing and Lebesgue measures

We start this section with the following two general results showing that despite Examples 5.2.6, 5.2.7 and 5.2.8 something positive can proved about Hausdorff and packing measures even in an entirely general conformal setting.

Theorem 4.5.1 *If m is a t-conformal measure on J, then the Hausdorff measure H_t restricted to J is absolutely continuous with respect to m and $||d\mathrm{H}_t/dm||_0 < \infty$. In particular, $\mathrm{H}_t(J)$ is finite.*

Proof. In view of (4.20) $\mathrm{diam}(\phi_\omega(X_{t(\omega)})) \leq D||\phi'_\omega||$ for every $\omega \in E^*$. Hence, by t-conformality of the measure m and the bounded distortion property, we get

$$m(\phi_\omega(X_{t(\omega)})) \geq K^{-t} m(X_{t(\omega)})||\phi'_\omega||^t \geq MK^{-t}||\phi'_\omega||^t,$$

where M is the constant coming from Lemma 4.2.10. Hence

$$\operatorname{diam}(\phi_\omega(X_{t(\omega)}))^t \le M^{-1}(DK)^t m(\phi_\omega(X_{t(\omega)})).$$

Let now A be a closed subset of J and for every $n \ge 1$ put $A_n = \{\omega \in E^n : \phi_\omega(J) \cap A \neq \emptyset\}$. Then the sequence of sets $\bigcup_{\omega\in A_n} \phi_\omega(X_{t(\omega)})$ is decreasing and $\bigcap_{n\ge1}(\bigcup_{\omega\in A_n} \phi_\omega(X_{t(\omega)})) = A$. Therefore

$$\begin{aligned}
\mathrm{H}_t(A) &\le \liminf_{n\to\infty} \sum_{\omega\in A_n} \left(\operatorname{diam}(\phi_\omega(X_{t(\omega)}))\right)^t \\
&\le \liminf_{n\to\infty} \left(M^{-1}(DK)^t \sum_{\omega\in A_n} m(\phi_\omega(X_{t(\omega)}))\right) \\
&= M^{-1}(DK)^t \liminf_{n\to\infty} \left(m(\bigcup_{\omega\in A_n} \phi_\omega(J))\right) = M^{-1}(DK)^t m(A).
\end{aligned}$$

Since J is a separable metric space, the measure m is regular and therefore the inequality $\mathrm{H}_t(A) \le M^{-1}(DK)^t m(A)$ extends to all Borel subsets of J. □

Let us now prove an analogous result for packing measures.

Theorem 4.5.2 *If m is a t-conformal measure for a CGDMS S and either the alphabet I is finite or $J \cap \operatorname{Int}(X) \neq \emptyset$, then m is absolutely continuous with respect to Π_t. Moreover, the Radon-Nikodym derivative $dm/d\Pi_t$ is uniformly bounded away from infinity. In particular $\Pi_h(J) > 0$.*

Proof. If I is finite, then the result follows from Theorem 4.2.11. So, suppose that $J \cap \operatorname{Int}(X_v) \neq \emptyset$ for some $v \in V$. Then there exists $q \ge 1$ and $\tau \in E^q$ with $i(\tau) = v$ and such that $\phi_\tau(X_{t(\tau)}) \subset \operatorname{Int}(X_v)$. Set $\gamma = \operatorname{dist}(\phi_\tau(X), \partial X)$. Let

$$R = \{\omega \in E^\infty : \omega|_{[n+1,n+q]} = \tau \text{ for infinitely many } n\text{'s}\}$$

and let R_0 be the set of those elements of E^∞ which contain no subword τ. Since $[\tau] \cap R_0 = \emptyset$, we get $\mu(R_0) < 1$, and since $\sigma^{-1}(E^\infty \setminus R_0) \subset E^\infty \setminus R_0$, it follows from the ergodicity of σ proved in Theorem 2.2.4 that $\mu(R_0) = 0$. As $E^\infty \setminus R = \bigcup_{n\ge0} \sigma^{-n}(R_0)$, we obtain $\mu(E^\infty \setminus R) = 0$. Therefore, using (4.29), we get $\mu(J \setminus \pi(R)) = \tilde\mu \circ \pi^{-1}(J \setminus \pi(R)) \le \tilde\mu(E^\infty \setminus R) = 0$. Take now $\omega \in R$ and an integer $n \ge 1$ such that $\omega|_{[n+1,n+q]} = \tau$. Put $x = \pi(\omega)$ and consider the ball $B(x, K^{-1}||\phi'_{\omega|_n}||\gamma)$. Since by (4.21) $B(x, K^{-1}||\phi'_{\omega|_n}||\gamma) \subset \phi_{\omega|_n}(B(\pi(\sigma^n(\omega)),\gamma))$ and since

$B(\pi(\sigma^n(\omega)),\gamma) \subset \text{Int}(X_v) \subset X_v$, using the conformality of m we get

$$m\big(B(x, K^{-1}||\phi'_{\omega|_n}||\gamma)\big) \le ||\phi'_{\omega|_n}||^t m(B(\pi(\sigma^n(\omega)),\gamma)) \le ||\phi'_{\omega|_n}||^t$$
$$= (K\gamma^{-1})^t (K^{-1}||\phi'_{\omega|_n}||\gamma)^t.$$

Since $m(J\backslash\pi(R)) = 0$, applying Theorem A2.0.13(1) we thus get $\Pi_t(E) \ge (K^{-1}\gamma)^t p_1(t)m(E)$ for every Borel subset E of J. □

The assumption $J \cap \text{Int}(X) \ne \emptyset$ is known in the literature concerning iterated function systems as the *strong open set condition* (abbreviated (SOSC)).

We shall now provide characterizations of positivity of the Hausdorff measure and finiteness of the packing measure of the limit set of a regular CGDMS. These characterizations reduce the procedure of comparing the ratios of measures of balls and their radii to the balls "containing" the first level sets. This is in the spirit of going to a "large scale". This idea is also employed in Theorem 4.6.2 and Theorem 6.3.1.

Theorem 4.5.3 *If S is a regular* CGDMS, *then the following conditions are equivalent.*

(a) $\text{H}_h(J) > 0$.

(b) There exists a constant $L > 0$ such that for every $i \in I$, every $r \ge \text{diam}(\phi_i(X_{t(i)}))$, and every $y \in \phi_i(V_{t(i)})$, $m(B(y,r)) \le Lr^h$.

(c) There are two constants $L > 0$, $\gamma \ge 1$ such that for every $i \in I$ and every $r \ge \gamma\text{diam}(\phi_i(X_{t(i)}))$ there exists $y \in \phi_i(V_{t(i)})$ such that $m(B(y,r)) \le Lr^h$.

Proof. (a)⇒(b). In order to prove this implication suppose that (b) fails. Then for every $L > 1/\text{dist}^h(X,\partial V)$ there exists $j \in I$ such that $m(B(x,r)) > Lr^h$ for some $x \in \phi_j(X_{t(j)})$ and some $r \ge \text{diam}(\phi_j(X_{t(j)}))$. Let J_1 be the image under π of all words of E^∞ that contain each element of I infinitely often. Consider $z \in J_1$, $z = \pi(\omega) \in E^\infty$ such that $\omega_{n+1} = j$ for some $n \ge 1$. Set $z_n = \pi(\sigma^n(\omega))$. Then $z = \phi_{\omega|_n}(z_n)$ and $z_n \in B(x,r)$. Since $r \le 1/L^{1/h} \le \text{dist}(X,\partial V)$, all the geometric consequences of the bounded distortion property (4e), (4f) and (4.19) – (4.24) are applicable to the ball $B(x,r)$. In particular, we get $|\phi_{\omega|_n}(z_n) - \phi_{\omega|_n}(x)| \le ||\phi'_{\omega|_n}||r$ and $B\big(\phi_{\omega|_n}(x), ||\phi'_{\omega|_n}||r\big) \supset \phi_{\omega|_n}(B(x,r))$. Therefore, $B(z, 2||\phi'_{\omega|_n}||r) \supset \phi_{\omega|_n}(B(x,r))$. By conformality and (4f) this implies that

$$m(B(z, 2||\phi'_{\omega|_n}||r)) \geq K^{-h}||\phi'_{\omega|_n}||^h m(B(x,r)) \geq K^{-h}L||\phi'_{\omega|_n}||^h r^h$$
$$= \left(2||\phi'_{\omega|_n}||r\right)^h \frac{L}{(2K)^h}.$$

Using Theorem A2.0.16, we get $\mathrm{H}_h(J_1) \leq C/L$, for some constant C independent of L. Now, letting $L \to \infty$ we conclude that $\mathrm{H}_h(J_1) = 0$. By Theorem 2.2.4, $m(J \backslash J_1) = 0$. This in turn, in view of Theorem 4.5.1, shows that $\mathrm{H}_h(J \setminus J_1) = 0$. Thus, $\mathrm{H}_h(J) = 0$ and therefore the proof of the implication (a)⇒(b) is finished.

The implication (b)⇒(c) is obvious.

(c)⇒(a). Set $\eta = \mathrm{dist}(X, \partial V)$. Increasing D or K if necessary, we may assume that $2KD\eta^{-1} \geq 1$. Take an arbitrary $x \in J$ and radius $r > 0$. Set $\tilde{r} = 2KD\eta^{-1}r$. For every $z \in B(x,r) \cap J$ consider a shortest word $\omega = \omega(z)$ such that $z \in \pi([\omega])$ and $\phi_\omega(X_{t(\omega)}) \subset B(z, \tilde{r})$. Then $\mathrm{diam}(\phi_{\omega|_{|\omega|-1}}(X_{t(\omega_{|\omega|-1})}) \geq \tilde{r}$. Let $R = \{\omega(z)|_{|\omega(z)|-1} : z \in J \cap B(x,r)\}$. Notice that R is finite since $\lim_{i \in I} \mathrm{diam}(\phi_i(X_{t(i)})) = 0$ and since $\lim_{n\to\infty} \sup\{\mathrm{diam}(\phi_\omega(X_{t(\omega)})) : \omega \in I^n\} = 0$. Therefore we can find a finite set $\{z_1, z_2, \ldots, z_k\} \subset J \cap B(x,r)$ such that the family $R^* = \{\omega(z_j)|_{|\omega(z_j)|-1} : j = 1, \ldots, k\} \subset R$ consists of mutually incomparable words and the family $\{\pi(\omega(z_j)|_{|\omega(z_j)|-1} : j = 1, \ldots, k\}$ covers $B(x,r) \cap J$. Now, temporarily fix an element $z \in \{z_1, z_2, \ldots, z_k\}$, set $\omega = \omega(z)$, $q = |\omega|$, and $\psi = \phi_{\omega|_{q-1}}$. Since $\mathrm{diam}(\psi(\mathrm{Int}(X_{t(\omega_{q-1})}))) \geq \tilde{r}$, it follows from Lemma 4.2.6 that $\#R^* \leq M$, a constant depending only on the system S. By the choice of ω, (4.23) and (4f) we have $D^{-1}K^{-1}||\psi'||\cdot||\phi'_{\omega_q}|| \leq 2\tilde{r}$, whence, by (4.20), $2KD^2||\psi'||^{-1}\tilde{r} \geq D||\phi'_{\omega_q}|| \geq \mathrm{diam}(\phi_{\omega_q}(X_{t(\omega_q)}))$. So, if $y \in \phi_{\omega_q}(X_{t(\omega_q)})$ is the point from the assumptions corresponding to the radius $4\gamma KD^2||\psi'||^{-1}\tilde{r} \geq 2\gamma \mathrm{diam}(\phi_{\omega_q}(X_{t(\omega_q)}))$, using (4.21), the inequality $2rK||\psi'||^{-1} \leq 2rKD\tilde{r}^{-1} = \eta$ and the relation $\psi^{-1}(z) \in \phi_{\omega_q}(X_{t(\omega_q)})$, we can write

$$\begin{aligned}
B(x,r) \cap \psi(X_{t(\omega_{q-1})}) &\subset B(z, 2r) \cap \psi(X_{t(\omega_{q-1})}) \\
&\subset \psi\big(B(\psi^{-1}(z), 2rK||\psi'||^{-1})\big) \\
&\subset \psi\big(B(y, 2\tilde{r}K||\psi'||^{-1} + 2\tilde{r}K||\psi'||^{-1}D^2)\big) \\
&\subset \psi\big(B(y, 4D^2K||\psi'||^{-1}\tilde{r})\big).
\end{aligned}$$

So, by the assumptions of the lemma,

$$\begin{aligned} m\big(B(x,r)\cap\psi(X_{t(\omega_{q-1})})\big) &\le m\big(\psi(X_{t(\omega_{q-1})})\cap B(y,4D^2K||\psi'||^{-1}\tilde r)\big)\\ &\le ||\psi'||^h m\big(B(y,4D^2K||\psi'||^{-1}\tilde r)\big)\\ &\le ||\psi'||^h L\big(4D^2K||\psi'||^{-1}\tilde r\big)^h\\ &= L(8D^3K^2\eta^{-1})^h r^h. \end{aligned}$$

Therefore $m(B(x,r)) \le \#R^*L(8D^3K^2\eta^{-1})^h r^h \le ML(8D^3K^2\eta^{-1})^h r^h$ and applying Theorem A2.0.12(2) finishes the proof of this implication and simultaneously the whole thorem. □

Remark 4.5.4 *It is obvious that it suffices the for conditions (b) and (c) of Theorem 4.5.3 to be satisfied for a cofinite subset of I.*

Theorem 4.5.5 *If $S=\{\phi_i : i\in I\}$ is a regular* CGDMS, *then the following conditions are equivalent.*

(a) $\Pi_h(J)<+\infty$.

(b) There are two constants $L>0$, $\xi>0$ such that for every $i\in I$, every $\mathrm{diam}(\phi_i(X))\le r\le\xi$ *and every $y\in\phi_i(W_{t(i)})$, $m(B(y,r))\ge Lr^h$.*

(c) There are three constants $L>0$, $\xi>0$, and $\gamma\ge 1$ such that for every $i\in I$ and every $\gamma\mathrm{diam}(\phi_i(X))\le r\le\xi$ *there exists $y\in\phi_i(W_{t(i)})$ such that $m(B(y,r))\ge Lr^h$.*

Proof. (a)⇒(b). In order to prove this implication suppose that (b) fails and (a) holds. Fix $L,\xi>0$. Then there are $i\in I$ and $\mathrm{diam}\phi_i(X)\le r\le\xi$ such that for some $x\in\phi_i(W_{t(i)})$, we have

$$m(B(x,r))\le Lr^h.$$

Since the system is regular, there is a Borel subset B of J with $m(B)=1$ and such that each point z of B has a unique code, ω, and $\pi(\sigma^n(\omega))$ is in the ball $B(x,r/2)$ for infinitely many n's. For such a point z and integer $n\ge 1$, we have by conformality of m and (4f)

$$\begin{aligned} m(\phi_{\omega|n}(B(\pi(\sigma^n(\omega)),r/2))) &\le \|\phi'_{\omega|n}\|^h m(B(\pi(\sigma^n(\omega)),r/2))\\ &\le \|\phi'_{\omega|n}\|^h m(B(x,r)) \le \|\phi'_{\omega|n}\|^h Lr^h. \end{aligned}$$

But, by (4.21),

$$\phi_{\omega|n}(B(\pi(\sigma^n(\omega)),r/2)) \supset B(z,\|\phi'_{\omega|n}\|K^{-1}r/2).$$

So, $m(B(z, \|\phi'_{\omega|n}\|r/2K)) \leq (\|\phi'_{\omega|n}\|r/2K)^h(2K)^hL$. Using Theorem A2.0.13(1), we get $\Pi^h(J) \geq \Pi^h(J \cap B) \geq (2K)^{-h}L^{-1}p_1(t)$. Now, letting $L \to 0$ we get $\Pi^h(J) = \infty$. This contradiction finishes the proof of the implication (a)$\Rightarrow$(b).

The implication (b)$\Rightarrow$(c) is obvious.

(c)$\Rightarrow$(a). First notice that decreasing L if necessary, the assumption of the lemma continues to be fulfilled if the number ξ is replaced by any other positive number, for example by $\eta/2$, where $\eta = \mathrm{dist}(X, \partial V)$. Fix $0 < r < \xi$, $x = \pi(\omega) \in J$, and take the maximal $k \geq 0$ such that

$$\phi_{\omega|_k}(W_{t(\omega_k)}) \supset B(x, D^{-2}r). \tag{4.39}$$

Abbreviate $\phi_{\omega|_{k+1}}$ by ψ. Then $\psi(W_{t(\omega_{k+1})})$ does not contain $B(x, D^{-2}r)$ and, as by (4.23), $\psi(V) \supset B(x, D^{-1}||\psi'||)$, we obtain $D^{-2}r > D^{-1}||\psi'||$. Hence, by (4.20), $B(x, r) \supset B(x, D||\psi'||) \supset \psi(W_{t(\omega_{k+1})})$ and therefore, using the conformality of m, we get $m(B(x, r)) \geq K^{-h}||\psi'||^h m(X_{t(\omega_{k+1})}))$ $\geq MK^{-h}||\psi'||^h$ where M comes from Lemma 4.2.10. If now $\gamma DK||\psi'|| \geq (2D^3)^{-1}\eta r$, then

$$\frac{m(B(x,r))}{r^h} \geq M\eta^h(2D^4K^2\gamma)^{-h}.$$

Otherwise,

$$\gamma DK||\psi'|| < (2D^3)^{-1}\eta r. \tag{4.40}$$

Set $g = \phi_{\omega_{k+1}}$ and let y be an arbitrary point in $g(W_{t(\omega_{k+1})})$. Since $\mathrm{diam}(g(W_{t(\omega_{k+1})})) \leq D||g'||$ and since $\gamma \geq 1$, it follows from (4.40) that

$$B\big(y, (2D^3)^{-1}\eta||\phi'_{\omega|_k}||^{-1}r\big) \subset B\big(\pi(\sigma^k(\omega)), D^{-3}\eta||\phi'_{\omega|_k}||^{-1}r\big). \tag{4.41}$$

From (4.19)

$$\begin{aligned} &\phi_{\omega|_k}\big(B(\pi(\sigma^k(\omega)), D^{-3}\eta||\phi'_{\omega|_k}||^{-1}r)\big) \\ &\quad \subset B\big(x, D^{-3}\eta r||\phi'_{\omega|_k}(g(z))||\big) \subset B(x, r). \end{aligned} \tag{4.42}$$

In view of (4.40) we have $(2D^3)^{-1}\eta||\phi'_{\omega|_k}||^{-1}r\gamma D||g'(z)|| \geq \gamma\mathrm{diam}$ $(g(W_{t(\omega_{k+1})}))$. By (4.39) and (4.20), $D^{-3}||\phi'_{\omega|_k}||^{-1}r \leq 1$; hence $(2D^3)^{-1}$ $\eta||\phi'_{\omega|_k}||^{-1}r\gamma D||g'(z)|| \leq \eta/2$. As the number $(2K)^{-1}||\phi'_{\omega|_k}(g(z))||^{-1}r$ does not depend on the choice of $y \in g(W_{t(\omega_{k+1})})$, we can assume that y satisfies the assumption of our lemma. So, using this assumption, it follows from (4.41) and (4.42) that

$$m(B(x,r)) \geq K^{-h}||\phi'_{\omega|_k}||^h L(2D^3)^{-h}||\phi'_{\omega|_k}||^{-h}r^h = L(2D^3K)^{-h}r^h.$$

□

Remark 4.5.6 *It is obvious that it suffices for the conditions (b) and (c) of Theorem 4.5.5 to be satisfied for a cofinite subset of I.*

As a consequence of Theorem 4.5.3, we get the following.

Corollary 4.5.7 *If $S = \{\phi_i : i \in I\}$ is a regular* CGDMS *and there exist a sequence of points $z_j \in S(\infty)$ and a sequence of positive reals $\{r_j : j \geq 1\}$ such that*

$$\limsup_{j\to\infty} \frac{m(B(z_j, r_j))}{r_j^h} = \infty,$$

then $\mathrm{H}_h(J) = 0$.

Proof. Indeed, by the definition of $S(\infty)$, for every $j \geq 1$ there exists $i(j) \in I$ such that $\phi_{i(j)}(X_{t(i(j))}) \subset B(z_j, r_j)$. Then for every $x \in \phi_{i(j)}(X_{t(i(j))})$ we have

$$\frac{m(B(x, 2r_j)}{(2r_j)^h} \geq \frac{m(B(z_j, r_j))}{(2r_j)^h}$$

and letting $j \nearrow \infty$ we see that condition (b) of Theorem 4.5.3 is not satisfied. □

As a consequence of Theorem 4.5.3, we get the following.

Corollary 4.5.8 *If $S = \{\phi_i : i \in I\}$ is a regular* CGDMS *and there exist a sequence of points $z_j \in S(\infty)$ and a sequence of positive reals $\{r_j : j \geq 1\}$ such that*

$$\liminf_{j\to\infty} \frac{m(B(z_j, r_j))}{r_j^h} = 0,$$

then $\Pi_h(J) = \infty$.

Proof. Indeed, by the definition of $S(\infty)$, for every $j \geq 1$ there exists $i(j) \in I$ such that $\phi_{i(j)}(X_{t(i(j))}) \subset B(z_j, r_j/4)$. Then for every $x \in \phi_{i(j)}(X_{t(i(j))})$, we have $\phi_{i(j)}(X_{t(i(j))}) \subset B(x, r_j/2) \subset B(z_j, r_j)$. Therefore

$$\frac{m(B(x, r_j/2)}{(r_j/2)^h} \leq \frac{m(B(z_j, r_j))}{(r_j/2)^h}$$

and letting $j \nearrow \infty$ we see that condition (b) of Theorem 4.5.5 is not satisfied. □

Now for each $n \geq 0$ put

$$X_n = \bigcup_{\omega \in E^n} \phi_\omega(X_{t(\omega)}).$$

We shall now prove two results concerning the d-dimensional Lebesgue measure λ_d of these sets, the Lebesgue measure of the limit set J, and an estimate on the Hausdorff dimension of J.

Proposition 4.5.9 *If S is a* CGDMS *and $\lambda_d\big(\mathrm{Int}(X) \setminus X_1\big) > 0$, then there exists $0 < \gamma < 1$ such that $\lambda_d(X_n) \leq \gamma^n \lambda_d(X)$ for all $n \geq 1$. In particular $\lambda_d(J) = 0$.*

Proof. For every $v \in V$ put $G = \mathrm{Int}(X) \setminus X_1$ and $\xi = K^{-d} \min\{\lambda_d(G_v)/\lambda_d(X_v)\} > 0$. In view of the bounded distortion property we have $\lambda_d(\phi_\omega(G_{t(\omega)})) \geq \xi \lambda_d(\phi_\omega(X_{t(\omega)}))$. In view of the open set condition we have $\phi_\omega(G_{t(\omega)}) \cap \phi_\tau(X_{t(\omega)}) = \emptyset$ if $\omega \neq \tau$ and $|\omega| = |\tau|$. Thus for all $n \geq 0$

$$\begin{aligned} X_{n+1} &= \bigcup_{\omega \in E^n} \phi_\omega(X_1) \subset \bigcup_{\omega \in E^n} \phi_\omega(X_{t(\omega)} \setminus G_{t(\omega)}) \\ &= \bigcup_{\omega \in E^n} \phi_\omega(X_{t(\omega)}) \setminus \bigcup_{\omega \in E^n} \phi_\omega(G_{t(\omega)}) = X_n \setminus \bigcup_{\omega \in E^n} \phi_\omega(G_{t(\omega)}). \end{aligned}$$

Therefore

$$\begin{aligned} \lambda_d(X_{n+1}) &= \lambda_d(X_n) - \lambda\Big(\bigcup_{\omega \in E^n} \phi_\omega(G_{t(\omega)})\Big) \\ &= \lambda_d(X_n) - \sum_{\omega \in E^n} \lambda_d(\phi_\omega(G_{t(\omega)})) \\ &\leq \lambda_d(X_n) - \xi \sum_{\omega \in E^n} \lambda_d(\phi_\omega(X_{t(\omega)})) \\ &\leq \lambda_d(X_n) - \xi \lambda_d(X_n) = (1 - \xi)\lambda_d(X_n). \end{aligned}$$

So, putting $\gamma = 1 - \xi$ finishes the proof. □

Theorem 4.5.10 *If S is a regular* CGDMS *and $\lambda_d(\mathrm{Int}(X) \setminus X_1) > 0$, then $h = \mathrm{HD}(J) < d$. Conversely, if $\lambda_d(X \setminus X_1) = 0$ and $\lambda_d(\partial X) = 0$, then S is regular, $\lambda_d(J) = \lambda_d(X) > 0$, in particular $\mathrm{HD}(J) = d$, and $\lambda_d/\lambda_d(X)$ is the only conformal measure.*

Proof. In order to prove the first part suppose to the contrary that $h = d$. Then for every $\omega \in E^*$ and every Borel set $A \subset X_{t(\omega)}$ with

$\lambda_d(A) > 0$ we have

$$\begin{aligned} m(\phi_\omega(A)) &\le ||\phi'_\omega||^d m(A) = K^{-d}||\phi'_\omega||^d \lambda_d(A)\frac{m(A)}{\lambda_d(A)}K^d \\ &\le \lambda_d(\phi_\omega(A))K^d\frac{m(A)}{\lambda_d(A)}. \end{aligned} \tag{4.43}$$

For every $n \ge 1$ and every $\omega \in E^n$ define $Y_\omega = \phi_\omega(X_{t(\omega)}) \cap \bigcup_{\tau\in E^n\setminus\{\omega\}} \phi_\tau(X_{t(\omega)})$. Then the sets $\phi_\omega(X_{t(\omega)}) \setminus Y_\omega$, $\omega \in E^n$, are mutually disjoint, $m(Y_\omega) = 0$ for all $\omega \in E^n$ in view of (4.29), and $\phi_\omega^{-1}(Y_\omega) \subset \partial X$ by the open set condition. Therefore, using (4.43), we get the following estimate:

$$\begin{aligned} m(X_n) &= \sum_{\omega\in E^n} m\big(\phi_\omega(X_{t(\omega)}) \setminus Y_\omega\big) = \sum_{\omega\in E^n} m\big(\phi_\omega(X_{t(\omega)} \setminus \phi_\omega^{-1}(Y_\omega))\big) \\ &\le \sum_{\omega\in E^n} \lambda_d\big(\phi_\omega(X_{t(\omega)} \setminus \phi_\omega^{-1}(Y_\omega))\big)K^d\frac{m(X_{t(\omega)} \setminus \phi_\omega^{-1}(Y_\omega))}{\lambda_d(X_{t(\omega)} \setminus \phi_\omega^{-1}(Y_\omega))} \\ &\le K^d\frac{m(X)}{\min\{\lambda_d(\mathrm{Int}(X_v)) : v \in V\}}\sum_{\omega\in E^n} \lambda_d\big(\phi_\omega(X_{t(\omega)}) \setminus Y_\omega\big) \\ &\le \frac{K^d}{\min\{\lambda_d(\mathrm{Int}(X_v)) : v \in V\}}\lambda_d(X_n). \end{aligned}$$

Thus, by Proposition 4.5.9, $m(J) = \lim_{n\to\infty} m(X_n) = 0$. This completes the proof of the first part.

Moving to the other part of the theorem notice first that for every $n \ge 0$ we have

$$\begin{aligned} X_n \setminus X_{n+1} &= \bigcup_{\omega\in E^n} \phi_\omega(X_{t(\omega)}) \setminus \bigcup_{\omega\in E^n} \phi_\omega(X_1) \\ &\subset \bigcup_{\omega\in E^n} \big(\phi_\omega(X_{t(\omega)}) \setminus \phi_\omega\big(X_1 \cap X_{t(\omega)}\big)\big) \\ &= \bigcup_{\omega\in E^n} \phi_\omega(X_{t(\omega)} \setminus X_1) \end{aligned}$$

Since $\lambda_d(X \setminus X_1) = 0$, we therefore obtain $\lambda_d(X_n \setminus X_{n+1}) = 0$ or equivalently $\lambda_d(X_n) = \lambda_d(X_{n+1})$. Hence $\lambda_d(J) = \lim_{n\to\infty} \lambda_d(X_n) = \lambda_d(X) > 0$. In particular $h = d$. Since obviously $\lambda_d(\phi_\omega(A)) = \int_A |\phi'_\omega|^d \, d\lambda_d$ for all $n \ge 0$ and all Borel subsets A of X, and since $\lambda_d\big(\phi_\omega(X_{t(\omega)}) \cap \phi_\tau(X_{t(\tau)})\big) = 0$ for all incomparable words $\omega, \tau \in E^*$ by the condition $\lambda_d(\partial X) = 0$, we conclude that $\lambda_d/\lambda_d(X)$ is a conformal measure for the system S. □

Let us finish this section with the following result concerning irregular systems.

Theorem 4.5.11 *If S is irregular, then each measure $\mathrm{H}_g(J)$ or $\Pi_g(J)$ is either zero or infinity for every gauge function g of the form $t^h L(t)$, where $L(t)$ is a slowly varying function. Additionally $\mathrm{H}_h(J) = 0$.*

Proof. Suppose that a measure $\mathrm{H}_g(J)$ or $\Pi_g(J)$ (call it G_g) is finite. Then the Jacobian (Radon-Nikodym derivative) of a map ϕ_ω, $\omega \in I^*$, with respect to the measure G_g is equal to $|\phi'_\omega|^h$. By the definition of pressure there exists $n_0 \geq 1$ such that $\sum_{\omega \in E^n} ||\phi'_\omega||^h < \exp(n\mathrm{P}(h)/2)$ for every $n \geq n_0$. Hence

$$G_g(J) \leq \sum_{\omega \in E^n} G_g(\phi_i(J)) \leq \sum_{\omega \in E^n} ||\phi'_\omega||^h G_g(J) < \exp(n\mathrm{P}(h)/2) G_g(J).$$

Thus letting $n \to \infty$ and noting that by Theorem 4.3.8 $\mathrm{P}(h) < 0$, we obtain $G_g(J) = 0$. The proof that $H_h(J) = 0$ is very similar but requires a slightly different argument as we do not know whether $\mathrm{H}_h(J)$ is finite. Indeed, if $n \geq n_0$ is as above, then

$$\sum_{\omega \in E^n} \left(\mathrm{diam}(\phi_i(X_{t(\omega)}))\right)^h \leq \sum_{\omega \in E^n} D^h ||\phi'_\omega||^h < D^h \exp(n\mathrm{P}(h)/2)$$

and letting $n \to \infty$ we conclude that $\mathrm{H}_h(J) = 0$. □

4.6 Porosity of limit sets

A bounded subset X of a Euclidean space is said to be *porous* if there exists a positive constant $c > 0$ such that each open ball B centered at a point of X and of an arbitrary radius $0 < r \leq 1$ contains an open ball of radius cr disjoint from X. If only balls B centered at a fixed point $x \in X$ are discussed, X is called porous at x.

Obviously the following, formally weaker, requirement also defines porosity. There exist positive constants $c, \kappa > 0$ such that each open ball B centered at a point of X and of an arbitrary radius $0 < \kappa r \leq 1$ contains an open ball of radius cr disjoint from X. For a fixed κ, c is called a porosity constant of X.

It is easy to see that each porous set has box counting dimension less than the dimension of the Euclidean space it is contained in. Further relations between porosity and dimensions can be found for example in [Ma2] and [Sal]. A much weaker property, also called porosity, was introduced in [De]. For a survey concerning this concept see [Zar]. Here we will only be interested in the notion of porosity described in the first paragraph of this section.

In this section, following the approach from [U2], we deal with the problem of porosity of limit sets of conformal graph directed Markov systems. We provide a necessary and sufficient condition for the limit sets of these systems to be porous and we show that the limit set of each non-trivial finite system is porous. In [U2] the reader may find some examples to which the results of this section apply. One of the most interesting is provided by the arithmetic characterization of subsets I of positive integers such that the set of all reals in $[0,1]$, all of whose entries in the continued fraction expansion belong to I, is porous. In [JJM], conformal iterated functions systems yielding limit sets with positive porosity are also characterized and applications are given to random fractals. We start this section with the following straightforward result.

Theorem 4.6.1 *If $S=\{\phi_i\}_{i\in I}$ is a* CGDMS *and* $\text{Int}X\setminus\overline{J}\neq\emptyset$, *then* $J\subset\mathbb{R}^d$ *is a nowhere-dense set.*

Proof. Consider an arbitrary point $x\in J$ and a radius $r>0$. By the definition of the limit set there exists $\omega\in E^*$ such that $\phi_\omega(X_{t(\omega)})\subset B(x,r)$. By the open set condition $\phi_\omega(\text{Int}(X_{t(\omega)})\setminus\overline{J})\subset B(x,r)$ is then an open set disjoint from J and we are done. □

The main result of this section is the following characterization of porosity of limit sets of conformal GDMSs in terms of the "large scale" behavior.

Theorem 4.6.2 *Let $S=\{\phi_i\}_{i\in I}$ be a* CGDMS *such that the cone condition is satisfied for every $x\in X$. Then the following three conditions are equivalent.*

(a) *The limit set J is porous.*

(b) $\exists c>0\,\exists\xi>0\,\forall i\in I\,\forall 0<r\leq\xi$
if $r\geq\text{diam}(\phi_i(X_{t(i)}))$ *then there exists* $x_i\in B(\phi_i(X_{t(i)}),r)\cap X_{i(i)}$ *such that*

$$B(x_i,cr)\cap J=\emptyset.$$

(c) $\exists\kappa\geq 1\,\exists c>0\,\exists\xi>0\,\exists\beta\geq 1\,\forall i\in I\,\forall 0<r\leq\xi$
if $r\geq\beta\text{diam}(\phi_i(X_{t(i)}))$ *then there exists* $x_i\in B(\phi_i(X_{t(i)}),\kappa r)\cap X_{i(i)}$ *such that*

$$B(x_i,cr)\cap J=\emptyset.$$

Proof. It is obvious that (a)⇒(b)⇒(c). So, suppose that condition (c) is satisfied. Decreasing $c > 0$ if necessary, we may assume that it holds with $\xi \geq 2KD^3\beta$. Fix an arbitrary $x = \pi(\omega) \in J$, $\omega \in E^\infty$, and a positive radius $r < 2KD^3\beta$. Let $n \geq 1$ be the least integer such that

$$\phi_{\omega|_n}(X_{t(\omega_n)}) \subset B\left(x, \frac{r}{2KD^2\beta}\right).$$

Suppose first that $n = 1$. Then $r \geq \beta\text{diam}(\phi_{\omega_1}(X_{t(\omega_1)}))$ and, as $r < 2KD^3\beta$, we conclude from (c) that $B(x_{\omega_1}, cr) \cap J = \emptyset$. Since also $B(x_{\omega_1}, cr) \subset B(x, cr + \kappa r) \subset B(x, (c+\kappa)r)$, we are done in this case with the porosity constant $\leq c/2$. So, suppose in turn that $n \geq 2$. Then

$$\text{diam}(\phi_{\omega|_n}(X)) \leq \frac{r}{KD^2\beta} \quad \text{and} \quad \text{diam}(\phi_{\omega|_{n-1}}(X)) \geq \frac{r}{2KD^2\beta}. \tag{4.44}$$

Therefore by (4.20) and (4f)

$$\begin{aligned}\text{diam}(\phi_{\omega_n}(X_{t(\omega_n)})) &\leq D||\phi'_{\omega_n}|| \leq DK\frac{||\phi'_{\omega|_n}||}{||\phi'_{\omega|_{n-1}}||} \\ &\leq DK||\phi'_{\omega|_{n-1}}||^{-1}D\text{diam}(\phi_{\omega|_n}(X_{t(\omega_n)})) \\ &\leq D^2Kr(\beta KD^2)^{-1}||\phi'_{\omega|_{n-1}}||^{-1} = \beta^{-1}r||\phi'_{\omega|_{n-1}}||^{-1}.\end{aligned}$$

Hence,

$$r||\phi'_{\omega|_{n-1}n}||^{-1} \geq \beta\text{diam}(\phi_{\omega_n}(X_{t(\omega_n)})). \tag{4.45}$$

Also by (4.44)

$$r||\phi'_{\omega|_{n-1}n}||^{-1} \leq Dr\text{diam}^{-1}(\phi_{\omega|_{n-1}}(X_{t(\omega_{n-1})})) \leq 2KD^3\beta. \tag{4.46}$$

Hence, condition (c) is applicable with $i = \omega_n$ and the radius $r||\phi'_{\omega|_{n-1}n}||^{-1}$. Using (4.45) we get

$$\begin{aligned}&\phi_{\omega|_{n-1}}(B(x_{\omega_n}, cr||\phi'_{\omega|_{n-1}}||^{-1})) \\ &\quad\subset \phi_{\omega|_{n-1}}(B(\phi_{\omega|_n}(X), cr||\phi'_{\omega|_{n-1}}||^{-1} + \kappa r||\phi'_{\omega|_{n-1}}||^{-1})) \\ &\quad\subset \phi_{\omega|_{n-1}}(B(\pi(\sigma^{n-1}(\omega)), \beta^{-1}r||\phi'_{\omega|_{n-1}}||^{-1} + cr||\phi'_{\omega|_{n-1}}||^{-1} \\ &\quad+ \kappa r||\phi'_{\omega|_{n-1}}||^{-1})) \subset B(x, (2+\kappa)r).\end{aligned}$$

Since $B(x_{\omega_n}, cr||\phi'_{\omega|_{n-1}}||^{-1})$ may not be contained in $X_{i(\omega_n)}$, we need the following reasoning to conclude the proof. In view of (4.46) and the cone condition (4d), we get for some $y \in \mathrm{Con}(x_{\omega_n}, \alpha, \min\{c, l(2KD^3\beta)^{-1}\} r||\phi'_{\omega|_{n-1}}||^{-1})$,

$$\begin{aligned}
&\phi_{\omega|_{n-1}}(B(x_{\omega_n}, cr||\phi'_{\omega|_{n-1}n}||^{-1})) \\
&\quad \supset \phi_{\omega|_{n-1}}(\mathrm{Con}(x_{\omega_n}, \alpha, \min\{cr||\phi'_{\omega|_{n-1}}||^{-1}, l\})) \\
&\quad \supset \phi_{\omega|_{n-1}}(\mathrm{Con}(x_{\omega_n}, \alpha, \min\{c, l(2KD^3\beta)^{-1}\}r||\phi'_{\omega|_{n-1}}||^{-1})) \\
&\quad \supset \phi_{\omega|_{n-1}}(B(y, c' \min\{c, l(2KD^3\beta)^{-1}\}r||\phi'_{\omega|_{n-1}}||^{-1})) \\
&\quad \supset B(\phi_{\omega|_{n-1}}(y), K^{-1}c' \min\{c, l(2KD^3\beta)^{-1}\}r),
\end{aligned}$$

where $0 < c' \leq 1$ is so small that each central cone $\mathrm{Con}(z, \alpha, k)$ contains an open ball of radius $c'k$. Since $\mathrm{Con}(x_{\omega_n}, \alpha, \min\{cr||\phi'_{\omega|_{n-1}}||^{-1}, l\}) \subset \mathrm{Int}(X_{i(\omega_n)})$ and $J \cap B(x_{\omega_n}, cr||\phi'_{\omega|_{n-1}}||^{-1}) = \emptyset$, we conclude that

$$J \cap B(\phi_{\omega|_{n-1}}(y), K^{-1}c' \min\{c, l(2KD^3\beta)^{-1}\}r) = \emptyset.$$

□

Theorem 2.3 from [U2] shows that one cannot replace the set I by any of its cofinite subsystems (i.e. those whose complements in I are finite) in Theorem 4.6.2. As an immediate consequence of Theorem 4.6.2 we get however the following.

Theorem 4.6.3 *If there exists a cofinite subset F of I such that one of the following conditions is satisfied, then the limit set J_I is porous.*

(a) $\exists c > 0\, \exists \xi > 0\, \forall i \in I\, \forall 0 < r \leq \xi$
if $r \geq \mathrm{diam}(\phi_i(X))$ *then there exists* $x_i \in B(\phi_i(X), r)$ *such that*

$$B(x_i, cr) \cap J = \emptyset.$$

(b) $\exists \kappa \geq 1\, \exists c > 0\, \exists \xi > 0\, \exists \beta \geq 1\, \forall i \in I\, \forall 0 < r < \xi$
if $r \geq \beta \mathrm{diam}(\phi_i(X))$ *then there exists* $x_i \in B(\phi_i(X), \kappa r)$ *such that*

$$B(x_i, cr) \cap J = \emptyset.$$

As an immediate consequence of Theorem 4.6.2 and Theorem 4.6.1 we get the following result showing that in case of a finite iterated function system "most" limit sets are porous.

Theorem 4.6.4 *If $S = \{\phi_i\}_{i \in I}$ is a finite conformal CIFS and* $\mathrm{Int}(X) \setminus \overline{J} \neq \emptyset$, *then the limit set J_S is porous.*

4.7 The associated iterated function system

For every vertex $v \in V$ we define the alphabet $I_v \subset E^*$ by induction as a union $\bigcup_{n=1}^{\infty} I_n$ as follows.

$$I_1 = \{e \in I : i(e) = t(e) = v\}.$$

Suppose now that all the sets $I_k \subset E^k$, $k = 1, 2, \dots, n$, have been defined. We then say that $\omega \in E^{n+1}$ belongs to I_{n+1} if $i(\omega) = t(\omega) = v$ and ω cannot be represented as a concatenation of words from $\bigcup_{k=1}^{n} I_k$. Notice that our construction is "economical" in the sense that no element from I_v is a concatenation of other elements from I_v.

Definition 4.7.1 *The system $S_v = \{\phi_\omega\}_{\omega \in I_v}$ is called the conformal iterated function system associated with the conformal graph directed Markov system S via vertex v.*

We would like to remark that the associated iterated function system S_v is usually much "bigger" than the original system S in the sense that I_v is "bigger" than I. It is however sometimes more convenient to deal with an iterated function system rather than a CGDMS, especially when the former shares many properties with the latter. We shall now prove several properties towards this end.

Proposition 4.7.2 *For every vertex $v \in V$ the limit set J_{I_v} is contained in J_v and $\overline{J_{I_v}} = \overline{J_v}$.*

Proof. The first part of this proposition is obvious. Hence $\overline{J_{I_v}} \subset \overline{J_v}$. In order to prove the opposite inclusion it suffices to demonstrate that each element of J_v can be represented as a limit of elements from J_{I_v}. And indeed, let $x = \pi(\omega)$, $\omega \in E^\infty$ with $i(\omega_1) = v$. Since the incidence matrix A is irreducible and the graph Γ is strongly connected, for every $n \geq 1$ there exists $\alpha^{(n)} \in E^*$ and $\tau^{(n)} \in I_v^\infty$ such that $\omega|_n \alpha^{(n)} \tau^{(n)} \in E^\infty$. Since $\tau^{(n)} \in I_v^\infty$ and $i(\omega_1) = v$, $\omega|_n \alpha^{(n)} \tau^{(n)} \in I_v^\infty$. Hence $\pi(\omega|_n \alpha^{(n)} \tau^{(n)}) \in J_{I_v}$ and obviously $\lim_{n\to\infty} \pi(\omega|_n \alpha^{(n)} \tau^{(n)}) = x$. □

Remark 4.7.3 *Notice that the equality in the first part of this proposition has to hold if there exists a word $\omega \in E^\infty$ without elements $e \in I$ with $i(e) = v$ or $t(e) = v$.*

Theorem 4.7.4 *If S is a* CGDMS, *then for every vertex $v \in V$,* $\mathrm{HD}(J_{I_v}) = \mathrm{HD}(J)$.

Proof. By Proposition 4.7.2, $\mathrm{HD}(J_{I_v}) \le \mathrm{HD}(J)$ for every $v \in V$. In view of Theorem 4.2.13 it suffices now to show that for every $v \in V$ and every $t \ge 0$, $\mathrm{P}(t) \le \mathrm{P}_v(t)$, where $\mathrm{P}_v(t)$ is the pressure associated with the system S_v. And indeed, let $\Lambda \subset E^q$ be the set witnessing the finite primitivity of A. Then for every $e \in I$ there exist $\alpha(e), \beta(e) \in \Lambda$ such that $i(\alpha(e)) = v$, $t(\beta(e)) = v$ and $\alpha(e)e\beta(e) \in E^*$. Fix $u > \mathrm{P}_v(t)$. Then

$$
\begin{aligned}
&\sum_{\omega \in E^*} ||\phi'_\omega||^t \mathrm{e}^{-u|\omega|} \\
&\quad \le K^{2t} \left(\max\{||\phi'_\alpha||^{-t} : \alpha \in \Lambda\}\right)^2 \sum_{\omega \in E^*} ||\phi'_{\alpha(\omega_1)\omega\beta(\omega_{|\omega|})}||^t \mathrm{e}^{-u|\omega|} \qquad (4.47)\\
&\quad \le K^{2t}\mathrm{e}^{2qu} \left(\max\{||\phi'_\alpha||^{-t} : \alpha \in \Lambda\}\right)^2 \sum_{\tau \in I_v^*} ||\phi'_\tau||^t \mathrm{e}^{-u|\tau|_v} < \infty,
\end{aligned}
$$

where we could write the second inequality sign since $\alpha(\omega_1)\omega\beta(\omega_{|\omega|}) \in I_v^*$, since the function $\omega \mapsto \alpha(\omega_1)\omega\beta(\omega_{|\omega|})$ is 1-to-1, and since $|\tau|_v \le |\tau|$ for every $\tau \in I_v^*$, where $|\tau|_v$ denotes the length of τ written as a concatenation of letters from the alphabet I_V^*. The last inequality sign follows from Theorem 3.1.1 since $u > \mathrm{P}_v(t)$. Consequently $\mathrm{P}_v(t) \ge \mathrm{P}(t)$ and the proof is complete. □

Suppose that $F = \{f^{(i)} : X_{t(i)} \to I\!R\}_{i \in I}$ is a Hölder family of functions of some order $\beta > 0$. Given $v \in V$ define F_v, a family of maps from $X_{t(v)}$ to $X_{t(v)}$, as follows. If $\omega \in I_v$, then set

$$f_v^{(\omega)} = S_\omega(F) - \mathrm{P}(F)|\omega| : X_v \to I\!R.$$

As an immediate consequence of Lemma 3.1.2 we get the following.

Proposition 4.7.5 *If $F = \{f^{(i)} : X_{t(i)} \to I\!R\}_{i \in I}$ is a Hölder family of functions of some order $\beta > 0$, then for every vertex $v \in V$, F_v is also a Hölder family of functions of order β.*

As a consequence of the definition of F-conformal measures, and the definitions of families F_v, we get the following.

Theorem 4.7.6 *If $F = \{f^{(i)} : X_{t(i)} \to I\!R\}_{i \in I}$ is a Hölder family of functions and m is the F-conformal measure, then $m(J_v)^{-1} m|_{J_v}$ is F_v-conformal for S_v for every vertex $v \in V$.*

Proof. Since the shift-invariant measure $\tilde{\mu}$ is ergodic and since $\tilde{\mu}$ is positive on all cylinders, we get for every $v \in V$ that

$$\tilde{\mu}(I_v^\infty) = \tilde{\mu}\left(\{\omega \in E^\infty : i(\omega_1) = v\}\right).$$

Hence $\tilde{m}(I_v^\infty) = \tilde{m}(\{\omega \in E^\infty : i(\omega_1) = v\})$ and consequently

$$m(J_{I_v}) = \tilde{m}(I_v^\infty) = \tilde{m}(\{\omega \in E^\infty : i(\omega_1) = v\}) = m(J_v).$$

That the measure $m(J_v)^{-1}m|_{J_v}$ is F_v-conformal follows immediately from the definition of Fconformality and the definition of families F_v. □

In particular we get the following.

Corollary 4.7.7 *If the system S is regular, then so is S_v for every $v \in V$. In addition, if m is the unique h-conformal measure for S, then $m(J_{I_v}) = m(J_v)$ and $m(J_v)^{-1}m|_{J_v}$ is a unique h-conformal measure for S_v.*

4.8 Refined geometry, F-conformal measures versus Hausdorff measures

In this section we establish relations between Gibbs states and appropriate *generalized Hausdorff measures*, cf. [DU!], [PUZ] and [U1]. Let

$$\psi = f + \kappa\zeta - \mathrm{P}(F),$$

where $\kappa = \mathrm{HD}(\mu_F)$ and, as in the previous sections,

$$\zeta(\omega) = -\log|\phi'_{\omega_1} \circ \pi \circ \sigma(\omega)|.$$

Throughout this section we assume that

$$\int |f|^{2+\gamma} d\tilde{\mu}_F < \infty \quad \text{and} \quad \int |\zeta|^{2+\gamma} d\tilde{\mu}_F < \infty \tag{4.48}$$

for some $\gamma > 0$. In view of Lemma 2.5.6 and since $L^*(\sigma)$ is a linear space, $\psi \in L^*(\sigma)$ and, in particular, $\sigma^2 = \sigma^2(\psi)$ exists. The following lemma has been proved in [DU1] as Lemma 4.3. We provide a formulation and short proof of it for the sake of completeness.

Lemma 4.8.1 *Let $\eta, \chi > 0$ and let $\rho : [(\chi+\eta)^{-1}, \infty) \to I\!R_+$ belong to the upper (lower) class. Let $\theta : [(\chi+\eta)^{-1}, \infty) \to I\!R_+$ be a function such that $\lim_{t\to\infty} \rho(t)\theta(t) = 0$. Then there exists an upper (lower) class function $\rho_+ : [1,\infty) \to I\!R_+$, ($\rho_- : [1,\infty) \to I\!R_+$) with the following properties.*

(a) $\rho(t(\chi+\eta)) + \theta(t(\chi+\eta)) \le \rho_+(t)$, $(t \ge 1)$
(b) $\rho(t(\chi-\eta)) - \theta(t(\chi-\eta)) \ge \rho_-(t)$, $(t \ge 1)$.

Proof. Since $\lim_{t\to\infty}\rho(t)\theta(t) = 0$, there exists a constant M such that $(\rho(t)+\theta(t))^2 \le \rho(t)^2 + M$. Let ρ belong to the upper class. Then $t \mapsto \rho(t/(\chi+\eta))$ also belongs to the upper class. Hence we may assume that $\chi + \eta = 1$. Define

$$\rho_+(t)^2 = \inf\{u(t)^2 : u \text{ is non-decreasing and } u(t) \ge \rho(t) + \theta(t)\}.$$

Then $\rho_+(t) \ge \rho(t) + \theta(t)$ for $t \ge 1$ and ρ_+ is non-decreasing. Since $\rho_+(t)^2 \le \rho(t)^2 + M$, we also get

$$\int_1^\infty \frac{\rho_+(t)}{t}\exp\bigl(-(1/2)\rho_+^2(t)\bigr)dt$$
$$\ge \exp(-M/2)\int_1^\infty \frac{\rho_+(t)}{t}\exp\bigl(-(1/2)\rho^2(t)\bigr)dt = \infty.$$

The proof in the case of a function of the lower class is similar. □

A function $h : [1,\infty) \to I\!\!R_+$ is said to be slowly growing if $h(t) = o(t^\alpha)$ for all $\alpha > 0$. Let $\chi = \chi_{\tilde\mu_F}(\sigma) = \int \zeta d\tilde\mu_F$. First we shall prove the main geometrical lemma.

Lemma 4.8.2 *(Refined volume lemma) Suppose that $\sigma^2 = \sigma^2(\psi) > 0$. If a slowly growing function h belongs to the upper class, then for μ_F-a.e. $x \in J$,*

$$\limsup_{r\to 0} \frac{\mu_F(B(x,r))}{r^\kappa \exp\bigl(\sigma\chi^{-1/2}h(-\log r)\sqrt{-\log r}\bigr)} = \infty.$$

If, on the other hand, h belongs to the lower class, then for every $\epsilon > 0$ there exists a Borel set $J_\epsilon \subset J$ such that $\mu_F(J_\epsilon) \ge 1-\epsilon$, and there exists a constant $k(\epsilon) \ge 1$ such that for all $x \in J_\epsilon$ and all $0 < r \le 1/k(\epsilon)$

$$\frac{\mu_F(J_\epsilon \cap B(x,r))}{r^\kappa \exp\bigl(\sigma\chi^{-1/2}h(-\log r)\sqrt{-\log r}\bigr)} \le \epsilon.$$

Proof. Given $x = \pi(\omega) \in J$ and $r > 0$ let $n = n(\omega, r)$ be the least integer such that $\phi_{\omega|_n}(X_{t(\omega_n)}) \subset B(x,r)$. Then $r \le \text{diam}(\phi_{\omega|_{n-1}}(X_{t(\omega_{n-1})}))$ and

$$m_F(B(x,r)) \ge m_F(\phi_{\omega|_n}(X)) = \int_{X_{t(\omega_n)}} \exp\bigl(S_{\omega|_n}(F) - \mathrm{P}(F)n\bigr)(z)\,dm_F(z).$$

Hence, using Lemma 3.2.4, Lemma 3.1.2 and (4.20), we get the following.

$$\frac{m_F(B(x,r))}{r^\kappa \exp\bigl(\sigma\chi^{-1/2}h(-\log r)\sqrt{-\log r}\bigr)}$$

$$\geq \frac{\int_{X_{t(\omega_n)}} \exp\bigl(S_{\omega|_n}(f) - \mathrm{P}(F)n\bigr)(z)dm_F(z)}{r^\kappa \exp\bigl(\sigma\chi^{-1/2}h(-\log r)\sqrt{-\log r}\bigr)}$$

$$\geq \frac{Q^{-1}\exp\bigl(S_{\omega|_n}(F) - \mathrm{P}(F)n\bigr)(x)}{r^\kappa \exp\bigl(\sigma\chi^{-1/2}h(-\log r)\sqrt{-\log r}\bigr)}$$

$$\geq \frac{MT(F)^{-1}\exp\bigl(S_{\omega|_n}(F) - \mathrm{P}(F)n\bigr)(x)}{\mathrm{diam}(\phi_{\omega|_{n-1}}(X))^\kappa \exp\bigl(\sigma\chi^{-1/2}h(-\log r)\sqrt{-\log r}\bigr)} \tag{4.49}$$

$$\geq \frac{MT(F)^{-1}\exp\bigl(\sum_{j=0}^{n-1} f\circ\sigma^j(\omega) - \mathrm{P}(F)n\bigr)}{D^\kappa \exp\bigl(-\kappa\sum_{j=0}^{n-2} g\circ\sigma^j(\omega)\bigr)\exp\bigl(\sigma\chi^{-1/2}h(-\log r)\sqrt{-\log r}\bigr)}$$

$$= \exp\left(\sum_{j=0}^{n-1}(f\circ\sigma^j(\omega) + \kappa g\circ\sigma^j(\omega)) - \mathrm{P}(F)n\right.$$

$$\left. - \sigma\chi^{-1/2}h(-\log r)\sqrt{-\log r} - \kappa\zeta(\sigma^{n-1}(\omega))\right) \Big/ T(F)M^{-1}D^\kappa.$$

In view of the Birkhoff ergodic theorem there exists a Borel set $Y_1 \subset E^\infty$ of $\tilde{\mu}_F$ measure 1 such that for every $\eta > 0$, every $\omega \in Y_1$ and every n large enough, say $n \geq n_1(\omega,\eta)$,

$$-\log r \leq -\log\bigl(\mathrm{diam}(\phi_{\omega|_n}(X_{t(\omega_n)}))\bigr) + \log 2 \leq (\chi+\eta)n. \tag{4.50}$$

In fact, in what follows we will need a better upper estimate on $-\log r$. In order to provide it, notice that in view of Lemma 2.5.6 and (4.48) the function ζ is a member of $L^*(\sigma)$. Let τ^2 denote the asymptotic variance of ζ. If $\tau^2 = 0$, then by [IL] g is cohomologous to the constant χ by an L^1 coboundary. It turns out that the following proof, where we assume $\tau^2 > 0$ becomes much simpler when $\tau^2 = 0$. Since the function $t \mapsto 2\sqrt{t\log\log t}$ belongs to the lower class there exists $Y_2 \subset Y_1$ of $\tilde{\mu}_F$ measure 1 such that for all $\omega \in Y_2$, $\tau_1 > \tau$, and all n large enough, say

$n \geq n_2(\omega) \geq n_1(\omega, \eta)$,

$$\begin{aligned}
-\log r &\leq -\log\bigl(\operatorname{diam}(\phi_{\omega|_n}(X_{t(\omega_n)}))\bigr) + \log 2 \\
&\leq \log D + \log 2 + \sum_{j=0}^{n-1} g(\sigma^j(\omega)) \\
&\leq \chi n + 2\tau\sqrt{n \log\log n} + \log D + \log 2 \\
&\leq \chi n + 2\tau_1\sqrt{n \log\log n}
\end{aligned} \tag{4.51}$$

It is a simple exercise in measure theory to check that if $t > 0$ and $\int |g|^t d\mu < \infty$, then for every $a > 0$, $\mu(|g| \geq a) \leq a^{-t} \int |g|^t d\mu$. Since by (4.48) $\int |\zeta|^{2+4\gamma} d\tilde{\mu}_F < \infty$ for some $0 < \gamma < 1/2$, for all $\eta > 0$ we get

$$\begin{aligned}
&\tilde{\mu}_F\bigl(\{\omega \in E^\infty : \kappa|\zeta(\omega)| \geq \sigma(h((\chi+\eta)n)\sqrt{n})^{1-\gamma}\}\bigr) \\
&\qquad\qquad \leq \bigl(\sigma(h((\chi+\eta)n)\sqrt{n})^{1-\gamma}\bigr)^{-2-4\gamma} \int |\zeta|^{2+4\gamma} d\tilde{\mu}_F.
\end{aligned}$$

Since for every $n \geq 1$, $h((\chi+\eta)n) \geq h(\chi+\eta) > 0$, $\frac{1}{2}(1-\gamma)(-2-4\gamma) < -1$ and the measure $\tilde{\mu}_F$ is σ-invariant, we get

$$\begin{aligned}
&\sum_{n=1}^{\infty} \tilde{\mu}_F\bigl(\{\omega \in E^\infty : \kappa|\zeta(\sigma^{n-1}(\omega))| \\
&\quad \geq \sigma(h((\chi+\eta)n)\sqrt{n})^{1-\gamma}\}\bigr) \\
&\quad \leq \sum_{n=1}^{\infty} \bigl(\sigma(h((\chi+\eta)n)\sqrt{n})^{1-\gamma}\bigr)^{-2-4\gamma} \int |\zeta|^{2+4\gamma} d\tilde{\mu}_F < \infty
\end{aligned}$$

Therefore, in view of the Borel–Cantelli lemma, there exists a set $Y_3 \subset Y_2$ of $\tilde{\mu}_F$ measure 1 such that for all $\omega \in Y_3$ there exists $n_3(\omega) \geq n_2(\omega)$ such that for all $n \geq n_3(\omega)$

$$\kappa|\zeta(\sigma^{n-1}(\omega))| \leq \sigma(h((\chi+\eta)n)\sqrt{n})^{1-\gamma}. \tag{4.52}$$

Combining this, (4.51), (4.50) and (4.49) we get for all $a < 0$

$$\frac{m_F(B(x,r))\mathrm{e}^{\sigma a n^{1/4}}}{r^\kappa \exp\bigl(\sigma\chi^{-1/2}h(-\log r)\sqrt{-\log r}\bigr)}$$
$$\geq \frac{M}{T(F)D^\kappa}\exp\Biggl(\sum_{j=0}^{n-1}\psi\circ\sigma^j(\omega)-\sigma\chi^{-1/2}h((\chi+\eta)n)$$
$$\times\sqrt{\chi n+2\tau_1\sqrt{n\log\log n}}-\sigma(h((\chi+\eta)n)\sqrt{n})^{1-\gamma}\Biggr)\mathrm{e}^{\sigma a n^{1/4}}$$
$$=\frac{M}{T(F)D^\kappa}\exp\Biggl(\sum_{j=0}^{n-1}\psi\circ\sigma^j(\omega)-\sigma\sqrt{n}\Biggl(h((\chi+\eta)n)$$
$$\times\sqrt{1+\frac{2\tau_1}{\chi}\sqrt{\frac{\log\log n}{n}}}-an^{-1/4}+h((\chi+\eta)n)\sqrt{n})^{1-\gamma}n^{-\gamma/2}\Biggr)\Biggr). \tag{4.53}$$

Now, consider the function

$$\theta(t)=h(t)\left(\sqrt{1+\frac{2\tau_1}{\chi}\sqrt{\frac{\log\log(t(\chi+\eta)^{-1})}{t(\chi+\eta)^{-1}}}}-1\right)$$
$$-a(t(\chi+\eta)^{-1})^{-1/4}+\frac{h(t)^{1-\gamma}}{(t(\chi-\eta)^{-1})^{\gamma/2}}.$$

Thus, $\theta(t)>0$ and since $h(t)$ is slowly growing, $\lim_{t\to\infty}h(t)\theta(t)=0$. Therefore it follows from Lemma 4.8.1(a) that there exists $h_+(t)$ in the upper class such that $h_+(t)\geq h(t(\chi+\eta))+\vartheta(t(\chi+\eta))$. Since, by (4.48), Theorem 2.2.9 and Theorem 4.4.2,

$$\int\psi d\tilde{\mu}_F=\int fd\tilde{\mu}_F+\frac{\mathrm{h}_{\tilde{\mu}_F}}{\chi}\chi-\mathrm{P}(F)=\int fd\tilde{\mu}_F+\mathrm{h}_{\tilde{\mu}_F}-\mathrm{P}(F)=0,$$

it follows from Theorem 2.5.5 that for infinitely many n's

$$0\leq\sum_{j=0}^{n-1}\psi\circ\sigma^j(\omega)-\sigma\sqrt{n}h_+(n)$$
$$\leq\sum_{j=0}^{n-1}\psi\circ\sigma^j(\omega)-\sigma\sqrt{n}\bigl(h(n(\chi+\eta))+\theta(n(\chi+\eta))\bigr)$$
$$\leq\sum_{j=0}^{n-1}\psi\circ\sigma^j(\omega)-\sigma\sqrt{n}\Biggl(h(n(\chi+\eta))\sqrt{1+\frac{2\tau_1}{\chi}\sqrt{\frac{\log\log n}{n}}}$$
$$-an^{-1/4}+h((\chi+\eta)n)^{1-\gamma}n^{-\gamma/2}\Biggr).$$

Combining this and (4.53) we see that

$$\frac{m_F(B(x,r))}{r^\kappa \exp\big(\sigma\chi^{-1/2}h(-\log r)\sqrt{-\log r}\big)} \geq M(T(F)D^\kappa)^{-1}\exp(-\sigma a n^{1/4})$$

for $\tilde{\mu}_F$-a.e. ω and infinitely many n's provided they are of the form $n(\omega, r)$. But since there exists $n_3(\omega)$ such that each $n \geq n_3(\omega)$ is of the form $n(\omega, r)$, fixing $a < 0$, we eventually get

$$\limsup_{r\to 0} \frac{m_F(B(x,r))}{r^\kappa \exp\big(\sigma\chi^{-1/2}h(-\log r)\sqrt{-\log r}\big)} = \infty.$$

for μ_F a.e. $x \in J$. Since the measures μ_F and m_F are equivalent with bounded Radon-Nikodym derivatives, the proof of the first part of Lemma 4.8.2 is complete.

Let us now prove the second part of the lemma. For every $\omega \in E^\infty$ and every $r > 0$ let $n = n(\omega, r) \geq 0$ be the least integer n such that $\mathrm{diam}(\phi_{\omega|_{n+1}}(X_{t(\omega_{n+1})})) < r$. Clearly $\lim_{r\to 0} n(\omega, r) = \infty$ and therefore there exists $r_1(\omega)$ such that for all $0 < r \leq r_1(\omega)$

$$\mathrm{diam}(\phi_{\omega|_n}(X_{t(\omega_n)})) \geq r. \tag{4.54}$$

Fix now $\omega \in E^\infty$ and $0 < r \leq r_1(\omega)$. Since $\tilde{m}_F$ is a Gibbs state, we have by (2.3) that $\tilde{m}_F([\omega|_n]) \leq Q\exp\Big(\sum_{j=0}^{n-1} f\circ\sigma^j(\omega) - \mathrm{P}(F)n\Big)$. Hence

$$\begin{aligned}
&\frac{\tilde{m}_F([\omega|_n])}{r^\kappa \exp\big(\sigma\chi^{-1/2}h(-\log r)\sqrt{-\log r}\big)} \\
&\leq \frac{Q\exp\Big(\sum_{j=0}^{n-1} f\circ\sigma^j(\omega) - \mathrm{P}(F)n\Big)}{\mathrm{diam}^\kappa(\phi_{\omega|_{n+1}}(X))\exp\big(\sigma\chi^{-1/2}h(-\log r)\sqrt{-\log r}\big)} \\
&\leq \frac{Q\exp\Big(\sum_{j=0}^{n-1} f\circ\sigma^j(\omega) - \mathrm{P}(F)n\Big)}{D^{-\kappa}|\phi'_{\omega|_{n+1}}(\pi(\sigma^{n+1}(\omega)))|\exp\big(\sigma\chi^{-1/2}h(-\log r)\sqrt{-\log r}\big)} \\
&= QD^\kappa \exp\left(\sum_{j=0}^{n-1} f\circ\sigma^j(\omega) + \kappa\sum_{j=0}^{n}\zeta\circ\sigma^j(\omega)\right. \\
&\qquad \left. - P(F)n - \sigma\chi^{-1/2}h(-\log r)\sqrt{-\log r}\right) \\
&= QD^\kappa \exp\left(\sum_{j=0}^{n-1} f\circ\sigma^j(\omega)\right. \\
&\qquad \left. - \sigma\chi^{-1/2}h(-\log r)\sqrt{-\log r} + \kappa\zeta\circ\sigma^n(\omega)\right).
\end{aligned} \tag{4.55}$$

In view of the Birkhoff ergodic theorem there exists a Borel set $Y_1 \subset E^\infty$ of $\tilde{\mu}_F$ measure 1 such that for every $\eta > 0$, every $\omega \in Y_1$ and every n large enough, say $n \geq n_1(\omega, \eta)$,

$$-\log r \geq -\log\bigl(\mathrm{diam}(\phi_{\omega|_n}(X_{t(\omega_n)}))\bigr) \geq (\chi - \eta)n. \tag{4.56}$$

In fact, in what follows we will need a better upper estimate on $-\log r$. In order to provide it recall that the function ζ is a member of $L^*(\sigma)$. Let τ^2 denote the asymptotic variance of ζ. If $\tau^2 = 0$, then by [IL] ζ is cohomologous to the constant χ by an L^1 coboundary. It turns out that the following proof, where we assume $\tau^2 > 0$, becomes much simpler when $\tau^2 = 0$. Since the function $t \mapsto 2\sqrt{t \log\log t}$ belongs to the lower class there exists $Y_2 \subset Y_1$ of $\tilde{\mu}_F$ measure 1 such that for all $\omega \in Y_2$, $\tau_1 > \tau$, and all n large enough, say $n \geq n_2(\omega) \geq n_1(\omega, \eta)$,

$$\begin{aligned}
-\log r &\geq -\log\bigl(\mathrm{diam}(\phi_{\omega|_n}(X_{t(\omega_n)})\bigr) \geq -\log(DK) + \sum_{j=0}^{n-1} \zeta(\sigma^j(\omega)) \\
&\geq \chi n - 2\tau\sqrt{n \log\log n} - \log(DK) \geq \chi n - 2\tau_1\sqrt{n \log\log n}.
\end{aligned} \tag{4.57}$$

The same argument as that leading to (4.52) shows that there exists a Borel set $Y_3 \subset Y_2$ of $\tilde{\mu}_F$ measure 1 such that for all $\omega \in Y_3$ there exists $n_3(\omega) \geq n_2(\omega)$ such that for all $n \geq n_3(\omega)$,

$$\kappa|\zeta(\sigma^n(\omega))| \leq \sigma(h((\chi - \eta)n)\sqrt{n})^{1-\gamma}.$$

Combining this, (4.55), (4.56) and (4.57) we get for all $a > 0$

$$\begin{aligned}
&\frac{\tilde{m}_F([\omega|_n])\mathrm{e}^{\sigma a n^{1/4}}}{r^\kappa \exp\bigl(\sigma\chi^{-1/2}h(-\log r)\sqrt{-\log r}\bigr)} \\
&\leq QD^\kappa \exp\Biggl(\sum_{j=0}^{n-1} f \circ \sigma^j(\omega) - \sigma\chi^{-1/2}h((\chi - \eta)n)\sqrt{\chi n - 2\tau_1\sqrt{n \log\log n}} \\
&+ \sigma(h((\chi - \eta)n)\sqrt{n})^{1-\gamma}\Biggr)\mathrm{e}^{\sigma a n^{1/4}} \\
&= QD^\kappa \exp\Biggl(\sum_{j=0}^{n-1} \psi \circ \sigma^j(\omega) - \sigma\sqrt{n}\Biggl(h((\chi - \eta)n)\sqrt{1 - \frac{2\tau_1}{\chi}\sqrt{\frac{\log\log n}{n}}} \\
&- a n^{-1/4} - h((\chi - \eta)n)\sqrt{n})^{1-\gamma} n^{-\gamma/2}\Biggr)\Biggr)
\end{aligned} \tag{4.58}$$

Now, consider the function

$$\theta(t) = h(t)\left(1 - \sqrt{1 - \frac{2\tau_1}{\chi}\sqrt{\frac{\log\log(t(\chi-\eta)^{-1})}{t(\chi-\eta)^{-1}}}}\right) + a(t(\chi-\eta)^{-1})^{-1/4} + \frac{h(t)^{1-\gamma}}{(t(\chi-\eta)^{-1})^{\gamma/2}}.$$

Thus, $\theta(t) > 0$ and since $h(t)$ is slowly growing, $\lim_{t\to\infty} h(t)\theta(t) = 0$. Therefore it follows from Lemma 4.8.1(b) that there exists $h_-(t)$ in the lower class such that $h_-(t) \le h(t(\chi-\eta)) - \theta(t(\chi-\eta))$. Since, by (4.48), Theorem 2.2.9, and Theorem 4.4.2,

$$\int \psi d\tilde{\mu}_F = \int f d\tilde{\mu}_F + \frac{\mathrm{h}_{\tilde{\mu}_F}}{\chi}\chi - \mathrm{P}(F) = \int f d\tilde{\mu}_F + \mathrm{h}_{\tilde{\mu}_F} - \mathrm{P}(F) = 0,$$

it follows from Theorem 2.5.5 that there exists a Borel set $Y_4 \subset Y_3$ of $\tilde{\mu}_F$ measure 1 such that for all $\omega \in Y_4$ and all n large enough, say $n \ge n_4(\omega) \ge n_3(\omega)$,

$$\begin{aligned} 0 &\ge \sum_{j=0}^{n-1} \psi\circ\sigma^j(\omega) - \sigma\sqrt{n}h_-(n) \\ &\ge \sum_{j=0}^{n-1} \psi\circ\sigma^j(\omega) - \sigma\sqrt{n}\big(h(n(\chi-\eta)) - \theta(n(\chi-\eta))\big) \\ &\ge \sum_{j=0}^{n-1} \psi\circ\sigma^j(\omega) - \sigma\sqrt{n}\Bigg(h(n(\chi-\eta))\sqrt{1-\frac{2\tau_1}{\chi}\sqrt{\frac{\log\log n}{n}}} \\ &\qquad - an^{-1/4} - h((\chi-\eta)n)^{1-\gamma}n^{-\gamma/2}\Bigg). \end{aligned}$$

Combining this and (4.58) we conclude that for every $\omega \in Y_4$ and every $n \ge n_4(\omega)$

$$\frac{\tilde{m}_F([\omega|_n])}{r^\kappa \exp\big(\sigma\chi^{-1/2}h(-\log r)\sqrt{-\log r}\big)} \le QD^\kappa \mathrm{e}^{-\sigma a n^{1/4}}.$$

In other words, for every $\omega \in Y_4$ and every $r > 0$ small enough, say $r \le r(\omega) \le r_1(\omega)$,

$$\frac{\tilde{m}_F([\omega|_{n(\omega,r)}])}{r^\kappa \exp\big(\sigma\chi^{-1/2}h(-\log r)\sqrt{-\log r}\big)} \le QD^\kappa \mathrm{e}^{-\sigma a n(\omega,r)^{1/4}}. \tag{4.59}$$

Fix now $\epsilon > 0$ and take q so large that $QD^\kappa \mathrm{e}^{-\sigma a q^{1/4}} \le \epsilon$. Then, since $\lim_{r\searrow 0} n(\omega, r) = \infty$, there exists $k(\omega) \ge 1$ such that for all $0 < r \le$

$1/k(\omega)$, (4.58) holds and $n(\omega, r) \geq q$. Since

$$Y_4 = \bigcup_{k=1}^{\infty} \{\omega \in Y_4 : k(\omega) \leq k\},$$

there exists $k(\epsilon)$ so large that if $\tilde{J}_\epsilon = \{\omega \in Y_4 : k(\omega) \leq k(\epsilon)\}$, then

$$\tilde{m}_F(J \setminus \tilde{J}_\epsilon) \leq \epsilon. \tag{4.60}$$

Moreover for every $\omega \in \tilde{J}_\epsilon$ and every $r < 1/k(\epsilon)$ (so $r \leq 1/k(\omega)$)

$$\begin{aligned} \frac{\tilde{m}_F([\omega_{n(\omega,r)}])}{r^\kappa \exp\left(\sigma\chi^{-1/2} h(-\log r)\sqrt{-\log r}\right)} &\leq QD^\kappa \mathrm{e}^{-a\sigma n(\omega,r)^{1/4}} \\ &\leq QD^\kappa \mathrm{e}^{-a\sigma q^{1/4}} \leq \epsilon. \end{aligned} \tag{4.61}$$

Let

$$J_\epsilon = \pi(\tilde{J}_\epsilon).$$

It then follows from (4.60) that

$$m_F(J_\epsilon) = \tilde{m}_F \circ \pi^{-1}(J_\epsilon) = \tilde{m}_F \circ \pi^{-1}(\pi(\tilde{J}_\epsilon)) \geq \tilde{m}_F(\tilde{J}_\epsilon) \geq 1 - \epsilon.$$

Now, in view of Lemma 4.2.6, there exists a universal constant $L \geq 1$ such that for every $x \in J_\epsilon$ and every $0 < r < 1/k(\epsilon)$ there exist points $\omega^{(1)}, \omega^{(2)}, \ldots, \omega^{(L)} \in \tilde{J}_\epsilon$ such that

$$J_\epsilon \cap B(x, r) \subset \bigcup_{j=1}^{L} \phi_{\omega^{(j)}|_{n(\omega^{(j)},r)}}(X_{t(\omega^{(j)}|_{n(\omega^{(j)},r)})}).$$

Therefore, by (4.61)

$$\begin{aligned} &\frac{m_F(J_\epsilon \cap B(x,r))}{r^\kappa \exp\left(\sigma\chi^{-1/2} h(-\log r)\sqrt{-\log r}\right)} \\ &\quad \leq \sum_{j=1}^{L} \frac{m_F\left(\phi_{\omega^{(j)}|_{n(\omega^{(j)},r)}}(X_{t(\omega^{(j)}|_{n(\omega^{(j)},r)})})\right)}{r^\kappa \exp\left(\sigma\chi^{-1/2} h(-\log r)\sqrt{-\log r}\right)}) \leq L\epsilon. \end{aligned}$$

Since the measures μ_F and m_F are equivalent with bounded Radon-Nikodym derivatives, looking at the last two displays we conclude that the proof of the second part of Lemma 4.8.2 is complete. □

For a function $h : [1, \infty) \to (0, \infty)$ define for sufficiently small $t > 0$

$$\tilde{h}(t) = t^\kappa \exp\left(\frac{\sigma}{\sqrt{\chi}} h(-\log t)\sqrt{-\log t}\right).$$

One of the results we announced at the beginning of this section is here.

Theorem 4.8.3 *Suppose that $\sigma^2(\psi) > 0$ and that $h : [1,\infty) \to (0,\infty)$ is a slowly growing function.*

(a) If h belongs to the upper class, then the measures μ_F and $\mathrm{H}_{\tilde{h}}$ on J are singular.
(b) If h belongs to the lower class, then μ_F is absolutely continuous with respect to the Hausdorff measure $\mathcal{H}^{\tilde{h}}$.

Proof. Suppose first that h belongs to the upper class. For every $n \geq 1$ and every $\epsilon > 0$, by Lemma 4.8.2 there exists a Borel set $E_n \subset J$ such that $\mu_F(E_n) > 1 - \epsilon 2^{-n}$ and such that for every $x \in E_n$ and some closed ball $B(x)$ centered at x and with diameter $< 1/n$, $\mu_F(B(x)) > n\tilde{h}(B(x))$. By Besicovitch's covering theorem there exists a universal constant $C > 0$ such that from the cover $\{B(x) : x \in E_n\}$ one can choose a countable subcover $\{B(x_j) : j \geq 1\}$ of multiplicity $\leq C$. Since for every $j \geq 1$, $\mathrm{diam}(B(x_j)) < 1/n$, we get

$$\mathrm{H}^{\tilde{h}}(E_n, 1/n) \leq \frac{1}{n}\sum_{j=1}^{\infty}\mu_F(B(x_j)) \leq \frac{C}{n}\mu_F(J) = \frac{C}{n}.$$

Setting $E_\epsilon = \bigcap_{n\geq 1} E_n$ we then have $\mathrm{H}^{\tilde{h}}(E_\epsilon) = 0$ and $\mu_F(E_\epsilon) \geq 1 - \epsilon$. Finally the set $E = \bigcup_{q\geq 1} E_{1/q}$ satisfies $\mathrm{H}^{\tilde{h}}(E) = 0$ and $\mu_F(E) = 1$. The proof of item (a) is therefore complete.

Suppose in turn that h belongs to the lower class and consider a Borel set $E \subset J$ with $\mu_F(E) > 0$. Take $\epsilon = \mu_F(E)/2$. Then by Lemma 4.8.2 $\mu_F(J_\epsilon \cap E) \geq \mu_F(E) - \epsilon = \epsilon$. Fix $0 < \delta \leq 1/k(\epsilon)$ and consider $\mathcal{B} = \{B(x_i, r_i)\}$, the cover of $J_\epsilon \cap E$ by balls centered at points of $J_\epsilon \cap E$ and with radii $\leq \delta$. Then by Lemma 4.8.2

$$\sum_i \tilde{h}(r_i) \geq \frac{1}{\epsilon}\sum_i \mu_F(J_\epsilon \cap B(x_i, r_i)) \geq \frac{1}{\epsilon}\mu_F(J_\epsilon \cap E) \geq 1.$$

Hence $\mathrm{H}_\delta^{\tilde{h}}(E) \geq B > 0$, where B is a universal constant (see Theorem A2.0.13). Thus $\mathrm{H}_{\tilde{h}}(E) \geq B > 0$ and we are done. □

Remark 4.8.4 *Taking $h := 0$ it follows from Theorem 4.8.3(a) that the measure μ_F is singular with respect to the κ-dimensional Hausdorff measure $\mathcal{H}^\kappa$ on J.*

We recall that two functions $f_1, f_2 : E^\infty \to I\!R$ are cohomologous in a class H if there exists a function $u \in H$ such that

$$f_2 - f_1 = u \circ \sigma - u.$$

As a complementary result to Theorem 4.8.3 we shall prove the following.

Theorem 4.8.5 *If $\sigma^2(\psi) = 0$, then $\kappa = \mathrm{HD}(\mu_F) = \mathrm{HD}(J) := h$, the functions $-hg$ and $f - \mathrm{P}(F)$ are cohomologous in the class of Hölder continuous bounded functions, the system $\{\phi_i : i \in I\}$ is regular and m_F is equivalent with the h-conformal measure on J, that is with m_{-hg}, with bounded Radon-Nikodym derivatives. In addition, the invariant measures $\tilde{\mu}_F$ and $\tilde{\mu}_{hg}$ are equal.*

In order to prove Theorem 4.8.5 we need some preparations. First, let α_- be the partition of the *two-sided shift space* $E^{\boldsymbol{Z}} = \{\{\omega_n\}_{n\in\boldsymbol{Z}}\}$ into elements of the form $\omega \times E^{\{1,2,\dots\}}$, where $\omega \in E^{\{\dots,-2,-1,0\}}$. Given $-\infty \le m \le n \le +\infty$ let $\omega|_m^n = \omega_m\omega_{m+1}\dots\omega_n$ and let

$$[\omega]_m^n = \{\tau \in E^{\boldsymbol{Z}} : \tau_k = \omega_k \text{ for all } m \le k \le n\}.$$

Finally let $\overline{\mu}_F$ be *Rokhlin's natural extension* of the invariant measure $\tilde{\mu}_F$ onto the two-sided shift space $E^{\boldsymbol{Z}}$. Let us recall that $\overline{\mu}_F$ is defined on a cylinder

$$C = \pi_{n_1}^{-1}(C_1) \cap \pi_{n_2}^{-1}(C_2) \cap \dots \cap \pi_{n_k}^{-1}(C_k)$$

with $n_1 \le n_2 \le \dots \le n_k$, by the formula

$$\overline{\mu}_F(C) = \tilde{\mu}_F\big(\sigma^{-(n_1+1)}(C)|_1^\infty\big),$$

where for every $k \in \boldsymbol{Z}$, $\pi_k : E^{\boldsymbol{Z}} \to I$ is the projection onto the kth coordinate given by the formula $\pi_k(\omega) = \omega_k$ and for every set $B \subset E^{\boldsymbol{Z}}$, $B|_1^\infty$ is the projection of B onto $E^{\{1,2,\dots\}}$ denoted also by $\pi(B)$. Let us recall that the measure $\overline{\mu}_F$ is shift-invariant. We shall prove the following.

Lemma 4.8.6 *If $\{\mu_{\alpha_-(\omega)} : \omega \in E^{\boldsymbol{Z}}\}$ is the Rokhlin canonical system of measures of the measure $\overline{\mu}_F$ on the partition α_-, then for $\overline{\mu}_F$-a.e. $\omega \in E^{\boldsymbol{Z}}$ the conditional measure $\mu_{\alpha_-(\omega)}$ considered on $E^{\mathbb{N}}$ is equivalent with $\tilde{\mu}_F$. Moreover, the Radon-Nikodym derivative $d\mu_{\alpha_-(\omega)}/d\tilde{\mu}_F$ is bounded from above and from below respectively by $Q^2T(F)^2$ and $Q^{-2}T(f)^{-2}$, where Q comes from the Gibbs property of the measure $\tilde{\mu}_F$.*

Proof. By the martingale theorem, for $\overline{\mu}_F$-a.e. $\omega \in E^{\boldsymbol{Z}}$ and every Borel set $B \subset E^{\boldsymbol{Z}}$,

$$\mu_{\alpha_-(\omega)}(B) = \lim_{n\to\infty} \frac{\overline{\mu}_F(B \cap [\omega]_{-n}^0)}{\overline{\mu}_F([\omega]_{-n}^0)}.$$

It therefore suffices to show that for every $\tau \in E^*$

$$Q^{-4} \le \frac{\mu_{\alpha_-(\omega)}([\omega|_{-\infty}^0 \tau])}{\mu_F([\tau])} \le Q^4.$$

And indeed, in view of the Gibbs property of the measure $\tilde{\mu}_F$, we get

$$\begin{aligned}\overline{\mu}_F([\omega]_{-n}^0) &= \tilde{\mu}_F(\sigma^{-(n+1)}([\omega]_{-n}^0)|_1^\infty) \\ &\le Q \exp\big(\sup\big(S_{\sigma^{-(n+1)}([\omega]_{-n}^0)|_1^{n+1}}(F)\big) - \mathrm{P}(F)(n+1)\big) \\ &= Q \exp\big(\sup\big(S_{\omega_{-n}\ldots\omega_0}(F)\big) - \mathrm{P}(F)(n+1)\big).\end{aligned}$$

Using in addition Lemma 3.1.2 and putting $k = |\tau|$, we get

$$\begin{aligned}&\overline{\mu}_F([\tau] \cap [\omega]_{-n}^0) \\ &\quad = \tilde{\mu}_F(\sigma^{-(n+1)}([\tau] \cap [\omega]_{-n}^0)|_1^\infty)) \\ &\quad \ge Q^{-1} \inf\big(\exp\big(S_{\sigma^{-(n+1)}([\tau]\cap[\omega]_{-n}^0)|_1^{n+1}}(F)\big) - \mathrm{P}(F)(n+1+k)\big) \\ &\quad = Q^{-1} \exp\big(\inf\big(S_{\omega_{-n}\ldots\omega_0\tau_1\ldots\tau_k}(F)\big) - \mathrm{P}(F)(n+1+k)\big) \\ &\quad \ge Q^{-1} \exp\big(\inf\big(S_{\omega_{-n}\ldots\omega_0}(F)\big) + \inf\big(S_{\tau_1\ldots\tau_k}(F)\big) - \mathrm{P}(F)(n+1+k)\big) \\ &\quad \ge Q^{-1}T(f)^{-2} \exp\big(\sup\big(S_{\omega_{-n}\ldots\omega_0}(F)\big) - \mathrm{P}(F)(n+1)\big) \\ &\quad \times \exp\big(\sup\big(S_{\tau_1\ldots\tau_k}(F)\big) - \mathrm{P}(F)k\big).\end{aligned}$$

Applying the Gibbs property of the measure $\tilde{\mu}_F$ again, we therefore obtain

$$\begin{aligned}\frac{\overline{\mu}_F([\tau] \cap [\omega]_{-n}^0)}{\overline{\mu}_F([\omega]_{-n}^0)} &\ge Q^{-1}T(f)^{-2} \exp\big(\sup S_\tau(F) - \mathrm{P}(F)k\big) \\ &\ge Q^{-2}T(f)^{-2}\mu_F([\tau]).\end{aligned}$$

Hence $\mu_{\alpha_-(\omega)}([\omega|_{-\infty}^0 \tau]) \ge Q^{-2}T(f)^{-2}\mu_F([\tau])$. Similar computations show that

$$\mu_{\alpha_-(\omega)}([\omega|_{-\infty}^0 \tau]) \le Q^2T(f)^2\mu_F([\tau]).$$

□

As an immediate consequence of this lemma we get the following.

Corollary 4.8.7 *If $\{\mu_{\alpha_-(\omega)} : \omega \in E^{\mathbf{Z}}\}$ is the Rokhlin canonical system of measures of the measure $\overline{\mu}_F$ on the partition α_-, then for $\overline{\mu}_F$-a.e. $\omega \in E^{\mathbf{Z}}$, supp$(\mu_{\alpha_-(\omega)}) = \alpha_-(\omega)$, where $\alpha_-(\omega)$ is the only atom of α_- containing ω.*

Lemma 4.8.8 *If $\eta : E^\infty \to I\!R$ is a Hölder continuous function of some order $\beta > 0$ such that $\int |\eta|^{2+\gamma} d\tilde{\mu}_F < \infty$, $\int \eta d\tilde{\mu}_F = 0$ and $\sigma^2(\eta) = 0$,* *then there exists a bounded Hölder continuous function u of order $\beta > 0$ such that $\eta = u - u \circ \sigma$. In particular η turns out to be bounded.*

Proof. It follows from Theorem 2.5.1 and [IL] that there exists $u \in L_2(\tilde{\mu}_F)$ such that

$$\eta = u - u \circ \sigma \tag{4.62}$$

$\tilde{\mu}_F$-a.e. Our aim is to show that u has a Hölder continuous version of order β. We first extend η and u on the two-sided shift space $E^{\mathbb{Z}}$ by declaring

$$\eta(\omega) = \eta(\omega|_1^\infty) \text{ and } u(\omega) = u(\omega|_1^\infty)$$

wherever $u(\omega|_1^\infty)$ is defined. The cohomological equation (4.62) remains satisfied since

$$u(\omega) - u \circ \sigma(\omega) = u(\omega|_1^\infty) - u(\sigma(\omega)|_1^\infty) = u(\omega|_1^\infty) - u(\sigma((\omega|_1^\infty))) = \eta(\omega). \tag{4.63}$$

In view of Luzin's theorem there exists a compact set $D \subset E^{\mathbb{Z}}$ such that $\overline{\mu}_F(D) > 1/2$ and the function $u|_D$ is continuous. In view of Birkhoff's ergodic theorem there exists a Borel set $B \subset E^{\mathbb{Z}}$ such that $\overline{\mu}_F(B) = 1$, for every $x \in B$, $\sigma^n(x)$ visits D with the asymptotic frequency $> 1/2$, u is well defined on $\bigcup_{n\in\mathbb{Z}} \sigma^{-n}(B)$ and (4.62) holds on $\bigcup_{n\in\mathbb{Z}} \sigma^{-n}(B)$. By the definition of conformal measures and by Lemma 4.8.6 there exists a Borel set $F \subset E^{\mathbb{Z}}$ such that $\overline{\mu}_F = 1$, for all $\omega \in F$, $\mu_{\alpha_-(\omega)}(B \cap \alpha_-(\omega)) = 1$, and $supp(\mu_{\alpha_-(\omega)}) = \alpha_-(\omega)$. In particular, for every $\omega \in F$, the set $B \cap \alpha_-(\omega)$ is dense in $\alpha_-(\omega)$. Fix one $\omega \in F$ and consider two arbitrary elements $\rho, \tau \in \alpha_-(\omega)$. Then there exists a continuous increasing unbounded sequence $\{n_j\}$ such that $\sigma^{-n_j}(\rho), \sigma^{-n_j}(\tau) \in D$ for all $j \geq 1$. Using (4.62) we get

$$\begin{aligned}
&|u(\rho) - u(\tau)| \\
&= \left| u(\sigma^{-n_j}(\rho)) - \sum_{k=1}^{n_j} \eta(\sigma^{-k}(\rho)) - \left(u(\sigma^{-n_j}(\tau)) - \sum_{k=1}^{n_j} \eta(\sigma^{-k}(\tau)) \right) \right| \\
&\leq |u(\sigma^{-n_j}(\rho)) - u(\sigma^{-n_j}(\tau))| + \sum_{k=1}^{n_j} |\eta(\sigma^{-k}(\rho)) - \eta(\sigma^{-n_k}(\tau))|.
\end{aligned} \tag{4.64}$$

Now, since $\lim_{j\to\infty} \mathrm{dist}(\sigma^{-n_j}(\rho), \sigma^{-n_j}(\tau)) = 0$, since both $\sigma^{-n_j}(\rho)$ and $\sigma^{-n_j}(\tau)$ belong to D and since $u|_D$ is uniformly continuous (as D is compact), we conclude that

$$\lim_{j\to\infty} |u(\sigma^{-n_j}(\rho)) - u(\sigma^{-n_j}(\tau))| = 0.$$

Since η is Hölder continuous of order β, we get

$$\sum_{k=1}^{n_j} |\eta(\sigma^{-k}(\rho) - \eta(\sigma^{-n_k}(\tau)| \le \sum_{k=1}^{n_j} V_\beta(\eta)\mathrm{e}^{-\beta k} d_\beta(\rho,\tau) \le \frac{V_\beta \mathrm{e}^{-\beta}}{1-\mathrm{e}^{-\beta}} d_\beta(\rho,\tau).$$

Therefore, it follows from (4.64) that

$$|u(\rho) - u(\tau)| \le \frac{V_\beta \mathrm{e}^{-\beta}}{1-\mathrm{e}^{-\beta}} d_\beta(\rho,\tau).$$

Hence, as $\alpha_-(\omega)\cap B$ is dense in $\alpha_-(\omega)$, u has a bounded Hölder continuous extension from $\alpha_-(\omega)\cap B$ on $\alpha_-(\omega) = \underline{\omega}\times E^{I\!N}$, where $\underline{\omega} = \omega|_{-\infty}^0$. Denote this extension by $\overline{u} : \alpha_-(\omega) \to I\!R$ and for every $\tau \in E^{I\!N}$ set

$$\overline{u}(\tau) = \overline{u}(\underline{\omega}\tau).$$

This obviously defines a bounded Hölder continuous function $\overline{u} : E^{I\!N} \to I\!R$. Define now the set B_ω to be determined by the condition

$$\underline{\omega}B_\omega = \alpha_-(\omega)\cap B.$$

The function $\overline{u} : E^{I\!N} \to I\!R$ is a version of u. Indeed, since $\mu_{\alpha_-(\omega)}(\omega B_\omega) = 1$, it follows from Lemma 4.8.6 that $\tilde{\mu}_F(B_\omega) = 1$ and additionally, for every $\tau \in B_\omega$, $\overline{u}(\tau) = \overline{u}(\underline{\omega}\tau) = u(\tau)$. Since the measure $\tilde{\mu}_F$ is shift-invariant, $\tilde{\mu}_F(B_\omega \cap \sigma^{-1}(B_\omega)) = 1$. Take now an arbitrary element $\rho \in B_\omega\cap\sigma^{-1}(B_\omega)$. Then $\sigma(\omega) \in B_\omega$ and we have $\eta(\rho) = u(\rho) - u(\sigma(\rho)) = \overline{u}(\rho) - \overline{u}(\sigma(\rho))$. But since $\mathrm{supp}(\tilde{\mu}_F) = E^{I\!N}$, the set $B_\omega\cap\sigma^{-1}(B_\omega)$ is dense in $E^{I\!N}$ and therefore $\eta = \overline{u} - \overline{u}\circ\sigma$ on $E^{I\!N}$. □

Proof of Theorem 4.8.5 First notice that in view of Theorem 4.4.2, Theorem 2.2.9 and Lemma 2.2.8

$$\begin{aligned}\int \psi d\tilde{\mu}_F &= \int f d\tilde{\mu}_F + \frac{\mathrm{h}_{\tilde{\mu}_F}}{\chi_{\tilde{\mu}_F}}\chi_{\tilde{\mu}_F} - \mathrm{P}(F)\\ &= \int f d\tilde{\mu}_F + \mathrm{h}_{\tilde{\mu}_F} - \mathrm{P}(F) = \mathrm{P}(F) - \mathrm{P}(F) = 0.\end{aligned}$$

Hence the assumptions of Lemma 4.8.8 are satisfied with $\eta = \psi$ and therefore there exists a bounded function $u \in \mathcal{H}_\beta$ such that

$$f - \mathrm{P}(F) + \kappa\zeta = u - u\circ\sigma,$$

that is the functions $-\kappa\zeta$ and $f - \mathrm{P}(F)$ are cohomologous in the class of bounded functions of $\mathcal{H}_\beta$. It follows from this equation that the constant R appearing in Theorem 2.2.7(2) (with $f := f - \mathrm{P}(F)$ and $g = -\kappa\zeta$) is equal to zero. Therefore, it follows from this theorem that $\mathrm{P}(-\kappa\zeta) = \mathrm{P}(f) - \mathrm{P}(F) = 0$. Hence, the system S is regular, $\kappa = h$ and it follows from Theorem 2.2.7(1) that $\tilde{\mu}_F = \tilde{\mu}_{-\kappa\zeta}$. The equivalence of measures m_F and $m_{-\kappa\zeta}$ with bounded Radon-Nikodym derivatives follows now from the fact that both these measures are Gibbs states of the functions $f - \mathrm{P}(F)$ and $-\kappa\zeta$ respectively. □

4.9 Multifractal analysis

The *multifractal* formalism arose from various considerations in physics and mathematics (see e.g. [Man], [FP], [Gr], [Ha]). In this last paper a formulation of the scenarios of multifractal theory was elaborated in which there were strong hints of parallels to the theory of statistical physics. Some of the first rigorous mathematical results concerning this formalism are in [CM] and [Ra]. Since then there have been many papers written verifying some aspects of this formalism (see for example [O1]–[O4], [PW], [Pat1], [Pat2]). Recently, Pesin presented a general formulation of the setting for multifractal theory [Pe]. Also, many more references concerning this topic may be found in his book. In this section dealing with multifractal analysis of conformal GDMSs we develop Section 7 of [HMU]. We would like to emphasize that our analysis is performed only for cylinders (and also under some other technical assumptions), whereas the question concerning the analysis of balls, successfully taken care of (see the papers cited above) in the case of finite iterated function systems, remains open for infinite iterated function systems.

In this section $S = \{\phi_e : X_{t(e)} \to X_{i(e)} : e \in I\}$ is a regular conformal GDMS such that

$$\phi_i(X_{t(i)}) \cap \phi_j(X_{t(j)}) \text{ is at most countable} \tag{4.65}$$

for all $i \neq j \in I$ and

$$F = \{f^{(i)} : X \to I\!R : i \in I\}$$

is a summable Hölder family of functions. Subtracting from each of the functions $f^{(i)}$ the topological pressure of F we may assume that $\mathrm{P}(F) = 0$. We consider a two-parameter family of Hölder continuous families of functions

$$G_{q,t} = \{g_{q,t}^{(i)} := qf^{(i)} + t \log |\phi_i'|\}.$$

Let .

$$\begin{aligned}\mathcal{F}in(F) &= \{q \in I\!R : \mathcal{L}_{qF}(1\!\!1) < \infty\}\\ &= \{q \in I\!R : \mathrm{P}(qF) < \infty\} \text{ and } \theta(F) = \inf \mathcal{F}in(F),\end{aligned}$$

where the second equality follows from Proposition 2.1.9. By the definition of summable Hölder families of functions, $1 \in \mathcal{F}in(F)$ and, in particular, $\{i : \sup f^{(i)} > 0\}$ is finite. Before dealing with smoothness properties we shall prove the following result, which will be needed in the next section.

Lemma 4.9.1 *The function $(q,t) \mapsto \mathrm{P}(q,t) := \mathrm{P}(G_{q,t})$ is decreasing with respect to both variables $q \geq 0$ and $t \geq 0$.*

Proof. Consider two pairs (q_1, t_1) and (q_2, t_2) such that $q_1 \leq q_2$ and $t_1 \leq t_2$. If $\mathrm{P}(q_1, t_1) = \infty$, there is nothing to be proved. So, suppose that $\mathrm{P}(q_1, t_1) < \infty$. Then by Proposition 2.1.9 G_{q_1,t_1} is a summable Hölder family of functions. Since the set $\{i : \sup f^{(i)} > 0\}$ is finite and since all the functions $\log|\phi_i'|$ are negative, this implies that G_{q_2,t_2} also forms a summable Hölder family of functions. It then follows from Theorem 2.1.8 that for every $\epsilon > 0$ there exists a Borel probability measure μ on E^∞ such that $\int -(q_2 f - t_2\zeta)d\mu < \infty$ (which implies that $\int -(q_1 f - t_1\zeta)d\mu < \infty$) and

$$\begin{aligned}P(q_2, t_2) &\leq \mathrm{h}_\mu + \int (q_2 f - t_2\zeta)d\mu + \epsilon\\ &= \mathrm{h}_\mu + \int (q_1 f - t_1\zeta)d\mu + (q_2 - q_1)\int f d\mu + (t_1 - t_2)\int \zeta d\mu + \epsilon\\ &\leq \mathrm{h}_\mu + \int (q_1 f - t_1\zeta)d\mu + \epsilon \leq \mathrm{P}(q_1, t_1) + \epsilon,\end{aligned}$$

where the last inequality follows from Theorem 2.1.8. Letting $\epsilon \searrow 0$ we thus get $\mathrm{P}(q_2, t_2) \leq \mathrm{P}(q_1, t_1)$. The proof is complete. □

Given $q \geq 0$ let

$$\begin{aligned}\mathcal{F}in(q) &= \inf\{t : \mathcal{L}_{G_{q,t}}(1\!\!1) < \infty\} = \inf\{t : \mathrm{P}(G_{q,t}) < \infty\}\\ &\leq \theta(S) \text{ and let } \theta(q) = \inf \mathcal{F}in(q).\end{aligned}$$

Notice that if $q \in \mathcal{F}in(F)$, then $0 \in \mathcal{F}in(q)$. We assume that for every $q \in \mathcal{F}in(F)$ there exists $u \in \mathcal{F}in(q)$ such that

$$0 < \mathrm{P}(G_{q,u}) < \infty. \tag{4.66}$$

We shall prove the following.

Lemma 4.9.2 *If $q \in \mathcal{F}in(F)$, then there exists a unique $t = T(q)$, called the temperature function associated with q, such that $\mathrm{P}(G_{q,T(q)}) = 0$. In addition $T(q) \in (\theta(q), \infty)$.*

Proof. Fix $q > \theta(F)$. Since for every $n \geq 1$ the function $t \mapsto \sum_{|\omega|=n} ||\exp(\sum_{j=1}^n \phi_{q,t}^{\omega_j} \circ \phi_{\sigma^j\omega})||$, $t \in \mathcal{F}in(q)$, is logarithmic convex, the function $t \mapsto \mathrm{P}(G_{q,t})$ is convex and hence continuous in $(\theta(q), \infty)$. Since $0 < \mathrm{P}(G_{q,u}) < \infty$ for some $u \in \mathcal{F}in(q)$, in order to conclude the proof it therefore suffices to show that the function $t \mapsto \mathrm{P}(G_{q,t})$ is strictly decreasing on $t \in (\theta(q), \infty)$ and $\lim_{t\to+\infty} \mathrm{P}(G_{q,t}) = -\infty$. But for every $t \geq u$

$$
\begin{aligned}
\mathrm{P}(G_{q,t}) &= \lim_{n\to\infty} \frac{1}{n} \log\left(\sum_{|\omega|=n} || \, |\phi_\omega'|^t \exp(S_\omega(qF))||\right) \\
&\leq \lim_{n\to\infty} \frac{1}{n} \log\left(\sum_{|\omega|=n} ||\phi_\omega'||^{t-u} || \, |\phi_\omega'|^u \exp(S_\omega(qF))||\right) \\
&\leq \liminf_{n\to\infty} \frac{1}{n} \log\left(\sum_{|\omega|=n} s^{n(t-u)} || \, |\phi_\omega'|^u \exp(S_\omega(qF))||\right) \\
&= (t-u)\log s + \mathrm{P}(G_{q,u})
\end{aligned}
$$

Hence $t \in \mathcal{F}in(q)$ and moreover, as $s < 1$, $\lim_{t\to+\infty} \mathrm{P}(G_{q,t}) = -\infty$. To prove that $\mathrm{P}(G_{q,t})$ is strictly decreasing consider $t > \theta(q)$ and $\delta > 0$. We then have

$$
\begin{aligned}
\mathrm{P}(G_{q,t+\delta}) &= \lim_{n\to\infty} \frac{1}{n} \log\left(\sum_{|\omega|=n} || \, |\phi_\omega'|^{t+\delta} \exp(S_\omega(q\phi))||\right) \\
&\leq \lim_{n\to\infty} \frac{1}{n} \log\left(\sum_{|\omega|=n} ||\phi_\omega'||^{\delta} || \, |\phi_\omega'|^t \exp(S_\omega(qF))||\right) \\
&\leq \lim_{n\to\infty} \frac{1}{n} \log\left(s^{n\delta} || \, |\phi_\omega'|^t \exp(S_\omega(qF))||\right) \\
&= \delta \log s + \mathrm{P}(G_{q,t}) < \mathrm{P}(G_{q,t}).
\end{aligned}
$$

□

Given $q \in \mathcal{F}in(F)$ and $t \in \mathcal{F}in(q)$, let

$$\mu_{q,t} = \mu_{G_{q,t}}, \ \tilde{\mu}_{q,t} = \tilde{\mu}_{G_{q,t}}, \ m_{q,t} = m_{G_{q,t}}, \ \tilde{m}_{q,t} = \tilde{m}_{G_{q,t}}$$

and

$$\mu_q = \mu_{q,T(q)}, \ \tilde{\mu}_q = \tilde{\mu}_{q,T(q)}, \ m_q = m_{q,T(q)}, \ \tilde{m}_q = \tilde{m}_{q,T(q)}$$

and let

$$\alpha(q) = \frac{\int f d\tilde{\mu}_q}{-\int \zeta d\tilde{\mu}_q}$$

if $\int |f| d\tilde{\mu}_q < \infty$. By (4.65), $\pi : E^\infty \to J$ is 1-to-1, so given $x = \pi(\omega)$ we can speak about x_n and $x|_n$ respectively as ω_n and $\omega|_n$. Given μ, a Borel probability measure on J, and $x \in J$ we define

$$\underline{D}_\mu(x) = \liminf_{n\to\infty} \frac{\log\big(\mu(\phi_{x|_n}(X_{t(x_n)}))\big)}{\log\big(\mathrm{diam}(\phi_{x|_n}(X_{t(x_n)}))\big)},$$

$$\overline{D}_\mu(x) = \limsup_{n\to\infty} \frac{\log\big(\mu(\phi_{x|_n}(X_{t(x_n)}))\big)}{\log\big(\mathrm{diam}(\phi_{x|_n}(X_{t(x_n)}))\big)},$$

and

$$\underline{d}_\mu(x) = \liminf_{r\to 0} \frac{\log\big(\mu(B(x,r))\big)}{\log r},$$

$$\overline{d}_\mu(x) = \limsup_{r\to 0} \frac{\log\big(\mu(B(x,r))\big)}{\log r},$$

If $\overline{D}_\mu(x) = \underline{D}_\mu(x)$ we denote the common value by $D_\mu(x)$, and if $\overline{d}_\mu(x) = \underline{d}_\mu(x)$ the common value is denoted by $d_\mu(x)$. Given $\alpha \geq 0$ we define

$$K_\mu(\alpha) = \{x \in J : D_\mu(x) = \alpha\}$$

and

$$f_\mu(\alpha) = \mathrm{HD}(K_\mu(\alpha)),$$

the Hausdorff dimension of the set $K_\mu(\alpha)$. Let k be a strictly convex map on an interval I, so that $k'' > 0$ wherever this second derivative exists. The *Legendre transform* of k is the function l of a new variable p defined by

$$l(p) = \max_I\{px - k(x)\}$$

everywhere where the maximum exists. It can be proved that the domain of l is either a point, an interval or a half-line. It is also easy to show that l is strictly convex and that the Legendre transform is an involution. We then say that the functions k and l form a Legendre transform pair. The following theorem (see [Ro] for example) gives a useful characterization of a *Legendre transform pair.*

Theorem 4.9.3 *Two strictly convex differentiable functions k and g form a Legendre transform pair if and only if $l(-k'(q)) = k(q) - qk'(q)$.*

Our main result in this section is the following.

Theorem 4.9.4 *Suppose that condition (4.66) is satisfied for all $q \in \mathcal{F}in(F)$. Suppose also that there exists an interval $\Delta_1 \subset \mathcal{F}in(F)$ such that $1 \in \Delta_1$ and for every $q \in \Delta_1$ and all t in some neighborhood of $T(q)$ (contained in $(\theta(q), \infty)$),*

$$\int (|f|^{2+\gamma} + |\zeta|^{2+\gamma}) d\tilde{\mu}_q < \infty \quad \text{and} \quad \int (|f| + |\zeta|) d\tilde{\mu}_{q,t} < \infty$$

for some $\gamma > 0$. Suppose finally that $\mathrm{h}_{\mu_q}(\sigma)/\chi_{\mu_q}(\sigma) > \theta(S)$ for all $q \in \Delta_2 \subset \Delta_1$ for some interval $\Delta_2 \subset \Delta_1$. Then

(a) The number $D_{\mu_F}(x)$ exists for μ_F-a.e. $x \in J$ and

$$D_{\mu_F}(x) = \frac{-\int f d\tilde{\mu}_F}{\int \zeta d\tilde{\mu}_F}.$$

(b) The function $T : \Delta_1 \to I\!R$ is real-analytic, $T(0) = \mathrm{HD}(J)$, and $T'(q) < 0$, $T''(q) \geq 0$ for all $q \in \Delta_1$.
(c) For every $q \in \Delta_2$, $f_{\mu_F}(-T'(q)) = T(q) - qT'(q)$.
(d) If $\tilde{\mu}_F \neq \tilde{\mu}_{-\mathrm{HD}(J)\zeta}$, then the function $\alpha \mapsto f_{\mu_F}(\alpha)$, $\alpha \in (\alpha_1, \alpha_2)$ is real-analytic, where the interval (α_1, α_2), $0 \leq \alpha_1 < \alpha_2 \leq \infty$ is the range of the function $-T'(q)$ defined on the interval Δ_2. Otherwise $T'(q) = \mathrm{HD}(J)$ for every $q \in (\theta(F), \infty)$.
(e) If $\tilde{\mu}_F \neq \tilde{\mu}_{-\mathrm{HD}(J)\zeta}$, then the functions $f_{\mu_F}(\alpha)$ and $T(q)$ form a Legendre transform pair.
(f) For every $q \in \Delta_1$ the number $T(q)$ is uniquely determined by the property that there exists a constant $C \geq 1$ such that for every $n \geq 1$

$$C^{-1} \leq \sum_{|\omega|=n} \mu_F^q([\omega]) \mathrm{diam}^{T(q)}(\phi_\omega(X)) \leq C.$$

Proof. Since $1 \in \Delta_1$ and $\int |f| d\tilde{\mu}_F < \infty$, part (a) is a combined consequence of Birkhoff's ergodic theorem (along with (4f), (4.20) and (4.23)), the Breimann–Shannon–McMillan theorem and the assumption that $\mathrm{P}(F) = 0$. We shall now prove part (b). And indeed, since by Proposition 2.6.13, $\frac{\partial P}{\partial t}|_{q,t} = -\int \zeta d\tilde{\mu}_{q,t} < 0$ for every $q \in \Delta_1$ and all t in a neighborhood of q, and since $T(q)$ is uniquely determined by the condition $\mathrm{P}(q, T(q)) = 0$, it follows from Theorem 2.6.12 and the implicit function theorem that the map $q \mapsto T(q)$ is real-analytic on Δ_1. Since the system F is regular, $\mathrm{P}(-\mathrm{HD}(J)\zeta) = 0$, which means that $T(0) = \mathrm{HD}(J)$. It follows from Proposition 2.6.13 that for every $q \in \Delta_1$

$$0 = \frac{d\mathrm{P}}{dq}(q, T(q)) = \frac{\partial \mathrm{P}}{\partial q}|_{(q,T(q))} + \frac{\partial \mathrm{P}}{\partial t}|_{(q,T(q))} T'(q) = \int \phi d\tilde{\mu}_q - \int g d\tilde{\mu}_q T'(q)$$

and therefore

$$T'(q) = \frac{\int f d\tilde{\mu}_q}{\int \zeta d\tilde{\mu}_q} = -\alpha(q). \tag{4.67}$$

Since $\mathrm{P}(f) = 0$ and $\int f d\tilde{\mu}_q < \infty$, we deduce from Theorem 2.1.8 that $\int f d\tilde{\mu}_q + \mathrm{h}_{\tilde{\mu}_q}(\sigma) \le 0$ and therefore it follows from (4.67) that $T'(q) \le -\mathrm{h}_{\tilde{\mu}_q}(\sigma)/\int \zeta d\tilde{\mu}_q \le 0$. Thus to prove that $T'(q) < 0$ it suffices to notice that $\mathrm{h}_{\tilde{\mu}_q}(\sigma) > 0$, which follows immediately from Theorem 2.5.2. Hence, to complete the proof of Theorem 4.9.4(b) it is left to show that $T''(q) \ge 0$ for all $q \in \Delta_1$. This is done in the following.

Lemma 4.9.5 *The function $q \mapsto T(q)$, $q \in \Delta_1$ is convex. It is not strictly convex if and only if $\tilde{\mu}_f$ is equal to $\tilde{\mu}_{-\mathrm{HD}(J)\zeta}$.*

Proof. Differentiating the formula

$$0 = \frac{\partial \mathrm{P}(q,t)}{\partial t}|_{(q,T(q))} \cdot T'(q) + \frac{\partial \mathrm{P}(q,t)}{\partial q}|_{(q,T(q))}$$

and using Proposition 2.6.13 we obtain

$$\begin{aligned} T''(q) &= -\frac{T'(q)^2 \frac{\partial^2 \mathrm{P}(q,t)}{\partial t^2} + 2T'(q)\frac{\partial^2 \mathrm{P}(q,t)}{\partial q \partial t} + \frac{\partial^2 \mathrm{P}(q,t)}{\partial q^2}}{\frac{\partial \mathrm{P}(q,t)}{\partial t}} \\ &= \frac{T'(q)^2 \frac{\partial^2 \mathrm{P}(q,t)}{\partial t^2} + 2T'(q)\frac{\partial^2 \mathrm{P}(q,t)}{\partial q \partial t} + \frac{\partial^2 \mathrm{P}(q,t)}{\partial q^2}}{\chi_{\tilde{\mu}_q}}, \end{aligned}$$

where, let us recall, $\chi_{\tilde{\mu}_q} = \int \zeta d\tilde{\mu}_q$ is the Lyapunov characteristic exponent of the measure $\tilde{\mu}_q$. Invoking Proposition 2.6.14 we see that

$$\frac{\partial^2 \mathrm{P}}{\partial t^2} = \sigma^2_{\tilde{\mu}_q}(-\zeta), \quad \frac{\partial^2 \mathrm{P}}{\partial q \partial t} = \sigma^2_{\tilde{\mu}_q}(\zeta, f), \quad \frac{\partial^2 \mathrm{P}}{\partial q^2} = \sigma^2_{\mu_q}(f).$$

Hence, we can write

$$
\begin{aligned}
&T'(q)^2\frac{\partial^2 \mathrm{P}}{\partial t^2} + 2T'(q)\frac{\partial^2 \mathrm{P}}{\partial q \partial t} + \frac{\partial^2 \mathrm{P}}{\partial q^2} \\
&= T'(q)^2\sum_{k=0}^{\infty}\big(\tilde{\mu}_q(\zeta\cdot\zeta\circ\sigma^k) - \chi_{\tilde{\mu}_q}^2\big) + T'(q)\sum_{k=0}^{\infty}\big(\tilde{\mu}_q(-\zeta\cdot f\circ\sigma^k) \\
&+ \chi_{\tilde{\mu}_q}\tilde{\mu}_q(f)\big) + T'(q)\sum_{k=0}^{\infty}\big(\tilde{\mu}_q(f(-\zeta\circ\sigma^k)) \\
&+ \chi_{\tilde{\mu}_q}\tilde{\mu}_q(f)\big) + \sum_{k=0}^{\infty}\big(\tilde{\mu}_q(f\cdot f\circ\sigma^k) - \tilde{\mu}_q(f)^2\big) \\
&= \sum_{k=0}^{\infty}\tilde{\mu}_q\big(-T'(q)\zeta(-T'(q)\zeta\circ\sigma^k + f\circ\sigma^k)\big) \\
&+ \sum_{k=0}^{\infty}\tilde{\mu}_q\big(f(-T'(q)\zeta(-T'(q)\zeta\circ\sigma^k + f\circ\sigma^k)\big) \\
&- \sum_{k=0}^{\infty}\big(-T'(q)\chi_{\tilde{\mu}_q} + \tilde{\mu}_q(f)\big)^2 \\
&= \sum_{k=0}^{\infty}\tilde{\mu}_q\big((-T'(q)\zeta + f)(-T'(q)\zeta + f)\circ\sigma^k\big) - (-T'(q)\chi_{\tilde{\mu}_q} + \tilde{\mu}_q(f))^2 \\
&= \sigma_{\tilde{\mu}_q}^2(-T'(q)\zeta + f).
\end{aligned}
$$

It follows then from (4.67) that $\int(-T'(q)\zeta + f)d\tilde{\mu}_q = 0$. We know that $\sigma_{\tilde{\mu}_q}^2(-T'(q)\zeta + f) \geq 0$, and by Lemma 4.8.8 $\sigma_{\tilde{\mu}_q}^2(-T'(q)\zeta + f) = 0$ if and only if the function $-T'(q)\zeta + f$ is cohomologous to 0 in the class of bounded Hölder continuous functions. Therefore $T'(q)\zeta$ is cohomologous to f and, as $\mathrm{P}(f) = 0$, also $\mathrm{P}(T'(q)\zeta) = 0$. Thus, by Theorem 4.2.13, $T'(q) = -\mathrm{HD}(J)$ and consequently f is cohomologous to the function $-\mathrm{HD}(J)\zeta$. This implies that $\tilde{\mu}_f = \tilde{\mu}_{-\mathrm{HD}(J)\zeta}$, the latter being the equilibrium (invariant Gibbs) state of the potential $-\mathrm{HD}(J)\zeta$. The proof is complete. □

So, item (b) of Theorem 4.9.4 is now an immediate consequence of Lemma 4.9.5. We shall now focus on a contribution toward the proof of parts (c)–(e). Given $\alpha \geq 0$ we define

$$
\tilde{K}(\alpha) = \left\{x \in J : \lim_{n\to\infty}\frac{\sum_{j=0}^{n-1} f\circ\sigma^j(x)}{\sum_{j=0}^{n-1} -\zeta\circ\sigma^j(x)} = \alpha\right\}.
$$

Lemma 4.9.6 *For every $\alpha \ge 0$, $\tilde{K}(\alpha) = K_{\mu_f}(\alpha)$.*

Proof. In order to prove this lemma it suffices to show that for all $x \in J$

$$\lim_{n\to\infty} \frac{\log\big(\mu_f(\phi_{x|_n}(X_{t(x_n)}))\big)}{\sum_{j=0}^{n-1} f\circ\sigma^j(x)} = 1 \text{ and } \lim_{n\to\infty} \frac{\log\big(\text{diam}(\phi_{x|_n}(X_{t(x_n)}))\big)}{\sum_{j=0}^{n-1} -\zeta\circ\sigma^j(x)} = 1.$$

And in order to prove the first equality it suffices to demonstrate that

$$\lim_{n\to\infty} \frac{\log\big(m_F(\phi_{x|_n}(X_{t(x_n)}))\big)}{\sum_{j=0}^{n-1} f\circ\sigma^j(x)} = 1$$

and this follows immediately from the F-conformality of the measure m_F and the fact that $\mathrm{P}(F) = 0$. The second inequality to be proved is an immediate consequence of the bounded distortion property. □

Lemma 4.9.7 *If $x \in \tilde{K}(\alpha)$ and*

$$\liminf_{n\to\infty} \frac{\log|\phi'_{x_n}(\sigma^n(x))|}{\log|\phi'_{x|_{n-1}}(\sigma^{n-1}(x))|} = 0,$$

then for every $q \in I\!R$

$$\liminf_{n\to\infty} \left(\frac{\log|\phi'_{x_n}(\sigma^n(x))|}{\log|\phi'_{x|_{n-1}}(\sigma^{n-1}(x))|} + \frac{qf(\sigma^{n-1}x)}{\log|\phi'_{x|_{n-1}}(\sigma^{n-1}(x))|} \right) \le 0$$

Proof. If $q \le 0$, then our inequality follows immediately from the fact that $q \in \mathcal{F}in(F)$, our assumption and the formula $\lim_{n\to\infty} \log||\phi'_{x|_{n-1}}|| = -\infty$. So, we may assume that $q > 0$. Let $\{n_k\}_{k=1}^\infty$ be an increasing infinite sequence such that

$$\lim_{k\to\infty} \frac{\log|\phi'_{x_{n_k}}(\sigma^{n_k}(x))|}{\log|\phi'_{x|_{n_k-1}}(\sigma^{n_k-1}(x))|} = 0. \tag{4.68}$$

In order to conclude the proof it suffices to sow that

$$\lim_{k\to\infty} \frac{f(\sigma^{n_k-1}x)}{\log|\phi'_{x|_{n_k-1}}(\sigma^{n_k-1}(x))|} \le 0.$$

So, suppose on the contrary that

$$\limsup_{k\to\infty} \frac{f(\sigma^{n_k-1}x)}{\log|\phi'_{x|_{n_k-1}}(\sigma^{n_k-1}(x))|} \ge 2b > 0$$

for some positive b. Passing to a subsequence of the sequence $\{n_k\}_{k=1}^\infty$ we may assume that the limit $\lim_{k\to\infty} \frac{f(\sigma^{n_k-1}x)}{\log|\phi'_{x|_{n_k-1}}(\sigma^{n_k-1}(x))|}$ exists and is greater than or equal to $2b$ (perhaps $+\infty$). This, (4.68) and the fact that $x \in \tilde{K}(\alpha)$ imply the existence of an integer $l_0 \geq 1$ such that for every $l \geq l_0$

$$\frac{\sum_{j=0}^{l} -f(\sigma^j x)}{\sum_{j=0}^{l} \zeta(\sigma^j x)} \geq \alpha - \frac{b}{3}, \quad \frac{f(\sigma^l x)}{\sum_{j=0}^{l} \zeta(\sigma^j x)} \geq b \text{ and } \frac{\sum_{j=0}^{l} \zeta(\sigma^j x)}{\sum_{j=0}^{l+1} \zeta(\sigma^j x)} \geq 1 - \delta,$$

where δ is so small that $(\alpha - \frac{b}{3})(1-\delta) \geq \alpha - \frac{b}{2}$. But then, taking k so large that $n_k - 2 \geq l_0$, we get

$$\begin{aligned}\frac{\sum_{j=0}^{n_k-1} -f(\sigma^j x)}{\sum_{j=0}^{n_k-1} \zeta(\sigma^j x)} &= \frac{\sum_{j=0}^{n_k-2} -f(\sigma^j x)}{\sum_{j=0}^{n_k-2} \zeta(\sigma^j x)} \cdot \frac{\sum_{j=0}^{n_k-2} \zeta(\sigma^j x)}{\sum_{j=0}^{n_k-1} \zeta(\sigma^j x)} + \frac{-f(\sigma^{n_k-1} x)}{\sum_{j=0}^{n_k-1} \zeta(\sigma^j x)} \\ &\geq (\alpha - \frac{b}{3})(1-\delta) + b \geq \alpha - \frac{b}{2} + b = \alpha + \frac{b}{2}.\end{aligned}$$

This however implies that

$$\liminf_{n\to\infty} \frac{\sum_{j=0}^{n} -f(\sigma^j x)}{\sum_{j=0}^{n} \zeta(\sigma^j x)} \geq \alpha + \frac{b}{2} > \alpha$$

which is a contradiction since $x \in \tilde{K}(\alpha)$. □

Lemma 4.9.8 *With the same assumptions as in Theorem 4.9.4*

(a) $\mu_q(K_{\mu_F}(\alpha(q))) = 1$ *for all* $q \in \Delta_1$.

(b) $\underline{d}_{\mu_q}(x) \leq T(q) + q\alpha(q)$ *for all* $q \in \Delta_1$ *and for every* $x \in K_{\mu_F}(\alpha(q))$ *but a set of Hausdorff dimension* $\leq \theta(S)$.

(c) $f_{\mu_F}(\alpha(q)) = T(q) + q\alpha(q)$ *for every* $q \in \Delta_2$.

Proof. Fix $q \in \Delta_1$. Since the functions $|f|$ and $|\zeta|$ are integrable with respect to the measure μ_q, part (a) follows immediately from Lemma 4.9.6 and Birkhoff's ergodic theorem. In order to prove part (b) fix $x \in K_{\mu_F}(\alpha(q))$ and $r > 0$. Let $n = n(x,r)$ be the least integer such that $\phi_{x|_n}(X_{t(x_n)}) \subset B(x,r)$. Then $\mu_q(B(x,r)) \geq \mu_q(\phi_{x|_n}(X_{t(x_n)})$ and $\phi_{x|_{n-1}}(X_{t(x_{n-1})})$ is not contained in $B(x,r)$. From the latter, $\mathrm{diam}(\phi_{x|_{n-1}}(X_{t(x_{n-1})})) \geq r$. Hence, by Lemma 3.1.2

$$\begin{aligned}\frac{\log(\mu_q(B(x,r)))}{\log r} &\leq \frac{\log(\mu_q(\phi_{x|_n}(X)))}{\log(\mathrm{diam}(\phi_{x|_{n-1}}(X)))} \\ &\leq \frac{T(q)\sum_{j=1}^{n} \log|\phi'_{x_j}(\sigma^j(x))| + q\sum_{j=0}^{n-1} f\circ\sigma^j(x) + M_1}{\sum_{j=1}^{n-1} \log|\phi'_{x_j}(\sigma^j(x))| + M_2}\end{aligned}$$

for some constants M_1 and M_2. Since the range of the function $r \mapsto n(x,r)$, $r \in (0,1]$, is of the form $\mathbb{N} \cap [A, \infty)$, it follows from the last inequality, Lemma 4.9.7 and Lemma 4.9.6 that if

$$\liminf_{n\to\infty} \frac{\log|\phi'_{x_n}(\sigma^n(x))|}{\log|\phi'_{x|_{n-1}}(\sigma^{n-1}(x))|} = 0,$$

then $\underline{d}_{\mu_q}(x) \le T(q) + q\alpha(q)$. Consider the set

$$\mathrm{Bad} = \left\{ x \in J : \liminf_{n\to\infty} \frac{\log|\phi'_{x_n}(\sigma^n(x))|}{\log|\phi'_{x|_{n-1}}(\sigma^{n-1}(x))|} > 0 \right\}.$$

We shall show that $\mathrm{HD}(\mathrm{Bad}) \le \theta(S)$. So, given $\gamma > 0$ define

$$\mathrm{Bad}(\gamma) = \left\{ x \in J : \exists q \ge 1\, \forall n \ge q\, \frac{\log|\phi'_{x_n}(\sigma^n(x))|}{\log|\phi'_{x|_{n-1}}(\sigma^{n-1}(x))|} \ge \gamma \right\}$$

and given $n \ge 1$ put

$$\mathrm{Bad}_n(\gamma) = \left\{ x \in J : \frac{\log|\phi'_{x_k}(\sigma^k(x))|}{\log|\phi'_{x|_{k-1}}(\sigma^{k-1}(x))|} \ge \gamma\ \forall k \ge n \right\}.$$

Fix $\eta > \theta(S)$. By the definition of $\theta(S)$ there exists $k \ge 1$ so large that for all $l \ge k$

$$\sum_{\{i\in I:||\phi_i'||_0 \le Ks^{\gamma l}\}} ||\phi_i'||_0^\eta \le \frac{1}{2}. \tag{4.69}$$

Fix $n \ge 1$. For every $l \ge p = \max\{n-1, k\}$ let $\Omega_l = \{\omega \in E^l : \phi_\omega(X_{t(\omega)}) \cap \mathrm{Bad}_n(\gamma) \ne \emptyset\}$. We shall prove by induction that for every $l \ge p$

$$\sum_{\omega\in\Omega_l} ||\phi_i'||_0^\eta \le \left(\frac{1}{2}\right)^{l-p} \sum_{\omega\in\Omega_p} ||\phi_i'||_0^\eta, \tag{4.70}$$

where as $\eta > \theta(S)$,

$$\sum_{\omega\in\Omega_p} ||\phi_i'||_0^\eta \le \sum_{\omega\in I^p} ||\phi_i'||_0^\eta < \infty. \tag{4.71}$$

Indeed, for $l = p$ we even have equality. So, suppose that (4.70) holds for some $l \ge p$. Fix $\omega \in \Omega_{l+1}$. Then $\omega|_l \in \Omega_l$ and there exists $x \in \phi_\omega(X_{t(\omega)}) \cap \mathrm{Bad}_n(\gamma)$. Since $x = \phi_{x|_{l+1}}(\sigma^{l+1}(x))$, it follows from (4.65) that $\omega = x|_{l+1}$. Since $l \ge n-1$ and $x \in \mathrm{Bad}_n(\gamma)$, we therefore get

$$\begin{aligned}||\phi'_{\omega_{l+1}}||_0 &\le K|\phi'_{\omega_{l+1}}(\sigma^{l+1}(x))| = K|\phi'_{x_{l+1}}(\sigma^{l+1}(x))| \\ &\le K|\phi_{x|_l}(\sigma^l(x))|^\gamma \le Ks^{\gamma l}.\end{aligned}$$

Thus, using (4.69) and (4.70) for l we can write

$$\begin{aligned}\sum_{\omega\in\Omega_{l+1}} ||\phi_i'||_0^\eta &\le \sum_{\omega\in\Omega_l} \sum_{\{i\in I: ||\phi_i'||_0\le Ks^{\gamma l}\}} ||\phi_\omega'||_0^\eta ||\phi_i'||_0^\eta \\ &= \sum_{\omega\in\Omega_l} ||\phi_\omega'||_0^\eta \sum_{\{i\in I: ||\phi_i'||_0\le Ks^{\gamma l}\}} ||\phi_i'||_0^\eta \\ &\le \frac{1}{2}\sum_{\omega\in\Omega_l} ||\phi_\omega'||_0^\eta = \left(\frac{1}{2}\right)^{l+1-p} \sum_{\omega\in\Omega_p} ||\phi_i'||_0^\eta.\end{aligned}$$

The inductive proof of (4.70) is finished. By (4.20) we therefore get for all $l \ge k$

$$\sum_{\omega\in\Omega_l} \mathrm{diam}^\eta(\phi_\omega(X_{t(\omega)})) \le D\left(\frac{1}{2}\right)^{l-p} \sum_{\omega\in\Omega_p} ||\phi_i'||_0^\eta$$

and using (4.71) we conclude that $\mathcal{H}^\eta(\mathrm{Bad}_n(\gamma)) = 0$. Thus $\mathrm{HD}(\mathrm{Bad}_n(\gamma)) \le \eta$ which implies that $\mathrm{HD}(\mathrm{Bad}_n(\gamma)) \le \theta(S)$. Since $\mathrm{Bad}(\gamma) = \bigcup_{n\ge1}\mathrm{Bad}_n(\gamma)$, $\mathrm{HD}(\mathrm{Bad}(\gamma)) \le \theta(S)$ and since $\mathrm{Bad} = \bigcup_{m\ge1}\mathrm{Bad}(1/m)$, $\mathrm{HD}(\mathrm{Bad}) \le \theta(S)$. The proof of (b) is complete. Since $\mu_q(K_{\mu_F}(\alpha(q))) = 1$, it follows from Theorem 4.4.2 that $f_{\mu_F}(\alpha(q)) = \mathrm{HD}(K_{\mu_F}(\alpha(q)) \ge \mathrm{HD}(\mu_q) = \mathrm{h}_{\mu_q}(\sigma)/\chi_{\tilde\mu_q}(\sigma)$. Since $\mathrm{P}(G_{q,T(q)}) = 0$, using Theorem 2.2.9, we continue writing

$$\begin{aligned}f_{\mu_F}(\alpha(q)) \ge \frac{\mathrm{h}_{\tilde\mu_q}(\sigma)}{\chi_{\tilde\mu_q}(\sigma)} &= \frac{-\int g_{q,T(q)}d\tilde\mu_q}{\chi_{\tilde\mu_q}(\sigma)} = \frac{\int(-T(q)\zeta + qf)d\tilde\mu_q}{-\chi_{\tilde\mu_q}(\sigma)} \\ &= \frac{-T(q)\chi_{\tilde\mu_q}(\sigma) + q\int fd\tilde\mu_q}{-\chi_{\tilde\mu_q}(\sigma)} = T(q) + q\alpha(q).\end{aligned} \tag{4.72}$$

This proves one half of (c). If now $q \in \Delta_2$, then our assumptions give $\mathrm{h}_{\mu_q}(\sigma)/\chi_{\mu_q}(\sigma) > \theta(S)$. Applying this along with (a) and (b), it follows from Theorem A2.0.16 that $f_{\mu_F}(\alpha(q)) = \mathrm{HD}(K\mu_F(\alpha(q))) \le T(q) + q\alpha(q)$. This proves the other part of (c). □

Part (c) of Theorem 4.9.4 is an immediate consequence of Lemma 4.9.8(c) and formula (4.69). Part (d) is a combined consequence of Lemma 4.9.5 and part (c) of Theorem 4.9.4. Part (e) of Theorem 4.9.4 follows from Lemma 4.9.5, part (c) of Theorem 4.9.4 and Theorem 4.9.3. We end the proof of Theorem 4.9.4 by demonstrating part (f). And indeed, since the diameters of the images $\phi_\omega(X_{t(\omega)})$ tend to zero uniformly (exponentially) fast with respect to the length of ω, we conclude that there exists at most one value $t \in I\!R$ such that for some $C \geq 1$ and every $n \geq 1$

$$C^{-1} \leq \sum_{|\omega|=n} \mu_F^q([\omega])\mathrm{diam}^t(\phi_\omega(X_{t(\omega)})) \leq C.$$

So, we only need to show that the display appearing in part (f) of Theorem 4.9.4 is true. And indeed, if $\omega \in I^*$, say $|\omega| = n$ and $\rho \in [\omega]$, then it follows from the definition of measures $\tilde{\mu}_q$ and $\tilde{\mu}_F$ that

$$\begin{aligned}
\tilde{\mu}_q([\omega]) &\asymp \exp\left(q\sum_{j=0}^{n-1} f\circ\sigma^j(\rho) - T(q)\sum_{j=0}^{n-1}\zeta\circ\sigma^j(\rho)\right) \\
&= \left(\exp\sum_{j=0}^{n-1} f\circ\sigma^j(\rho)\right)^q\left(\exp\left(-\sum_{j=0}^{n-1}\zeta\circ\sigma^j(\rho)\right)\right)^{T(q)} \\
&\asymp \tilde{\mu}_F^q([\rho|_n])\mathrm{diam}^{T(q)}(\phi|_{\rho|_n}(X_{t(\rho_n)})) \\
&= \tilde{\mu}_F^q([\omega])\mathrm{diam}^{T(q)}(\phi|_\omega(X_{t(\omega)})).
\end{aligned}$$

Since $\sum_{|\omega|=n}\tilde{\mu}_q([\omega]) = 1$, summing the above display over all $\omega \in I^n$ we obtain the desired inequalities. The proof of Theorem 4.9.4 is complete. □

Let us recall that in [MU2] we have introduced the class of absolutely regular conformal iterated function systems S by the requirement that $\theta(S) = 0$. The same definition extends to conformal GDMSs. For these systems we can rewrite Theorem 4.9.4, relaxing the assumption $\mathrm{h}_{\tilde{\mu}_q}(\sigma)/\chi_{\tilde{\mu}_q} > \theta(S)$ since we already know (see the paragraph preceeding Lemma 4.9.5) that the entropy $\mathrm{h}_{\tilde{\mu}_q}(\sigma)$ is always positive. It then reads as follows.

Theorem 4.9.9 *Suppose that condition (4.66) is satisfied for all $q \in \mathcal{F}in(F)$. Suppose also that there exists an interval $\Delta_1 \subset \mathcal{F}in(F)$ such that $1 \in \Delta_1$ and for every $q \in \Delta_1$ and all t in some neighborhood of $T(q)$ (contained in $(\theta(q), \infty)$),*

$$\int(|f|^{2+\gamma} + |\zeta|^{2+\gamma})d\tilde{\mu}_q < \infty \quad \text{and} \quad \int(|f| + |\zeta|)d\tilde{\mu}_{q,t} < \infty$$

for some $\gamma > 0$. Suppose finally that the system S is absolutely regular. Then

(a) *The number* $D_{\mu_F}(x)$ *exists for* μ_F*-a.e.* $x \in J$ *and*

$$D_{\mu_F}(x) = \frac{-\int f d\tilde{\mu}_F}{\int \zeta d\tilde{\mu}_F}.$$

(b) *The function* $T : \Delta_1 \to I\!R$ *is real-analytic,* $T(0) = \mathrm{HD}(J)$*, and* $T'(q) < 0$, $T''(q) \geq 0$ *for all* $q \in \Delta_1$.

(c) *For every* $q \in \Delta_2$, $f_{\mu_F}(-T'(q)) = T(q) - qT'(q)$.

(d) *If* $\tilde{\mu}_F \neq \tilde{\mu}_{-\mathrm{HD}(J)\zeta}$*, then the function* $\alpha \mapsto f_{\mu_F}(\alpha)$, $\alpha \in (\alpha_1, \alpha_2)$ *is real-analytic, where the interval* (α_1, α_2), $0 \leq \alpha_1 < \alpha_2 \leq \infty$ *is the range of the function* $-T'(q)$ *defined on the interval* Δ_2*. Otherwise* $T'(q) = \mathrm{HD}(J)$ *for every* $q \in (\theta(F), \infty)$.

(e) *If* $\tilde{\mu}_F \neq \tilde{\mu}_{-\mathrm{HD}(J)\zeta}$*, then the functions* $f_{\mu_F}(\alpha)$ *and* $T(q)$ *form a Legendre transform pair.*

(f) *For every* $q \in \Delta_1$ *the number* $T(q)$ *is uniquely determined by the property that there exists a constant* $C \geq 1$ *such that for every* $n \geq 1$

$$C^{-1} \leq \sum_{|\omega|=n} \mu_F^q([\omega]) \mathrm{diam}^{T(q)}(\phi_\omega(X)) \leq C.$$

5
Examples of GDMSs

5.1 Examples of GDMSs in other fields of mathematics

In this section we mainly provide some classes of conformal GDMS naturally generated in other areas of mathematics. We would especially like to call the reader's attention to the class of *Kleinian groups of Schottky type*. We start however with the following two operations on iterated function systems which lead to graph directed Markov systems.

Example 5.1.1 *(Gluing)*

Suppose that we are given finitely many iterated function systems $\{S_v\}_{v\in V}$ in the same Euclidean space, say $I\!R^d$, and let S be a GDMS with vertices V and edges E. We form a new GDMS $\hat{S}$ by adding to S all the contractions $\{\phi_i^{(v)} : X_v \to X_v\}_{i\in I_v}$, where I_v is the alphabet of the system S_v, and by declaring that the new incidence matrix $\hat{A}$ contains the old matrix and has additional entries $\hat{A}_{a,b}$ equal to 1 if either $t(a) = v$ and $b \in I_v$ or $a \in I_v$ and $i(b) = v$.

Example 5.1.2 *(Restrictions)*

Let $S = \{\phi_i : X \to X\}_{i\in I}$ be a conformal iterated function system (e.g. generated by a continued fraction algorithm with restricted entries see Example 5.1.4 for more details) and let $A : I \times I \to \{0, 1\}$. The system $\hat{S}$ generated by taking the set of vertices to be a singleton $\{v\}$ with $X_v = X$, the set of edges equal to I and the incidence matrix equal to A is a GDS.

Example 5.1.3 *(Expanding maps)*

Each distance expanding map $f : X \to X$ (see [Ru], cf. [PU]) has Markov partitions $R = \{X_t\}_{t\in T}$ with arbitrarily small diameters. It gives rise to a GDS with T, the set of vertices, the contractions formed by continuous inverse branches of f defined on X_t, $t \in T$, and the incidence matrix determined by the Markov partition R.

Example 5.1.4 *(Continued fractions)*

This system is given by the maps $\phi_n : [0,1] \to [0,1]$ defined by the formulae

$$\phi_n(x) = \frac{1}{x+n}.$$

It is easy to see that

$$\phi(\omega) = \cfrac{1}{\omega_1 + \cfrac{1}{\omega_2 + \cfrac{1}{\omega_3}}}$$

The continued fraction system with restricted entries (see Example 5.1.2) has been thoroughly explored from the geometric viewpoint in [MU2]. Compare also [HeU] and [U2].

Example 5.1.5 *(Kleinian groups of Schottky type)*

Fix finitely many, say $q \geq 1$, closed balls $B_1, B_2, \dots, B_q \subset \mathbb{R}^d$. For each $j = 1, 2, \dots, q$ let g_j be the inversion with respect ∂B_j, the boundary of B_j. The group $G = \langle g_1, g_2, \dots, g_q \rangle$ is the Kleinian group of Schottky type generated by the inversions $g_1, g_2, \dots, g_q$. Recall that $L(G)$, the limit set of a Kleinian group G is the set of limit points $\lim_{n\to\infty} g_n(z)$, where $g_n \in G$ are mutually distinct. The value of this limit does not depend on the point $z \in \overline{\mathbb{R}^d}$. Our goal is to represent $L(G)$, the limit set of the Schottky group G as the limit set on an appropriate conformal GDS. Let us first construct this GDS. Set $V = \{1, 2, \dots, q\}$, $E = V \times V \setminus \{(i,i) : i \in \{1, 2, \dots, q\}\}$ and $X_v = B_v$ for all $v \in V$. Since $g_i(\overline{\mathbb{R}^d}) = \mathrm{Int}(B_i)$, we see that for every $(i,j) \in E$ the map $g_{(i,j)} = g_i|_{B_j} : B_j \to B_i$ is well defined. The associated incidence matrix is defined as follows.

$$A_{(i,j),(k,l)} = \begin{cases} 1 & \text{if } k = j \\ 0 & \text{if } k \neq j. \end{cases}$$

The system $S_G = \langle V, E, \{g_e\}_{e\in E} \rangle$ obviously satisfies all the requirements of a conformal GDS except that the maps $\{g_e\}_{e\in E}$ need not be uniform contractions. But since the diameters of the sets $g_\omega(B_{t\omega})$ converge to zero uniformly with respect to the length of the word ω, the bounded distortion property (see [PU] for details) implies that all maps from a sufficiently high iterate of the system S_G are uniformly contracting. And this is precisely what we need. We shall prove the following.

Theorem 5.1.6 *If G is a Kleinian group of Schottky type, then $L(G) = J_{S_G}$.*

Proof. The inclusion $J_{S_G} \subset L(G)$ is obvious. In order to prove the opposite inclusion, fix a sequence $\{g_n\}_{n=1}^\infty$ of mutually distinct elements of G such that $\lim_{n\to\infty} g_n(z)$ exists for some (and equivalently all) $z \in \overline{I\!R^d}$. With appropriate $n_{k_j} \in \{1, 2, \dots, q\}$ write

$$g_n = g_{n_{k_n}} \circ g_{n_{k_n-1}} \circ \cdots \circ g_{n_2} \circ g_{n_1}$$

in its unique irreducible form, that is with $g_{n_j} \neq g_{n_{j+1}}$ for all $1 \le j \le k_n - 1$. Passing to a subsequence we may assume that $n_1 \neq i$ for all $n \ge 1$ and some $i \in \{1, 2, \dots, q\}$. Fix $z \in B_i$. We shall show by induction that $g_{n_l} \circ g_{n_{l-1}} \circ \cdots \circ g_{n_1}(z) \in B_{n_l}$ for all $l = 1, 2, \dots, k_n$. And indeed, for $l = 1$ this follows from the fact that $z \notin B_{n_1}$ and $g_{n_1}(\overline{I\!R^d} \setminus B_{n_1}) = \text{Int}(B_{n_1})$. So, suppose that $g_{n_l} \circ g_{n_{l-1}} \circ \cdots \circ g_{n_1}(z) \in B_{n_l}$ for some $1 \le l \le k_n - 1$. Since, by irreducibility of the word $g_{n_{k_n}} \circ g_{n_{k_n-1}} \circ \cdots \circ g_{n_2} \circ g_{n_1}$, $B_{n_{l+1}} \cap B_{n_l} = \emptyset$ and since $g_{n_{l+1}}(\overline{I\!R^d} \setminus B_{n_{l+1}}) = \text{Int}(B_{n_{l+1}})$, we conclude that $g_{n_{l+1}} \circ g_{n_l} \circ \cdots \circ g_{n_1}(z) \in B_{n_{l+1}}$. The inductive proof is finished. Hence, we can write

$$g_n(z) = g_{(n_{k_n}, n_{k_n-1})} \circ g_{(n_{k_n-1}, n_{k_n-2})} \circ \cdots g_{(n_2, n_1)} \circ g_{(n_1, i)}(z).$$

Thus $\lim_{n\to\infty} g_n(z) \in \overline{J}_{S_G} = J_{S_G}$, where we could write the equality sign since the system S_G is finite. □

We shall now generalize the above considerations by establishing natural relations between the limit set of each subgroup of a Kleinian group of Schottky type and the limit set of a naturally associated conformal GDS, perhaps with infinitely many edges. Indeed, let $\Gamma \subset G = \langle g_1, \dots, g_q \rangle$ be a subgroup of a Schottky group. We first look at the set $\Gamma_0 \subset \Gamma$ formed by those elements $g = g_{i_n} \circ \cdots \circ g_{i_1}$ ($i_j \in \{1, 2, \dots, q\}$) written in irreducible form as elements of G, with no proper initial block belonging to Γ. In order to associate with Γ the conformal GDS S_Γ we take as before $V = \{1, 2, \dots, q\}$ and $X_v = B_v$ for all $v \in V$. The elements of the system S_Γ are formed by the maps

$$\phi_{g,j} = g_{i_n} \circ \cdots \circ g_{i_1} : B_j \to B_{i_n}$$

for every element $g = g_{i_n} \circ \ldots \circ g_{i_1} \in \Gamma_0$ and every $j \in \{1, 2, \ldots, q\} \setminus \{i_1\}$. The incidence matrix is determined by the requirement that the compositions $\phi_{g,j} \circ \phi_{\gamma,i}$ are allowed if and only if $j = i_m$, where $\gamma = g_{i_m} \circ \ldots \circ g_{i_1} \in \Gamma_0$ is represented in its irreducible form. As in the case of a (full) Schottky group, the system S_Γ is a conformal GDS but, recalling the discussion in the case of a (full) Schottky group, we see that the only point which may require some explanation is the open set condition. So, suppose that $\phi_{g,j} \neq \phi_{\gamma,i}$, both belonging to S_Γ. If $g = \gamma$, then $j \neq i$ and $\phi_{g,j}(B_j) \cap \phi_{\gamma,i}(B_i) = g(B_j) \cap g(B_i) = \emptyset$ since g is 1-to-1 and $B_j \cap B_i = \emptyset$. If $g \neq \gamma$, then $\phi_{g,j}$ and $\phi_{\gamma,i}$ are two incomparable elements of S_G, and therefore $\phi_{g,j}(B_j) \cap \phi_{\gamma,i}(B_i) = \emptyset$. Thus, we have proved that S_Γ is a conformal GDS. Repeating essentially the proof of the previous theorem, we shall demonstrate the following.

Theorem 5.1.7 *If Γ is a subgroup of a Kleinian group of a Schottky type, then $J_{S_\Gamma} \subset L(\Gamma)$, $L(\Gamma) = \overline{J_{S_\Gamma}}$ and $L(\Gamma) = J_{S_\Gamma}$ if Γ_0 is finite.*

Proof. The inclusion $J_{S_\Gamma} \subset L(\Gamma)$ is obvious. In order to prove that $L(\Gamma) \subset \overline{J_{S_\Gamma}}$ fix a sequence $\{\gamma_n\}_{n=1}^\infty$ of mutually distinct elements of Γ such that $\lim_{n\to\infty} \gamma_n(z)$ exists for some (and equivalently all) $z \in \overline{I\!R^d}$. Write

$$\gamma_n = g_{n_{k_n}} \circ g_{n_{k_n - 1}} \circ \cdots \circ g_{n_2} \circ g_{n_1},$$

in its unique irreducible form in G and decompose this representation of γ_n into blocks $\gamma_{n_l} \circ \ldots \circ \gamma_{n_1}$ of elements from Γ_0. Passing to a subsequence we may assume that $t(n_1) \neq i$ for all $n \geq 1$ and some $i \in \{1, 2, \ldots, q\}$. Fix $z \in B_i$. Then, as in the proof of Theorem 5.1.6, we see that

$$\gamma_n(z) = \phi_{\gamma_{n_l}, i(\gamma_{n_{l-1}})} \circ \phi_{\gamma_{n_{l-1}}, i(\gamma_{n_{l-2}})} \circ \cdots \circ \phi_{\gamma_{n_2}, i(\gamma_{n_1})} \circ \phi_{\gamma_{n_1}, i}(z)$$

and therefore $\lim_{n\to\infty} \gamma_n(z) \in \overline{J_{S_\Gamma}}$. Thus the proof of the inclusion $L(\Gamma) \subset \overline{J_{S_\Gamma}}$ is complete. If now Γ_0 is finite, then S_Γ is also finite and consequently $J_{S_G} = \overline{J_{S_\Gamma}} = L(G)$, which finishes the proof. □

5.2 Examples with special geometric features

In this section we provide a number of examples of infinite conformal iterated function systems showing how flexible they are, and how large a variety of fractal features one can already meet find among them, not to mention general GDMSs. We begin with an extremely simple example of a conformal graph directed Markov system whose limit set cannot be represented as the limit set of a conformal iterated function system.

Example 5.2.1 *A conformal finite irreducible graph directed Markov system whose limit set cannot be represented as the limit set of a conformal iterated function system.*

Let the set of vertices consist of two elements v and w and let $E = \{vw, vv, wv, ww\}$. We put $X_v = [2,4] \times [1,3] \subset I\!\!R^2$ and $X_w = [0,2] \times [3,5] \subset I\!\!R^2$. Let $\phi_{vw} : X_w \to X_v$, $\phi_{ww} : X_w \to X_w$, $\phi_{wv} : X_v \to X_w$, $\phi_{vv} : X_v \to X_v$ be similarity maps with scaling factor $1/2$ mapping respectively the set X_w onto the square $[2,3]\times[2,3]$, X_w onto $[1,2]\times[3,4]$, X_v onto $[1,2] \times [4,5]$ and X_v onto $[3,4] \times [2,3]$. Obviously this graph directed system is irreducible and its limit set J is

$$[2,4] \times \{3\} \cup \{2\} \times [3,5].$$

Since J has Hausdorff dimension is equal to 1 and it is not an analytic arc, it follows from the second paragraph following Theorem 6.4.1 that J cannot be represented as the limit set of a conformal iterated function system.

Example 5.2.2 *The limit set J is an $F_{\sigma\delta}$ but not a G_δ.*

Denote by Q the set of all rational numbers in $[0,1]$. Let $X = [0,1]\times[0,1]$ and let $\Delta = \{(x,x) \in X\}$ be the diagonal of X. Consider a conformal iterated function system $\{\phi_i : X \to X : i \in Q \cup \{-1\}\}$ consisting of linear mappings and such that

(a) $\phi_i(X) \cap \Delta = \{\phi_i(0,1)\} = \{(i,i)\}$ for all $i \in Q$
(b) $\phi_{-1}(x,y) = (x/2, (y+1)/2)$
(c) The sets $\phi_i(X)$, $i \in Q \cup \{-1\}$, are mutually disjoint.

Then $J \cap \Delta = Q$ is not G_δ, so neither is J. Let us also note that this system is not locally finite. □

Example 5.2.3 *An iterated function system for which* $\mathrm{PD}(J) \geq \underline{\mathrm{BD}}(J) > \mathrm{HD}(J)$.

Take any sequence of positive numbers $\{r_i : i \geq 1\}$ (for example of the form b^i, $0 < b < 1$) such that the equation $\sum_{i\geq 1} r_i^t = 1$ has a (unique) solution and this solution is less than 1. Consider a family $\{\phi_i : \{z \in C\!\!\!\!I : |z| \leq 1\} \to \{z \in C\!\!\!\!I : |z| \leq 1\} : i \geq 1\}$ of similarity maps satisfying the open set condition and such that $||\phi_i'|| = r_i$ and $X(\infty) = \{z : |z| = 1\}$. Then by Theorem 4.2.14, $\mathrm{PD}(J) \geq \underline{\mathrm{BD}}(J) \geq \underline{\mathrm{BD}}(X(\infty)) \geq 1$ and by Theorem 4.2.13, $\mathrm{HD}(J) < 1$. □

Example 5.2.4 *Irregular system.*

A model for such a system has been described in Example 4.5 of [MW1]. Since this is a very short and important example we repeat its construction here. Let $I = \{(n,k) : n \geq 1 \text{ and } 1 \leq k \leq 2^{n^2-1}\}$, let $X = [0,1]$, and let $S = \{\phi_{n,k} : X \to X : (n,k) \in I\}$ be a system consisting of similarity maps $\phi_{n,k}$ such that $||\phi'_{n,k}|| = 2^{-(n^2+n)}$ and such that the intervals $\phi_{n,k}(X)$ are mutually disjoint. This last requirement can be satisfied since $\sum_{(n,k)\in I} ||\phi'_{n,k}|| = \sum_{n\geq 1} 2^{-(n^2+n)}2^{n^2-1} = 1/2 < 1$. Notice that by this computation we have shown that $\psi(1) = 1/2 < 1$. Observe also that $\psi(t) = \sum_{n\geq 1} 2^{n^2-1}2^{-(n^2+n)t} = \sum_{n\geq 1} 2^{n^2(1-t)-nt-1} = \infty$ for all $0 < t < 1$. Thus, in view of Theorem 4.3.8, S is irregular, in view of Theorem 4.3.9 $h = \mathrm{HD}(J) = 1$, and in view of Theorem 4.5.11, J is dimensionless in the restricted sense. □

Example 5.2.5 *Linear, regular but not hereditarily regular iterated function system.*

This example is very similar to Example 5.2.4. The only difference in its definition is that now we take $I = \{(n,k) : n \geq 1 \text{ and } 1 \leq k \leq 2^{n^2}\}$. Then the same computations as in Example 5.2.4 above show that $\psi(1) = 1$, whence $\mathrm{P}(1) = 0$, and $\psi(t) = \infty$ for all $0 < t < 1$. Hence, in view of Theorem 4.5.10, S is regular, the only conformal measure is the Lebesgue measure λ_1, and $h = \mathrm{HD}(J) = 1$. Moreover, in view of Theorem 4.3.4, S is not hereditarily regular. □

Notice that Example 5.2.5 provides a number of irregular examples. In fact every cofinite subsystem of S is irregular.

Example 5.2.6 *A hereditarily regular linear system with* $0 < \Pi_h(J) < \infty$ *and* $\mathrm{H}_h(J) = 0$.

Let $X = [0,1]$ and let $S = \{\phi_n : X \to X : n \geq 1\}$ be the CIFS consisting of similarities $\phi_n(x) = \frac{x}{3n^2} + \frac{1}{n} - \frac{1}{3n^2}$ so that $\phi_n(0) = \frac{1}{n} - \frac{1}{3n^2}$ and $\phi_n(1) = \frac{1}{n}$. Thus $||\phi'_n|| = \frac{1}{3n^2}$ and $\psi(t) = \sum_{n\geq 1} ||\phi'_n||^t = \sum_{n\geq 1} 3^{-t}n^{-2t}$. Hence $h = \mathrm{HD}(J) > 1/2$ and by Theorem 4.3.4 S is hereditarily regular. Let m be the corresponding conformal measure. Then for every $n \geq 1$

$$m(B(0,1/n)) = \sum_{k\geq n} \left(\frac{1}{3k^2}\right)^h \geq 3^{-h}\int_n^\infty x^{-2h}\,dx = 3^{-h}\frac{1}{2h-1}\left(\frac{1}{n}\right)^{2h-1}.$$

Taking now for any $0 < r \leq 1$ the unique integer $n \geq 1$ such that $1/(n+1) < r \leq 1/n$, we get $m(B(0,r)) \geq Cr^{2h-1}$, where

$C = ((2h-1)3^h 2^{2h-1})^{-1}$. Since $h-1<0$ it now follows from Corollary 4.5.7 that $H_h(J) = 0$. Positivity of $\Pi_h(J)$ is guaranteed by Theorem 4.5.2. We now show that the assumptions of Lemma 4.5.5 are satisfied with $\gamma = 3$ if for every $n \geq 1$ the point y is chosen to be $1/n$. Indeed, fix $n \geq 1$ and take $1/n^2 \leq r \leq 1$. Suppose first that $r \leq 1/(2n)$. Then $n \geq 2$ and $\frac{1}{n} - r > \frac{1}{2n}$. Let $I(r) = \{k \geq 1 : \frac{1}{k} \leq \frac{1}{n} \text{ and } \frac{1}{k+1} \geq \frac{1}{n} - r\}$. Notice that $\#I(r) \geq (1/n - r)^{-1} - n = n^2 r/(1-nr) \geq n^2 r$. Therefore

$$m(B(1/n, r)) \geq \sum_{k \in I(r)} \left(\frac{1}{3k^2}\right)^h \geq \left(\frac{1}{3(2n)^2}\right)^h \#I(r) \geq (12)^{-h} n^{-2h} n^2 r \geq$$

$$= (12)^{-h} \left(\frac{1}{n^2}\right)^{h-1} r \geq (12)^{-h} r^{h-1} r = (12)^{-h} r^h.$$

Now suppose that $1/(2n) \leq r \leq 2/n$. Then $1/n^2 \leq r/4 \leq 1/(2n)$ and in view of the previous case $m(B(1/n, r)) \geq m(B(1/n, r/4)) \geq (12)^{-h}(r/4)^h = (48)^{-h} r^h$. Finally suppose that $r \geq 2/n$. Then $B(1/n, r) \supset B(0, r/2) \geq C(r/2)^{2h-1} = 2C4^{-h} r^h r^{h-1}$. Thus the assumptions of Lemma 4.5.5 are satisfied and therefore $\Pi_h(J) < \infty$. □

We should mention here that in the next section the CIFS induced by complex continued fractions will be considered, which is also hereditarily regular and whose limit set has h-dimensional Hausdorff measure 0 and h-dimensional finite packing measure. The idea for proving these properties will be the same there as in Example 5.2.6.

Example 5.2.7 *Hereditarily regular linear system with* $\Pi_h(J) = \infty$ *and* $\mathrm{H}_h(J) > 0$.

Let $X = [0, 1]$ and let $S = \{\phi_n : X \to X : n \geq 1\}$ be the CIFS consisting of similarities $\phi_n(x) = 2^{-2n} x + 2^{-n} - 2^{-2n}$ so that $\phi_n(0) = 2^{-n} - 2^{-2n}$ and $\phi_n(1) = 2^{-n}$. Thus $||\phi_n'|| = 2^{-2n}$ and $\psi(t) = \sum_{n \geq 1} ||\phi_n'||^t = \sum_{n \geq 1} 2^{-2nt}$. Hence $h = 1/2$ and by Theorem 4.3.4 S is hereditarily regular. Let m be the corresponding conformal measure. Then for every $n \geq 1$ we have $m(B(0, 2^{-n})) = \sum_{k \geq n} 2^{-2kh} = 2(2^{-2nh})$. Taking now for any $0 < r \leq 1/2$ the unique integer $n \geq 1$ such that $2^{-(n+1)} < r \leq 2^{-n}$, we get

$$m(B(0, r)) \leq 4r^{2h}.$$

Thus, $\Pi_h(J) = \infty$ by Corollary 4.5.8. Finiteness of $\mathrm{H}_h(J)$ is guaranteed by Lemma 4.2. We now show that the assumptions of Theorem 4.5.3 are satisfied with $\gamma = 1$ if for every $n \geq 1$ the point y is chosen to be

2^{-n}. Indeed, fix $n \ge 1$ and take $2^{-2n} \le r \le 1/2$. If $r \ge 2^{-n}$, then $m(B(2^{-n}, r)) \le m(B(0, 2r)) \le 4(2r)^{2h} = 8r^{2h}$. In the general case $r^{1/2} \ge 2^{-n}$ and then $m(B(2^{-n}, r)) \le m(B(2^{-n}, r^{1/2})) \le 8(r^{1/2})^{2h} = 8r^h$. Thus the assumptions of Theorem 4.5.3 are satisfied and therefore $\mathrm{H}_h(J) > 0$. □

Example 5.2.8 *Hereditarily regular linear system with* $\Pi_h(J) = \infty$ *and* $\mathrm{H}_h(J) = 0$.

This example is made up by gluing together Examples 5.2.6 and 5.2.7. Namely let $X = [0, 2]$ and $S = \{\phi_{n,0}, \phi_{n,1} : n \ge 2\}$, where $\phi_{n,0}(x) = \frac{1}{3n^2}\frac{x}{2} + \frac{1}{n} - \frac{1}{3n^2}$ and $\phi_{n,1}(x) = 2^{-2n}\frac{x}{2} + 2^{-n} - 2^{-2n} + 1$. Then $\phi_{n,0}([0, 2]) = [\frac{1}{n} - \frac{1}{3n^2}, \frac{1}{n}] \subset [0, 1/2]$ and $\phi_{n,1}([0, 2]) = [2^{-n} - 2^{-2n} + 1, 2^{-n} + 1] \subset [1, 2]$ and $\psi(t) = \sum_{n\ge2}(||\phi'_{n,0}||^t + ||\phi'_{n,1}||^t) = \sum_{n\ge2} 6^{-t}n^{-2t} + 2^{-t}2^{-2nt}$. So, the interval of convergence of $\psi(t)$ is $(1/2, \infty)$. Thus, in view of Theorem 3.20, S is hereditarily regular and $h = \mathrm{HD}(J) > 1/2$. We see that $X(\infty) = \{0, 1\}$ and, if m is the corresponding h-conformal measure, then as in Example 5.2.6 we get $m(B(0, 1/n)) \ge (2h - 1)^{-1}(1/n)^{2h-1}$. In view of Theorem 4.3.10 this implies that $\mathrm{H}_h(J) = 0$ and as in Example 5.2.7 we get $m(B(1, 2^{-n})) \le 4^h(4^h - 1)^{-1}(2^{-n})^{2h}$ and this in view of Corollary 4.5.8 implies that $\Pi_h(J) = \infty$. □

Example 5.2.9 *One-dimensional systems.*

Here we want to describe how every compact subset F of the interval $X = [0, 1]$ canonically gives rise to a linear CIFS on X such that

$$S(\infty) = (\partial F)^d = \{x \in X : x \text{ is an accumulation point of } \partial F\},$$

the Cantor–Bendixson derived set of ∂F. Indeed, let R be the family of all connected components of $X \setminus F$ and for every $C \in R$ let $\phi_C : X \to X$ be the unique linear map such that $\phi_C(0)$ is the left endpoint of the closure of C and $\phi_C(1)$ is the right endpoint of the closure of C. The system $S = \{\phi_C : X \to X : C \in R\}$ has the property required. □

6
Conformal Iterated Function Systems

In this chapter we deal with conformal iterated function systems CIFS. Recall this means that we assume S to be a CGDMS such that the set of vertices is a singleton, the corresponding spaces are denoted by X and W, and all the entries of the incidence matrix are equal to 1. We would like to note at the very beginning that in the context of CIFSs, indeed in the wider context of conformal-like iterated function systems, the name S-invariant becomes meaningful since each measure μ_F, where F is a summable Hölder family of functions, enjoys the following two properties expressed for a Borel probability measure ν supported on J_S.

$$\nu(\bigcup_{i\in I} \phi_i(A)) = \nu(A) \tag{6.1}$$

and

$$\nu(\phi_i(X) \cap \phi_j(X)) = 0 \tag{6.2}$$

for all $i \in I$, $j \in I \setminus \{i\}$, and all Borel sets $A \subset X$. Any measure ν satisfying (6.1) and (6.2) will be in the sequel called S-invariant.

6.1 The Radon-Nikodym derivative $\rho = \frac{d\mu}{dm}$

In this section we study analytic properties of the Radon-Nikodym derivative $\rho = \frac{d\mu}{dm}$ where m is the h-conformal measure of a regular CIFS S and μ is its S-invariant version. We first introduce an auxiliary Perron–Frobenius operator $F : C(X) \to C(X)$ and then we show that ρ has a unique continuous extension on the whole set X such that $F(\rho) = \rho$ on X. This will play a crucial role in Section 6.2. Our ultimate aim in the present section, playing an important role in Section 6.4 and Section 6.7 is to show that in dimension $d \geq 2$ (and also in dimension $d = 1$ if all the contractions forming the system S are real-analytic) the density ρ has a real-analytic extension on a neighborhood of $\overline{J_S}$. So, notice first that if the system S is regular then by Lemma 4.2.12 the series $\sum_{i\in I} ||\phi_i'||^h$ converges, and therefore the operator $F : C(X) \to C(X)$ defined by the formula

$$F(g) = \sum_{i\in I} |\phi_i'|^h g \circ \phi_i$$

acts continuously on $C(X)$. Let us recall from [Lj] that a bounded operator $L : B \to B$ defined on a Banach space B is said to be *almost periodic* if for every $x \in B$ the orbit $\{L^n x\}_{n=0}^\infty$ is relatively compact in B. We start with the following result establishing almost periodicity of the operator $F : C(X) \to C(X)$.

Lemma 6.1.1 *If S is a regular CIFS, then the operator $F : C(X) \to C(X)$ is almost periodic and the sequence $\{F^n(1\!\!1)\}_{n=0}^\infty$ is uniformly bounded between K^h and K^{-h}.*

Proof. The uniform bound from above of the sequence $\{F^n(1\!\!1)\}_{n=0}^\infty$ is an immediate consequence of Lemma 4.2.12. The lower bound follows from this lemma combined with (4f). Although the proof of almost periodicity of the operator $F : C(X) \to C(X)$ is similar to the proof of Lemma 2.4.1, following [MU1], we provide it here for the sake of completeness and the convenience of the reader. And indeed, fix $g \in C(X)$ and $\epsilon > 0$. Since g is uniformly continuous, there exists $\eta_1 > 0$ such that $|g(y) - g(x)| < \epsilon$ if $x, y \in X$ and $||y - x|| \le \eta_1$. Since it follows from the proof of Proposition 4.2.7 that the family $\{\log|\phi_\omega'|\}_{\omega\in I^*}$ is equicontinuous, there exists $\eta_2 > 0$ so small that

$$|\log|\phi_\omega'(y)| - \log|\phi_\omega'(x)|| \le \min\{\frac{1}{2h}, \epsilon\}$$

for all $\omega \in I^*$ and all $x, y \in X$ with $||y - x|| \le \eta_2$. Put $\eta = \min\{\eta_1, \eta_2\}$ and consider two points $x, y \in X$ with $||y - x|| \le \eta$. Then using Lemma 4.2.12 we obtain for every $n \ge 1$

$$\begin{aligned}
|F^n(g)(y) - F^n(g)(x)| &= \left|\sum_{\omega\in I^n} g(\phi_\omega(y))|\phi_\omega'(y)|^h - \sum_{\omega\in I^n} g(\phi_\omega(x))|\phi_\omega'(x)|^h\right| \\
&\le \sum_{\omega\in I^n} |g(\phi_\omega(y))|\phi_\omega'(y)|^h - g(\phi_\omega(x))|\phi_\omega'(x)|^h| \\
&\le \sum_{\omega\in I^n} |g(\phi_\omega(y))|\,\big||\phi_\omega'(y)|^h - |\phi_\omega'(x)|^h\big| \\
&\quad + \sum_{\omega\in I^n} |\phi_\omega'(x)|^h |g(\phi_\omega(y)) - g(\phi_\omega(x))| \\
&\le ||g||_0 \sum_{\omega\in I^n} \big||\phi_\omega'(y)|^h - |\phi_\omega'(x)|^h\big| + \epsilon K^h.
\end{aligned}$$

Looking at the Taylor's series expansion of e^t about 0, we deduce that there exists $M > 0$ such that $|e^t - 1| \le M|t|$ if only $|t| \le 1/2$. Hence

$$\begin{aligned}
\big||\phi'_\omega(y)|^h - |\phi'_\omega(x)|^h\big| &= |\phi'_\omega(y)|^h \left|1 - \frac{|\phi'_\omega(x)|^h}{|\phi'_\omega(y)|^h}\right| \\
&= |\phi'_\omega(y)|^h \left|1 - \exp\big(h\big(\log|\phi'_\omega(x)| - \log|\phi'_\omega(y)|\big)\big)\right| \\
&\le |\phi'_\omega(y)|^h M h \left|\log|\phi'_\omega(x)| - \log|\phi'_\omega(y)|\right| \\
&\le |\phi'_\omega(y)|^h M h \epsilon.
\end{aligned}$$

Consequently, using Lemma 4.2.12 again, we get

$$\sum_{\omega \in I^n} \big||\phi'_\omega(y)|^h - |\phi'_\omega(x)|^h\big| \le M\epsilon \sum_{\omega \in I^n} |\phi'_\omega(y)|^h \le K^h M\epsilon.$$

Thus, finally

$$|F^n(g)(y) - F^n(g)(x)| \le K^h(M||g||_0 + 1)\epsilon.$$

Due to the Ascoli–Arzela theorem along with Lemma 4.2.12 this demonstrates that the family $\{F^n(g)\}_{n=1}^\infty$ is relatively compact in the sup norm on $C(X)$. □

We are now in position to prove the following result, which can be found in [MU4]. Perhaps its most important and rather unexpected part is that the density function ρ has a canonical extension to the whole space X.

Theorem 6.1.2 *Suppose that S is a regular* CIFS *and m is the corresponding conformal measure. Then*

(a) There exists a unique continuous function $\rho : X \to [0, \infty)$ such that

$$F\rho = \rho \text{ and } \int \rho dm = 1.$$

(b) $K^{-h} \le \rho \le K^h$.

(c) The sequence $\{F^n(1\!\!1)\}_{n=1}^\infty$ converges uniformly to ρ on X.

(d) $\rho|_J = \frac{d\mu}{dm}$, where μ is the S-invariant version of the conformal measure m.

Proof. Suppose that $\rho : X \to [0, \infty)$ is a continuous function such that $F\rho = \rho$ and $\int \rho dm = 1$. Since $F\rho(\pi\omega) = \mathcal{L}_{h\zeta}(\rho \circ \pi)(\omega)$, where $\mathcal{L}_{h\zeta}$ is the operator considered in Chapter 2, we thus get $\mathcal{L}_{h\zeta}(\rho \circ \pi) = \rho \circ \pi$. It therefore follows from Proposition 2.4.7 and Theorem 2.4.6 that

$\rho \circ \pi = \frac{d\tilde{\mu}}{d\tilde{m}}$ and consequently $\rho|_J = \frac{d\mu}{dm}$. So, item (d) is proved and if $\rho_1, \rho_2 : X \to [0, \infty)$ are two functions satifying the requirements of item (a), then $\rho_1|_{\overline{J}} = \rho_2|_{\overline{J}}$. Denote this common restriction by $\hat{\rho}$. Fix $\epsilon > 0$ and consider $\eta > 0$ so small that for each $i = 1, 2$, $|\rho_i(y) - \rho_i(x)| < \epsilon$ if $x, y \in X$ and $||y - x|| \leq \eta$. Take an arbitrary $n \geq 1$ so large that $Ds^n \leq \eta$. Finally fix an arbitrary $z \in X$ and consider an $\omega \in I^n$. Then $\text{diam}(\phi_\omega(X)) \leq Ds^n \leq \eta$. Choose $x \in J \cap \phi_\omega(X)$. Then

$$\begin{aligned}|\rho_2(\phi_\omega(z)) - \rho_1(\phi_\omega(z))| &\leq |\rho_2(\phi_\omega(z)) - \hat{\rho}(x)| + |\hat{\rho}(x) - \rho_1(\phi_\omega(z))| \\ &\leq \epsilon + \epsilon = 2\epsilon.\end{aligned}$$

Hence, using Lemma 4.2.12, we get

$$\begin{aligned}|\rho_2(z) - \rho_1(z)| &= |F^n\rho_2(z) - F^n\rho_1(z)| = |F^n(\rho_2 - \rho_1)(z)| \\ &\leq \sum_{|\omega|=n} |\rho_2(\phi_\omega(z)) - \rho_1(\phi_\omega(z))| \cdot |\phi_\omega'(z)|^h \\ &\leq \sum_{|\omega|=n} 2\epsilon||\phi_\omega'||^h \leq 2K^h\epsilon.\end{aligned}$$

Therefore, letting $\epsilon \searrow 0$ we conclude that $\rho_2(z) = \rho_1(z)$ and the uniqueness part of item (a) is proved.

Since by Lemma 6.1.1 the sequence $\{F^n(1\!\!1_X)\}_{n=1}^\infty$ is uniformly bounded between $K^{-\delta}$ and K^δ and is equicontinuous, the sequence $\frac{1}{n}\sum_{j=0}^{n-1} F^j(1\!\!1_X)$ has the same properties. Let ρ be an accumulation point of this sequence of averages. Then obviously, ρ is continuous and $F\rho = \rho$, $\int \rho dm = 1$ and $K^{-h} \leq \rho \leq K^h$. Thus items (a) and (b) are also proved. It remains to demonstrate item (c). And indeed, since by Lemma 6.1.1 the Perron–Frobenius operator $F : C(X) \to C(X)$ is almost periodic, it follows from a Lyubich's result (see [Lj]) that

$$C(X) = E_0 \oplus E_u,$$

where $E_0 = \{f : ||F^n(f)|| \mapsto 0\}$ and E_u is the closed span of $\{f : F(f) = \lambda f$ for some λ with $\lambda| = 1\}$. We shall demonstrate first that

$$E_u = \{c\rho : c \in \mathbb{C}\}. \tag{6.3}$$

Indeed, suppose $F(\psi) = \lambda\psi$ with $|\lambda| = 1$. Since F is a positive operator on the Banach lattice, $C(X)$, it follows from Lemma 18, Theorem 4.9 and Exercise 2 in [Sc] (p. 326–327) that the spectrum of F meets the unit circle in a cyclic compact group. Therefore, the group is finite and there is some positive integer r such that $\lambda^r = 1$. Thus, $F^r(\psi) = \psi$ and $F^r(\text{Re}\psi) = Re\psi$ $F^r(\text{Im}\psi) = \text{Im}\psi$. Let us suppose $Re\psi \neq 0$. Fix

$M \in R$ so large that $\text{Re}\psi + M\rho > 0$. But, by lemma 2.2.4, σ^r is ergodic with respect to $\tilde{\mu}$. This means that $\rho \circ \pi\tilde{m}$ is the only invariant measure for σ^r equivalent to $\tilde{m}$. Therefore, there is a constant $c > 0$ such that $\text{Re}\psi \circ \pi + M\rho \circ \pi = c\rho \circ \pi$. So, $\text{Re}\psi \circ \pi = (c - M)\rho \circ \pi$ and $\int \text{Re}\psi dm = c - M$. Repeating this argument for $\text{Im}\psi$, we obtain $\psi \circ \pi = (\int \psi dm)\rho \circ \pi$. Since $F\rho = \rho$, this implies that $\lambda = 1$ and since $\pi(E^\infty)$ is dense in $\overline{J}$, $\frac{\psi}{\int \psi dm}|_{\overline{J}} = \rho|_{\overline{J}}$. But repeating now the argument of the proof of uniqueness in item (a), we conclude that $\psi = \int \psi dm\, \rho$ on X. The proof of (6.3) is complete.

Representing the function $1\!\!1$ as a unique sum of an element from E_u and E_0, it follows from (6.3) that there exists $c \in \mathbb{C}$ such that $1\!\!1 - c\rho \in E_0$. But since the operator F preserves integration with respect to the measure m, $\int g dm = 0$ for every $g \in E_0$. Consequently $c = 1$. Therefore $||F^n(1\!\!1) - \rho|| = ||F^n(1\!\!1 - \rho)|| \to 0$ when $n \to \infty$. We are done. □

The main result of this section is contained in the following.

Theorem 6.1.3 *If $d \geq 2$ and the system S is regular, then the Radon-Nikodym derivative $\rho = \frac{d\mu}{dm}$ has a real-analytic extension on an open connected neighborhood U of X in W.*

Proof. We shall deal first with the case $d \geq 3$. In view of (4.1), there then exist $\lambda_\omega > 0$, a linear isometry A_ω, an inversion (or the identity map) $i_\omega = i_{a_\omega, r_\omega}$ and a vector $b_\omega \in \mathbb{R}^d$ such that $\phi_\omega = \lambda_\omega A_\omega \circ i_\omega + b_\omega$. Then

$$|\phi'_\omega(z)| = \frac{\lambda_\omega r_\omega^2}{||z - a_\omega||^2} \text{ if } i_\omega \neq \text{Id}$$

and

$$|\phi'_\omega(z)| = \lambda_\omega \text{ if } i_\omega = \text{Id}$$

Since $\phi_\omega(W) \subset W$, $a_\omega \notin W$. Fix $\xi \in X$ and consider the function $\rho_\omega : \mathbb{C}^d \to \mathbb{C}$ given by the formula

$$\rho_\omega(z) = \frac{||\xi - a_\omega||^2}{\sum_{j=1}^d \left(z_j - (a_\omega)_j\right)^2} \text{ or } \rho_\omega(z) = 1 \text{ if } i_\omega = \text{Id}.$$

We shall show that there exist a constant $B > 0$ and a neighborhood $\tilde{U}$ of X in $\mathbb{C}^d$ such that

$$|\rho_\omega(z)| \leq B \tag{6.4}$$

for every $\omega \in I^*$ and every $z \in \tilde{U}$. Indeed, otherwise there exist sequences $\omega^{(n)} \in I^*$ and $z^{(n)} \in \mathbb{C}^d$, $n \geq 1$ such that $\lim_{n\to\infty} \text{dist}(z^{(n)}, X) =$

0 and $|\rho_{\omega^{(n)}}(z^{(n)})| \geq n$ for every $n \geq 1$. Passing to a subsequence we may assume that the limit $w = \lim_{n\to\infty} z^{(n)}$ exists. Then $w \in X \subset \mathbb{R}^d$ and for every $n \geq 1$ we have

$$\begin{aligned}\sum_{j=1}^d \big(z_j^{(n)} - (a_{\omega^{(n)}})_j\big)^2 &= \sum_{j=1}^d \big((z_j^{(n)} - w_j) + (w_j - (a_{\omega^{(n)}})_j)\big)^2 \\ &= \sum_{j=1}^d \big((z_j^{(n)} - w_j)^2 + 2\sum_{j=1}^d (z_j^{(n)} - w_j)(w_j - (a_{\omega^{(n)}})_j) \\ &\quad + \sum_{j=1}^d (w_j - (a_{\omega^{(n)}})_j)^2.\end{aligned} \tag{6.5}$$

Now, since $a_{\omega^{(n)}} \notin W$ and $w \in X$, we get

$$\sum_{j=1}^d (w_j - (a_{\omega^{(n)}})_j)^2 = ||w - a_{\omega^{(n)}}||^2 \geq \operatorname{dist}^2(X, \partial W). \tag{6.6}$$

Fix now $q \geq 1$ so large that for every $n \geq q$

$$\sum_{j=1}^d |z_j^{(n)} - w_j|^2 \leq \frac{1}{4}\operatorname{dist}^2(X, \partial W) \tag{6.7}$$

and

$$\sum_{j=1}^d |z_j^{(n)} - w_j| \leq \min\left\{\frac{1}{8d}\operatorname{dist}^2(X, \partial W), \frac{1}{8d}\right\}. \tag{6.8}$$

So, if $|w_j - (a_{\omega^{(n)}})_j| \leq 1$, then by (6.8)

$$2|z_j^{(n)} - w_j||w_j - (a_{\omega^{(n)}})_j| \leq \frac{1}{4d}\operatorname{dist}^2(X, \partial W) \tag{6.9}$$

and if $|w_j - (a_{\omega^{(n)}})_j| \geq 1$, then by the other part of (6.8)

$$\begin{aligned}2|z_j^{(n)} - w_j||w_j - (a_{\omega^{(n)}})_j| &\leq 2|z_j^{(n)} - w_j||w_j - (a_{\omega^{(n)}})_j|^2 \\ &\leq \frac{1}{4d}|w_j - (a_{\omega^{(n)}})_j|^2 \\ &\leq \frac{1}{4d}||w - a_{\omega^{(n)}}||^2.\end{aligned} \tag{6.10}$$

Applying now (6.5) along with (6.6), (6.7), (6.9), and (6.10), we get for every $n \geq q$

$$\begin{aligned}\left|\sum_{j=1}^{d}\left(z_j^{(n)} - (a_{\omega^{(n)}})_j\right)^2\right| &\geq \sum_{j=1}^{d}(w_j - (a_{\omega^{(n)}})_j)^2 - \sum_{j=1}^{d}|z_j^{(n)} - w_j|^2 \\ &\quad - 2\sum_{j=1}^{d}|z_j^{(n)} - w_j||w_j - (a_{\omega^{(n)}})_j| \\ &\geq ||w - a_{\omega^{(n)}}||^2 - \frac{1}{4}||w - a_{\omega^{(n)}}||^2 \\ &\quad - d\frac{1}{4d}||w - a_{\omega^{(n)}}||^2 = \frac{1}{2}||w - a_{\omega^{(n)}}||^2.\end{aligned}$$

And therefore, using also (4f) and (4.2), we get

$$n \leq |\rho_{\omega^{(n)}}(z^{(n)})| = \frac{||\xi - a_{\omega^{(n)}}||^2}{\sum_{j=1}^{d}\left(z_j^{(n)} - (a_{\omega^{(n)}})_j\right)^2} \leq 2\frac{||\xi - a_{\omega^{(n)}}||^2}{||w - a_{\omega^{(n)}}||^2} \leq 2K.$$

This contradiction finishes the proof of (6.4). Decreasing $\tilde{U}$ if necessary, we may assume that this set is connected. Now, for every $n \geq 1$ define the function $b_n : \tilde{U} \to \mathbb{C}$ by setting

$$b_n(z) = \sum_{|\omega|=n} \rho_\omega^h(z)|\phi'_\omega(\xi)|^h.$$

Since each term of this series is an analytic function and since, by Lemma 6.1.1 and (6.4),

$$\sum_{|\omega|=n} |\rho_\omega^h(z)||\phi'_\omega(\xi)|^h \leq B^h \sum_{|\omega|=n} ||\phi'_\omega||^h \leq B^h K^h,$$

we conclude that all the functions $b_n : \tilde{U} \to \mathbb{C}$ are analytic and $||b_n||_\infty \leq B^h K^h$ for every $n \geq 1$. Hence, in view of *Montel's theorem*, we can choose a subsequence $\{b_{n_k}\}_{k=1}^{\infty}$ converging on a connected neighborhood $\tilde{U}_1$ of X (with closure $\tilde{U}_1$ contained in $\tilde{U}$) to an analytic function $b : \tilde{U}_1 \to \mathbb{C}$. Since for every $n \geq 1$ and every $z \in X$, $b_n(z) = \sum_{|\omega|=n} |\phi'_\omega(z)|^h = \mathcal{L}^n(1\!\!1)$, it therefore follows from Theorem 6.1.2(c) that $b|_X = \rho = \frac{d\mu}{dm}$. Hence, putting $U = \Pr(\tilde{U}_1)$, where Pr: $\mathbb{C}^d \to \mathbb{R}^d$ is the orthogonal projection from $\mathbb{C}^d$ to $\mathbb{R}^d$, completes the proof in the case $d \geq 3$.

Following [MPU] we shall deal now with the case $d = 2$. We start by defining the sequence of functions $b_n : V \to (0, \infty)$ by setting

$$b_n(z) = \sum_{|\omega|=n} |\phi'_\omega(z)|^h = \sum_{|\omega|=n} |\psi'_\omega(z)|^h, \tag{6.11}$$

where $\psi_\omega = \phi_\omega$ if ϕ_ω is holomorphic and $\psi_\omega = \overline{\phi_\omega}$ if ϕ_ω is antiholomorphic. In view of Lemma6.1.1 $|b_n(z)| = b_n(z) \le K^h$ for all $z \in X$ and all $n \ge 1$. Hence, applying the Koebe distortion theorem, we conclude that there exists $T > 0$ such that for each point $w \in X$ there exists a radius $r = r(w) > 0$ such that $B(w, 2r) \subset V$ and for all $z \in B(w, 2r)$ and all $n \ge 1$

$$|b_n(z)| = b_n(z) \le T. \tag{6.12}$$

Now equate $\mathbb{C}$, where our contractions ϕ_i, $i \in I$, act, to $\mathbb{R}^2$ with coordinates x, y, the real and complex part of z. Embed this into $\mathbb{C}^2$ with x, y complex. Denote the above $\mathbb{C} = \mathbb{R}^2$ by $\mathbb{C}_0$. We may assume that $w = 0$ in $\mathbb{C}_0$. Given $\omega \in I^*$ define the function $\rho_\omega : B_{\mathbb{C}_0}(0, 2r) \to \mathbb{C}$ by setting

$$\rho_\omega(z) = \frac{\phi'_\omega(z)}{\phi'_\omega(0)}.$$

Since $B_{\mathbb{C}_0}(0, 2r) \subset \mathbb{C}_0$ is simply connected and ρ_ω nowhere vanishes, all the branches of the $\log \rho_\omega$ are well-defined on $B_{\mathbb{C}_0}(0, 2r)$. Choose the branch that maps 0 to 0 and denote it also by $\log \rho_\omega$. By Koebe's distortion theorem $|\rho_\omega|$ and $|\arg\rho_\omega|$ are bounded on $B(0, r)$ by universal constants K_1, K_2 respectively. Hence $|\log \rho_\omega| \le K = \log K_1 + K_2$. We write

$$\log \rho_\omega = \sum_{m=0}^{\infty} a_m z^m$$

and note that by Cauchy's inequalities

$$|a_m| \le K/r^m. \tag{6.13}$$

We can write for $z = x + iy$ in $\mathbb{C}_0$

$$\begin{aligned}
\operatorname{Re} \log \rho_\omega &= \operatorname{Re} \sum_{m=0}^{\infty} a_m (x + iy)^m \\
&= \sum_{p,q=0}^{\infty} \operatorname{Re}\left(a_{p+q} \binom{p+q}{q} i^q\right) x^p y^q := \sum c_{p,q} x^p y^q.
\end{aligned}$$

In view of (6.13) we can estimate $|c_{p,q}| \le |a_{p+q}| 2^{p+q} \le K r^{-(p+q)} 2^{p+q}$.

Hence $\mathrm{Re}\log\rho_\omega$ extends, by the same power series expansion $\sum c_{p,q}x^py^q$, to a complex-valued function on the polydisk $\mathbb{D}_{\mathbb{C}^2}(0,r/2)$ and

$$|\mathrm{Re}\log\rho_\omega| \le 4K \quad \text{on } \mathbb{D}_{\mathbb{C}^2}(0,r/4). \tag{6.14}$$

Now each function b_n, $n \ge 1$, extends to the function

$$B_n(z) = \sum_{|\omega|=n} |\phi'_\omega(0)|^h \mathrm{e}^{h\mathrm{Re}\log\rho_\omega(z)}, \tag{6.15}$$

whose domain, like the domains of the functions $\mathrm{Re}\log\rho_\omega$, contains the polydisk $\mathbb{D}_{\mathbb{C}^2}(0,r/2)$. Finally, using (6.12) and (6.14) we get for all $n \ge 0$ and all $z \in \mathbb{D}_{\mathbb{C}^2}(0,r/4)$

$$\begin{aligned}|B_n(z)| &\le \sum_{|\omega|=n} |\phi'_\omega(0)|^h \mathrm{e}^{\mathrm{Re}(h\mathrm{Re}\log\rho_\omega(z))} \le \sum_{|\omega|=n} |\phi'_\omega(0)|^h \mathrm{e}^{h|\mathrm{Re}\log\rho_\omega(z)|} \\ &\le \mathrm{e}^{Kh} \sum_{|\omega|=n} |\phi'_\omega(0)|^h \le \mathrm{e}^{Kh}T.\end{aligned}$$

Now by Cauchy's integral formula (in $\mathbb{D}_{\mathbb{C}^2}(0,r/4)$) for second derivatives we prove that the family B_n is equicontinuous on, say, $\mathbb{D}_{\mathbb{C}^2}(0,r/5)$. Hence we can choose a uniformly convergent subsequence whose limit function G is complex and analytic and extends ρ on $X \cap B(0,r/5)$, in the manner described in Theorem 6.1.2. Thus we have proved that for every $w \in X$, the density ρ extends to a complex and analytic function in an open ball contained in $\mathbb{C}^2$ and centered at the point $w \in X$. Since $X = \overline{\mathrm{Int}X}$, these extensions restricted to $\mathbb{C}_0$, coincide on intersections of these balls. This finishes the proof in the dimension $d = 2$. □

As an immediate consequence of this theorem we get the following.

Corollary 6.1.4 *Suppose that $d = 1$, the system S is regular and all contractions are real-analytic. Moreover, suppose that there exists an open connected set $U \subset \mathbb{R}^2$ containing X and invariant under all elements of S. Then the Radon-Nikodym derivative $\rho = \frac{d\mu}{dm}$ has a real-analytic extension on an open connected neighborhood of X in $\mathbb{R}$.*

For every $\omega \in I^*$ denote by $D_{\phi_\omega} = \frac{d\mu\circ\phi_\omega}{d\mu}$ the *Jacobian* of the map $\phi_\omega : J \to J$ with respect to the measure μ. As an immediate consequence of Theorem 6.1.3, the following computation

$$\frac{d\mu\circ\phi_\omega}{d\mu} = \frac{d\mu\circ\phi_\omega}{dm\circ\phi_\omega}\cdot\frac{dm\circ\phi_\omega}{dm}\cdot\frac{dm}{d\mu} = \left(\frac{d\mu}{dm}\circ\phi_\omega\right)\cdot|\phi'_\omega|^\delta\cdot\frac{dm}{d\mu}$$

and the observation that $|\phi'_\omega|^\delta$ is real-nalytic on V, we get the following.

Corollary 6.1.5 *For every $i \in I$ the Jacobian D_{ϕ_i} has a real-analytic extension $\tilde{D}_{\phi_i}$ on the neighborhood U of X produced either in Theorem 6.1.3 or Corollary 6.1.4.*

6.2 Rate of approximation of the Hausdorff dimension by finite subsystems

Let us recall that in view of Proposition 4.2.7, for every $t \geq 0$ the collection $\zeta_t = \{t \log |\phi_i'|\}_{i \in I}$ is a Hölder family of functions. If $t > \theta$, then ζ_t is a summable Hölder family of functions. Denote the corresponding measure m_{ζ_t} by m_t and the pressure $\mathrm{P}(\zeta_t)$ by $\mathrm{P}(t)$. As in the previous section we define the operator $F_t : C(X) \to C(X)$ by the formula

$$F_t(g) = \sum_{i \in I} |\phi_i'|^t g \circ \phi_i.$$

F_t acts continuously on $C(X)$. In this section we provide an effective estimate on the perturbation of topological pressure if we form iterated function systems consisting of bigger and bigger finite subsets of I. From Theorem 4.2.13 we know that the limit of Hausdorff dimensions of the limit sets of these finite systems converges to the Hausdorff dimension of the limit set of the original system. In this section, as our main result, we estimate (see Theorem 6.2.3) the rate of this convergence, thus making possible numerical calculations of this Hausdorff dimension. Proceeding in a manner similar to the proof of Theorem 6.1.2 we can demonstrate the following generalization of that theorem.

Theorem 6.2.1 *If S is a CIFS and $t > \theta_S$, then*

(a) There exists a unique continuous function $\rho_t : X \to [0, \infty)$ such that

$$F_t \rho_t = e^{\mathrm{P}(t)} \rho_t \text{ and } \int \rho_t dm = 1.$$

(b) $K^{-t} \leq \rho_t \leq K^t$.

(c) The sequence $\{F_t^n(1\!\!1)\}_{n=1}^{\infty}$ converges uniformly to ρ_t on X.

(d) $\rho_t|_J = \frac{d\mu_t}{dm_t}$, *where μ_t is the S-invariant version of the measure m_t.*

(e)

$$C(X) = C(X)_t^0 \oplus \mathbb{C}\rho_t,$$

where $C(X)_t^0 = \{f \in C(X) : ||F_t^n(f)|| \mapsto 0\} = \{f \in C(X) : \int f dm_t = 0\}$.

If instead of S we consider a system $S_G = \{\phi_i\}_{i\in G}$ for some subset G of I, we add the subscript G to the objects F_t, $\mathrm{P}(t)$, m_t, ρ_t, and C_t^0 resulting respectively in $F_{G,t}$, $\mathrm{P}_G(t)$, $m_{G,t}$, $\rho_{G,t}$, and $C_{G,t}^0$. We are now in a position to prove the following.

Theorem 6.2.2 *If* $S = \{\phi_i\}_{i\in I}$ *is a CIFS, then for all* $G \subset I$ *and* $t \in (\theta_S, d]$ *we have*

$$|e^{\mathrm{P}(t)} - e^{\mathrm{P}_G(t)}| \le K^{3d}(2 + K^d)||F_t - F_{G,t}||.$$

Proof. Consider a function $\psi = r\rho_t + u$, $r \in I\!R$, $u \in C_t^0(X)$ and assume that $||\psi|| = 1$. Then

$$1 = ||\psi|| \ge |\int \psi dm_t| = |r| \int \rho_t dm_t = |r|.$$

Hence, using also Theorem 6.2.1(b), we conclude that $||u|| \le ||\psi|| + |r| \cdot ||\rho_t|| \le 1 + K^d$. Thus

$$||r\rho_t + u|| = 1 = \frac{1}{2+K^d}(1 + K^d + 1) \ge \frac{1}{2+K^d}(||u|| + |r|). \tag{6.16}$$

If we now relax the assumption $||\psi|| = 1$ but still keep $\psi \ne 0$, then it follows from (6.16) that

$$\left\| \frac{\psi}{||\psi||} \right\| \ge \frac{1}{2+K^d} \left(\left\| \frac{|u}{||\psi||} \right\| + \left| \frac{r}{||\psi||} \right| \right)$$

and therefore $||\psi|| \ge \frac{1}{2+K^d}(||u|| + |r|)$, so that (6.16) remains true in this case. If $\psi = 0$, then $r = 0$, $u = 0$, and (6.16) is obviously satisfied in this case. In view of Theorem 6.2.1(e) there is a unique representation

$$\rho_{G,t} = r\rho_t + u$$

with appropriate $r \in I\!R$ and $u \in C_t^0(X)$. In view of Theorem 6.2.1(b) we get

$$K^{-d} \le \int \rho_{G,t} dm_t = \int (r\rho_t + u) dm_t = r \int \rho_t dm_t = r. \tag{6.17}$$

Write $\lambda_t = \mathrm{e}^{\mathrm{P}(t)}$ and $\lambda_{G,t} = \mathrm{e}^{\mathrm{P}_G(t)}$. Applying Theorem 6.2.1(b), Theorem 6.2.1(a), $\mathcal{L}_t$-invariantness of the space $C_t^0(X)$, (6.16) and (6.17), we get

$$\begin{aligned}||F_t - F_{G,t}||K^d &\geq ||(F_t - F_{G,t})\rho_{G,t}|| = ||rF_t(\rho_t) + F_t(u) - \lambda_{G,t}\rho_{G,t}|| \\ &= ||(r\lambda_t\rho_t - r\lambda_{G,t}\rho_t) + F_t(u) - \lambda_{G,t}u|| \\ &\geq \frac{1}{2+K^d}\big(r|\lambda_t - \lambda_{G,t}|\cdot||\rho_t|| + ||F_t(u) - \lambda_{G,t}u||\big) \\ &\geq \frac{1}{2+K^d}K^{-d}K^{-d}|\lambda_t - \lambda_{G,t}|.\end{aligned}$$

Thus $|\lambda_t - \lambda_{G,t}| \leq K^{3d}(2+K^d)||F_t - F_{G,t}||$ and the proof is complete. □

Let

$$\chi = \sup_{n\geq 1}\inf_{\omega\in I^n}\left\{-\frac{1}{n}\log||\phi'_\omega||\right\}.$$

The main result of this section, coming mainly from [HeU] and culminating the approach begun in [GM] and continued in [MU1], establishing the rate of approximation of the Hausdorff dimension $h = \mathrm{HD}(J_S)$ of the limit set J_S of the system S by the Hausdorff dimensions of the limit sets of its finite subsystems, is included in the following.

Theorem 6.2.3 *If $\gamma > \theta_S$ and $h_G \geq \gamma$ for some finite set $G \subset I$, then*

$$0 \leq h - h_G \leq \chi^{-1}K^{3d}(2+K^d)\sum_{i\in I\setminus G}||\phi'_i||^\gamma.$$

Proof. Applying Proposition 2.6.13 along with Birkhoff's ergodic theorem for the function $f(\omega) = \log|\phi_{\omega_1}(\pi(\sigma(\omega)))|$, $\omega \in I^\infty$, we get

$$\frac{d\mathrm{P}(t)}{dt} = \int \log|\phi_{\omega_1}(\pi(\sigma(\omega)))|d\tilde{\mu}_t \leq -\chi,$$

where $\tilde{\mu}_t$ is the lift to the coding space I^∞ of the measure $\rho_t m_t$. Hence

$$\begin{aligned}e^{P(h_G)} - 1 = e^{P(h_G)} - e^{P(h)} &= \int_h^{h_G}\frac{d}{dt}e^{P(t)}dt = \int_h^{h_G}\mathrm{P}'(t)e^{P(t)}dt \\ &= -\int_{h_G}^{h} -\mathrm{P}'(t)e^{P(t)}dt \geq \int_{h_G}^{h}\chi dt = \chi(h - h_G).\end{aligned} \tag{6.18}$$

On the other hand, using Theorem 6.2.2, we obtain

$$\begin{aligned} e^{P(h_G)} - 1 = e^{P(h_G)} - e^{P_G(h_G)} &\le K^{3d}(2+K^d)||F_{h_G} - F_{G,h_G}|| \\ &= K^{3d}(2+K^d) \sum_{i\in I\setminus G} ||\phi_i'||^{h_G} \\ &\le K^{3d}(2+K^d) \sum_{i\in I\setminus G} ||\phi_i'||^{\gamma}. \end{aligned}$$

Combining this inequality and (6.18) completes the proof. □

Notice that the assumptions $\gamma > \theta_S$ and $h_G \ge \gamma$ imply that $h \ge h_G > \theta_S$, and consequently the system S is strongly regular. On the other hand, if the the system S is strongly regular, then there exist a finite $G \subset I$ with these properties.

6.3 Uniform perfectness

A compact set $L \subset \mathbb{C}$ is called *uniformly perfect* if there exists a constant $0 < c < 1$ such that for every $x \in L$ and every radius $0 < r \le 1$

$$L \cap \{w \in \mathbb{C} : cr \le |w - z| < r\} \ne \emptyset.$$

This notion was introduced by Ch. Pommerenke in [Po1]. The main consequence of uniform perfectness of the compact set L is that all its points are regular in the sense of Dirichlet. The concept of uniform perfectness has been widely used in harmonic analysis. Some applications for studying harmonic measure on the limits sets of CIFS and selected literature can be found in [UZd]. Let us recall that

$$S(\infty) = \overline{\lim}_{i\to\infty}\phi_i(X) = \bigcap_F \overline{\bigcup_{i\in I\setminus F} \phi_i(X)},$$

where the intersection is taken over all finite subsets of I. We start with the following sufficient condition for the limit set of a CGDMS to be uniformly perfect.

Theorem 6.3.1 *Suppose that the following condition (UP) holds. $S(\infty)$ is finite and for each index $i \in I$ there exists an infinite countable set $\{i_n\}_{n\ge 1}$ of elements in I such that $i_1 = i$ and*

$$\sup_{i\in I}\sup_{n\ge 1} \left\{ \frac{\max\{\text{diam}(\phi_{i_n}(X)), \text{diam}(\phi_{i_{n+1}}(X)), \text{dist}(\phi_{i_{n+1}}(X), \phi_{i_n}(X))\}}{\min\{\text{diam}(\phi_{i_n}(X)), \text{diam}(\phi_{i_{n+1}}(X))\}} \right\} < \infty.$$

Then $\overline{J}$, the closure of the limit set J, is uniformly perfect.

Proof. It follows from (4.20) and (4.23) that for every $\tau \in I^*$

$$\text{diam}(\phi_\tau(X)) \le D||\phi'_\tau|| \le D^2\text{diam}(\phi_\tau(J)). \tag{6.19}$$

Let C be the maximum of 1 and the supremum appearing in the condition (UP) above. It is sufficient to demonstrate that there exists a constant $0 < c < 1$ such that for each positive radius small enough and each point $z \in J$ the annulus $A(z, cr, r) := \{w \in \mathbb{C} : cr \le |w - z| < r\}$ intersects J. Let us consider first the set $X_a(\infty) \subset S(\infty)$ of all those points $w \in S(\infty)$ for which there exists an infinite sequence $\{j_n\}_{n \ge 1}$ for which the supremum appearing in condition (UP) is bounded from above by C and

$$w \in \overline{\lim}_{n \to \infty} \phi_{j_n}(X) := \bigcap_{k \ge 1} \overline{\bigcup_{n \ge k} \phi_{j_n}(X)}.$$

We shall show first the uniform perfectness of $\overline{J}$ at the points w of $X_a(\infty)$. By the cone condition (4d) and the bounded distortion property (4f), the set $\{n \ge 1 : w \in \phi_{j_n}(X)\}$ is finite. Let $n_w \ge 1$ be the least element in the complement of this set. Set

$$R_w = \text{dist}(w, \phi_{j_{n_w}}(X))$$

and consider any radius $0 < r < R_w$. Since $w \in \overline{\lim}_{n \to \infty} \phi_{j_n}(X)$ and $\lim_{n \to \infty} \text{diam}(\phi_{j_n}(X)) = 0$, there exists an element $k \ge n_w$ such that $\phi_{j_k}(X) \subset B(w, r)$. Then $j > n_w$. Let p be a least such index k. If $\text{diam}(\phi_{j_p}(J)) \ge r/8D^2C$, then using the fact that $\phi_{j_p}(J) \subset B(w, r)$, we conclude that $A(z, r/16D^2C, r) \cap \phi_{j_p}(J) \ne \emptyset$. But since $\phi_{j_p}(J) \subset J$, we get

$$A(z, r/16D^2C, r) \cap J \ne \emptyset$$

and we are done in this case with any constant $c \le 1/16D^2C$. So suppose that $\text{diam}(\phi_{j_p}(J)) \le r/8D^2C$. Then by (6.19), $\text{diam}(\phi_{j_p}(X)) \le r/8C$. So, by the definition of w,

$$\text{dist}(\phi_{j_p}(X), \phi_{j_{p-1}}(X)) < \frac{r}{8} \text{ and } \text{diam}(\phi_{j_{p-1}}(X)) \le \frac{r}{8}.$$

Since $\phi_{j_{p-1}}(X) \cap (\mathbb{C} \setminus B(w, r)) \ne \emptyset$, we deduce that

$$\text{dist}(\phi_{j_p}(X), \partial B(w, r)) < \frac{r}{8} + \frac{r}{8} = \frac{r}{4}.$$

Since $\phi_{j_p}(X) \subset B(w, r)$ and $\text{diam}(\phi_{j_p}(X)) \le r/8C < r/4$, we conclude that $\phi_{j_p}(X) \subset A(w, r - \frac{r}{4} - \frac{r}{4}, r) = A(w, r/2, r)$. Since $J \cap \phi_{j_p}(X) \ne \emptyset$, we are done in this case with any constant $c \leq 1/2$. Put

$$R = \min\{R_w : w \in X_a(\infty)\} > 0 \text{ and } c_1 = \min\{1/2, 1/16D^2C\}.$$

Consider now an arbitrary point $z \in \phi_i(J)$ for some $i \in I$ and $\frac{4}{c_1}$ $\mathrm{diam}(\phi_i(X)) < r < R$. Let $\{i_n\}_{n\geq 1}$ be the sequence claimed by our hypothesis. Suppose that

$$\phi_{i_n}(X) \cap \left(\mathbb{C} \setminus B\left(z, \frac{c_1}{4}r\right)\right) \neq \emptyset$$

for some $n \geq 1$. Then $n \geq 2$. Let $q \geq 2$ be the least index n with this property. If $\mathrm{diam}(\phi_{i_{q-1}}(J)) \geq c_1 r/32D^2C$, then using the fact that $\phi_{i_{q-1}}(J) \subset B(z, c_1 r/4)$, we conclude that $A(z, c_1 r/64D^2C, c_1 r/4) \cap \phi_{i_{q-1}}(J) \neq \emptyset$. But since $\phi_{i_{q-1}}(J) \subset J$, we get

$$A(z, c_1 r/64D^2C, c_1 r/4) \cap J \neq \emptyset$$

and we are done in this case with any constant $c \leq c_1/64D^2C$. So suppose that $\mathrm{diam}(\phi_{i_{q-1}}(J)) \leq c_1 r/32D^2C$. Then by (6.19), $\mathrm{diam}(\phi_{i_{q-1}}(X)) \leq c_1 r/32C$. But then

$$\mathrm{dist}(\phi_{i_{q-1}}(X), \phi_{i_q}(X)) < \frac{c_1 r}{32} \text{ and } \mathrm{diam}(\phi_{i_q}(X)) \leq \frac{c_1 r}{32}.$$

Since $\phi_{i_q}(X) \cap (\mathbb{C} \setminus B(z, c_1 r/4)) \neq \emptyset$, we deduce then that

$$\mathrm{dist}(\phi_{i_{q-1}}(X), \partial B(z, c_1 r/4)) < \frac{c_1 r}{32} + \frac{c_1 r}{32} = \frac{c_1}{16} r.$$

Since $\phi_{i_{q-1}}(X) \subset B(z, c_1 r/4)$ and $\mathrm{diam}(\phi_{i_{q-1}}(X)) \leq c_1 r/32D^2C < c_1 r/16$, we conclude that $\phi_{i_{q-1}}(X) \subset A(z, \frac{c_1}{4}r - \frac{c_1}{16} - \frac{c_1}{16}, \frac{c_1}{4}r) \subset A(z, \frac{c_1}{8}r, r)$. Since $J \cap \phi_{i_{q-1}}(X) \neq \emptyset$, we are done in this case with any constant $c \leq c_1/8$. So, suppose in turn that $\phi_{i_n}(X) \subset B(z, c_1 r/4)$ for all $n \geq 1$. Let $w \in \mathbb{C}$ be an arbitrary point of $\overline{\lim}_{n\to\infty} \phi_{i_n}(X)$. Then $w \in X_a(\infty) \cap \overline{B}(z, c_1 r/4)$ and, as $r/2 < R \leq R_w$, we conclude from what has been already proved that $A(w, c_1 r/2, r/2) \cap J \neq \emptyset$. Then $A(w, c_1 r/2, r/2) \subset \overline{B}(z, (r/2) + (c_1 r/4)) \subset B(z, r)$. Take an arbitrary point $x \in A(w, c_1 r/2, r/2) \cap J$. Then $|x - z| \geq |x - w| - |z_w| \geq c_1 r/2 - c_1 r/4 = c_1 r/4$ which implies that $A(z, c_1 r/4, r) \cap J \neq \emptyset$ and we are done in this case with the constant $c_2 = c_1/4 = \min\{1/8, 1/64D^2C\}$. Obviously, taking $c_2 > 0$ appropriately smaller, if necessary, we have local uniform perfectness at the point z for every r satisfying $\frac{4}{c_1}\mathrm{diam}(\phi_i(X)) < r < K^{-1}D$.

Passing to the last step of this proof, fix an arbitrary point $z = \pi(\tau) \in J$, $\tau \in I^\infty$, and a positive radius $r < \mathrm{diam}(X)$. Let $n \geq 1$ be the least integer such that

$$\phi_{\tau|_n}(X) \subset B(z, K^{-2}D^{-2}c_2r). \tag{6.20}$$

Consider the ball

$$B(\pi(\sigma^n(\tau)), K^{-1}||\phi'_{\tau|_{n-1}}||^{-1}r)$$

(note that $\pi(\sigma^n(\tau)) = \phi^{-1}_{\tau|_{n-1}}(z)$ and that if $n = 1$, then $\phi_{\tau|_{n-1}}$ is the identity map). Since $r \le \operatorname{diam}(\phi_{\tau|_{n-1}}(X))$, using (4.20), we get

$$\begin{aligned} K^{-1}||\phi'_{\tau|_{n-1}}||^{-1}r &\le K^{-1}||\phi'_{\tau|_{n-1}}||^{-1}\operatorname{diam}(\phi_{\tau|_{n-1}}(X)) \\ &\le K^{-1}||\phi'_{\tau|_{n-1}}||^{-1}D||\phi'_{\tau|_{n-1}}|| = K^{-1}D. \end{aligned} \tag{6.21}$$

Using (6.20) and (4.23) we obtain

$$\begin{aligned} K^{-1}||\phi'_{\tau|_{n-1}}||^{-1}r &= K^{-1}||\phi'_{\tau|_{n-1}}||^{-1}(K^2D^2c_2^{-1}(K^{-2}D^{-2}c_2)r) \\ &\ge KD^2c_2^{-1}||\phi'_{\tau|_{n-1}}||^{-1}\operatorname{diam}(\phi_{\tau|_n}(X)) \\ &\ge KD^2c_2^{-1}||\phi'_{\tau|_{n-1}}||^{-1}D^{-1}||\phi'_{\tau|_n}|| \ge D^2c_2^{-1}||\phi'_{\tau_n}|| \\ &\ge c_2^{-1}\operatorname{diam}(\phi_{\tau_n}(X)) \ge \frac{4}{c_1}\operatorname{diam}(\phi_{\tau_n}(X)). \end{aligned}$$

This inequality and (6.21) enable us to apply the previous case, and as a consequence, we obtain an annulus $A(\pi(\sigma^n(\tau)), c_2K^{-1}||\phi'_{\tau|_{n-1}}||^{-1}r, K^{-1}||\phi'_{\tau|_{n-1}}||^{-1}r)$ having a non-empty intersection with J. Hence

$$\phi_{\tau|_{n-1}}(A(\pi(\sigma^n(\tau)), c_2K^{-1}||\phi'_{\tau|_{n-1}}||^{-1}r, K^{-1}||\phi'_{\tau|_{n-1}}||^{-1}r)) \cap J \ne \emptyset.$$

Assuming K to be so large that $K^{-1}D < \operatorname{dist}(X, \partial V)$, using the bounded distortion property, the mean value inequality, and (4.21), we get

$$\begin{aligned} &\phi_{\tau|_{n-1}}(A(\pi(\sigma^n(\tau)), c_2K^{-1}||\phi'_{\tau|_{n-1}}||^{-1}r, K^{-1}||\phi'_{\tau|_{n-1}}||^{-1}r)) \\ &\quad \subset A(z, c_2K^{-2}r, r). \end{aligned}$$

Hence $A(z, c_2K^{-2}r, r) \cap J \ne \emptyset$ and the proof is complete. □

As an immediate consequence of this theorem and Example 5.1.4 we get the following.

Corollary 6.3.2 *If $I = \{a_n\}_{n=1}^\infty \subset I\!N$ is represented as a non-decreasing sequence, then the closure of the set of numbers in $[0, 1]$ whose continued* *fraction expansion entries are all contained in I, is uniformly perfect if and only if*

$$\limsup_{n\to\infty}(a_{n+1} - a_n) < \infty.$$

As the following theorem shows, the limit sets of finite CIFS are all uniformly perfect. The idea of the proof is however completely different.

Theorem 6.3.3 *The limit set of a finite CIFS is uniformly perfect.*

Proof. In view of Theorem 4.2.11 the system S is regular. Let m be the corresponding h-conformal measure. Fix $x \in J_S$ and $0 < r \leq 1$. Applying Theorem 4.2.11 we then get

$$\begin{aligned} m(A(x,(2C^2)^{-1/h}r,r)) &= m(B(x,r)) - m(B(x,(2C^2)^{-1/h}r)) \\ &\geq C^{-1}r^h - C((2C^2)^{-1/h}r)^h = C^{-1}r^h - \frac{1}{2}C^{-1}r^h \\ &= \frac{1}{2}C^{-1}r^h > 0. \end{aligned} \tag{6.22}$$

In particular $A(x,(2C^2)^{-1/h}r,r) \cap J_S \neq \emptyset$ and the proof is complete. □

6.4 Geometric rigidity

In this section, following [MMU] we explore the structure of limit sets J of infinite conformal iterated function systems whose closure is a continuum (compact connected set). Under a natural easily verifiable technical condition (always satisfied if the system is finite), we demonstrate the following dichotomy. Either the Hausdorff dimension of J exceeds 1 or else $\overline{J}$ is a proper compact segment of either a geometric circle or a straight line if $d \geq 3$ or an analytic interval if $d = 2$ (cf. Theorem 6.4.1). From the viewpoint of conformal dynamics, this result can be thought of as a far going generalization of results originated in [Su1] and [B2], which are formulated in the plane case. The proofs contained there use the Riemann mapping theorem and can be carried out only in the plane. The proof presented in our paper is different and holds in any dimension. The reader is also encouraged to notice an analogy between our result and a series of other papers (see e.g. ([B2], [FU], [MU2], [Ma], [Pr], [Ru], [Su1], [U1], [UV], [Z1], [Z2]) which are aimed toward establishing a similar dichotomy. However, to our knowledge, all these results like those in [B2] and [Su1], were formulated in the plane and used the

Riemann mapping theorem, except those in [MU2]. The current result is however much stronger than that in [MU2] and in particular with our present approach the main result of [MU2] can be strengthened as described at the end of this section. Another corollary of our result is the following: if a continuum C in $I\!R^d$ is the self-conformal set generated by finitely many conformal mappings satisfying the open set condition, the Hausdorff 1-measure of C is finite and one of the mappings is a similarity, then the continuum is a line-segment. In particular, this holds if all the maps are similarities, a result obtained early on by Mattila [Ma3].

The main result of this section is the following.

Theorem 6.4.1 *If $d \geq 3$, $S = \{\phi_i\}_{i \in I}$ is a CIFS, $\overline{J}$ is a continuum (compact connected set) and $\dim_H(S(\infty)) < \dim_H(J)$, then either*

(a) $\dim_H(J) > 1$ or
(b) $\overline{J}$ is a proper compact segment of either a geometric circle or a straight line.

In addition, if any one of the maps ϕ_i is a similarity mapping, then $\overline{J}$ is a line segment.

We note that the technical condition in Theorem 6.4.1 is necessary. A modification of Example 5.2 of [MU1] shows that the dichotomy of Theorem 6.4.1 in general fails if $\dim_H(S(\infty)) \geq \dim_H(J)$. Here is such an example. Take any sequence of positive reals $\{r_i : i \geq 1\}$ (for example of the form b^i, $0 < b < 1$) such that the equation $\sum_{i \geq 1} r_i^t = 1$ has a (unique) solution and this solution is less than or equal to 1. Consider a family $\{\phi_i : \{z \in I\!R^3 : ||z|| \leq 1\} \to \{z \in I\!R^3 : ||z|| \leq 1\} : i \geq 1\}$ of similarity maps satisfying the open set condition and such that $||\phi_i'|| = r_i$ and $S(\infty) = \{z \in I\!R^3 : ||z|| = 1\}$. Then $\mathrm{HD}(J_S) \leq 1 \leq \mathrm{HD}(S(\infty))$ and the closure $\overline{J_S}$ is connected. The set $\overline{J_S}$ is obviously not an arc.

We would also like to remark that in the case $d = 2$, for every $i \in I$, ϕ_{ii} is a holomorphic map bi-holomorphically conjugate with the linear map $\psi(z) = x_{ii} + \phi'(x_{ii})(z - x_{ii})$ on some neighborhood W of x_{ii}. Proceeding then similarly as in the proof of Theorem 6.4.1 we could demonstrate the same statement with the segment of the line or the circle replaced by an analytic arc.

Since in the finite case the set $S(\infty)$ is empty, we get immediately from Theorem 6.4.1 the following.

Corollary 6.4.2 *If $d \geq 3$, $S = \{\phi_i\}_{i \in I}$ is a finite CIFS and $\overline{J}$ is a continuum, then either*

(a) $\dim_H(J) > 1$ *or*

(b) $\overline{J}$ *is a proper compact segment of either a geometric circle or a straight line.*

In addition, if any one of the maps ϕ_i is a similarity mapping, then $\overline{J}$ is a line segment.

The proof of Theorem 6.4.1 will consist of several steps. First of all we assume from now on throughout the entire section that the assumptions of Theorem 6.4.1 are satisfied and $\dim_H(J) = 1$. Our goal is to show that then item (b) is satisfied. Since $\dim_H(S(\infty)) < \dim_H(J) = 1$ and $\overline{J}$ is a continuum, using Lemma 1.0.1, we conclude that $\mathrm{H}_1(J) > 0$. It therefore follows from Theorem 4.5.11 that the system S is regular. Let m be the corresponding 1-dimensional measure. By Theorem 4.5.1 and since $\dim_H(S(\infty)) < \dim_H(J) = 1$ the 1-dimensional Hausdorff measure H_1 on $\overline{J}$ is absolutely continuous with respect to m and in consequence $\mathrm{H}_1|_{\overline{J}}$ is a multiple of m. So, $\overline{J}$ is a continuum whose H_1 measure is finite. Therefore, the following fact follows from [EH] and [Wh].

Lemma 6.4.3 $\overline{J}$ *is a locally arcwise connected continuum.*

Given $x \in I\!R^d$, $\theta \in I\!P I\!R^d$, the $(d-1)$-dimensional projective space, and $\gamma > 0$, we put

$$\mathrm{Con}(x, \theta, \gamma) = \mathrm{Con}(x, \eta, \gamma) \cup \mathrm{Con}(x, -\eta, \gamma),$$

where $\eta \in I\!R^d$ is a representative of $\theta \in I\!P I\!R^d$. We recall that a set Y has a *tangent* in the direction $\theta \in I\!P I\!R^d$ at a point $x \in Y$ if for every $\gamma > 0$

$$\lim_{r \to 0} \frac{\mathrm{H}_1\big(Y \cap (B(x,r) \setminus \mathrm{Con}(x, \theta, \gamma))\big)}{r} = 0.$$

Since we will consider only tangents of 1-sets (the set $\overline{J}$ above), this definition coincides with the definition given on p. 31 of [Fa1]. We say that a set Y has a *strong tangent* in the direction $\theta \in I\!P I\!R^d$ at a point x provided for each $0 < \beta \leq 1$ there is some $r > 0$ such that $Y \cap B(x,r) \subset \mathrm{Con}(x, \theta, \beta)$. We shall prove the following.

Theorem 6.4.4 *If Y is locally arcwise connected at a point x and Y has a tangent θ at x, then Y has strong tangent θ at x.*

Proof. Suppose there is some $0 < \beta < 1$ and points x_n in Y such that for each n, $|x_n - a - < x_n - a, \theta > \theta| > \beta |x_n - x|$. For each n, let $\alpha_n : [0,1] \to Y$ be an arc from x to x_n with $\operatorname{diam}(\alpha_n) \to 0$. For each n, note that since Y has a tangent $\theta \in \mathbb{P}\mathbb{R}^d$ at x, there is some t, $0 < t < 1$ such that $\alpha_n(t) \in \operatorname{Con}(x, \theta, \beta/2)$. Let t_n be the largest number such that $\alpha_n(t_n) \in \operatorname{Con}(x, \theta, \beta/2)$ and let s_n be the first number larger than t_n such that $\alpha_n(s_n) \in \partial\operatorname{Con}(x, \theta, \beta)$. Consider y_n, a point of the arc α_n from $\alpha_n(t_n)$ to $\alpha_n(s_n)$ at maximum distance from x, and take z_n to be a point of this same subarc at minimum distance from x. If $\|z_n - a\| \le \|y_n - a\|/2$, then considering the projection of this subarc on the line through a and y_n,

$$\frac{\mathrm{H}_1(X \cap B(x, \|y_n - x\|))}{\operatorname{Con}(x, \theta, \beta/2)} \ge \|y_n - x\|/2.$$

If $\|z_n - x\| > \|y_n - x\|/2$, then, considering the projection of this subarc on the sphere with center a and radius $\|y_n - a\|/2$, we get

$$\frac{\mathrm{H}_1(X \cap B(x, \|y_n - x\|))}{\operatorname{Con}(x, \theta, \beta/2)} \ge (\pi/2)(\beta/2)\|y_n - a\|.$$

Thus, Y does not have a tangent at x. □

We call a point $\tau \in I^\infty$ transitive if $\omega(\tau) = I^\infty$, where $\omega(\tau)$ is the ω-limit set of τ under the shift transformation $\sigma : I^\infty \to I^\infty$. We denote the set of these points by I_t^∞ and put

$$J_t = \pi(I_t^\infty).$$

We call the J_t the set of transitive points of J and notice that for every $\tau \in I_t^\infty$, the set $\{\pi(\sigma^n \tau) : n \ge 0\}$ is dense in J (or $\overline{J}$ if this is the space under consideration).

Lemma 6.4.5 *If $\overline{J}$ has a strong tangent at a point $x = \pi(\tau)$, $\tau \in I^\infty$, then $\overline{J}$ has a strong tangent at every point of $\overline{\pi(\omega(\tau))}$.*

Proof. Suppose on the contrary that $\overline{J}$ does not have a strong tangent at some point $y \in \overline{\pi(\omega(\tau))}$. Let $\theta \in \mathbb{P}\mathbb{R}^d$ be the tangent direction of $\overline{J}$ at x and let $\{n_k\}_{k=1}^\infty$ be an increasing sequence of positive integers such that $\lim_{k\to\infty} \pi(\sigma^{n_k}\tau) = y$. Passing to a subsequence, we may assume that

$$\lim_{k\to\infty} \frac{\left(\phi_{\omega|_{n_k}}^{-1}\right)'(x)}{|\left(\phi_{\omega|_{n_k}}^{-1}\right)'(x)|}\theta = \xi$$

for some $\xi \in I\!\!PI\!\!R^d$. Since $\overline{J}$ does not have a strong tangent at y, there exists $0 < \beta \le 1$ such that for every $r > 0$

$$\overline{J} \cap B(y,r) \setminus \overline{J} \cap \mathrm{Con}(y,\xi,\beta) \neq \emptyset.$$

Then

$$\overline{J} \cap B(\pi(\sigma^{n_k}\tau), r) \setminus \overline{J} \cap \mathrm{Con}(\pi(\sigma^{n_k}\tau), \xi_k, \beta/2) \neq \emptyset \tag{6.23}$$

for all k large enough where

$$\xi_k = \frac{\left(\phi^{-1}_{\omega|_{n_k}}\right)'(x)}{|\left(\phi^{-1}_{\omega|_{n_k}}\right)'(x)|}\theta.$$

But in view of Proposition 4.2.3, for all $r > 0$ small enough we have

$$\begin{aligned}\phi_{\omega|_{n_k}}\big(B(\pi(\sigma^{n_k}\tau), r)\big) \setminus \mathrm{Con}(\pi(\tau), \xi_k, \beta/2) \subset &\\ \subset B\big(x, r||\phi'_{\omega|_{n_k}}||\big) \setminus \mathrm{Con}&\left(x, \frac{\phi'_{\omega|_{n_k}}(\pi(\sigma^{n_k}\tau))}{|\phi'_{\omega|_{n_k}}(\pi(\sigma^{n_k}\tau))|}\xi_k, \frac{\beta}{4}\right)\\ = B\big(x, r||\phi'_{\omega|_{n_k}}||\big) \setminus \mathrm{Con}&(x, \theta, \beta/4).\end{aligned}$$

Since in view of (6.23), $\overline{J} \cap \phi_{\omega|_{n_k}}\big(B(\pi(\sigma^{n_k}\tau), r)\big) \neq \emptyset$, we conclude that for every k large enough, $\overline{J} \cap \Big(B\big(x, r||\phi'_{\omega|_{n_k}}||\big) \setminus \mathrm{Con}(x,\theta,\beta/4)\Big) \neq \emptyset$. Since $\lim_{k\to\infty} ||\phi'_{\omega|_{n_k}}|| = 0$, this implies that θ is not the strong density direction of $\overline{J}$ at x. This contradiction finishes the proof. □

Corollary 6.4.6 *The continuum $\overline{J}$ has a strong tangent at every point.*

Proof. Since $\mathrm{H}_1(\overline{J}) < \infty$, Corollary 3.15 from [Fa1] shows that $\overline{J}$ has a tangent at H_1-a.e. point in $\overline{J}$, and therefore at a set of points of positive m measure. Since $m(J_t) = 1$, there thus exists at least one transitive point x in J having a tangent of J. By Theorem 6.4.4 and Lemma 6.4.3 $\overline{J}$ has a strong tangent at x, and it then follows from Lemma 6.4.5 that $\overline{J}$ has a strong tangent at every point. □

Now, the following lemma finishes the proof of Theorem 6.4.1.

Lemma 6.4.7 *Suppose that $\phi : \overline{I\!\!R^d} \to \overline{I\!\!R^d}$, $d \ge 3$, is a conformal diffeomorphism that has an attracting fixed point a ($\phi(a) = a$, $|\phi'(a)| < 1$). Suppose that a compact connected set M has a strong tangent at a, that $\phi(M) \subset M$ and that $\lim_{n\to\infty} \phi^n(x) = a$ for all $x \in M$. Then M*

is a segment of a ϕ-invariant line or circle. If ϕ is affine ($\phi(\infty) = \infty$), then the former possibility holds.

Proof. Since a is an attracting fixed point of ϕ, there exists a radius $r > 0$ so small that $\phi^{-1}(\overline{R}^d \setminus B(a,r)) \subset \overline{R}^d \setminus B(a,r)$, where $\overline{R}^d$ is the Alexandrov compactification of $I\!R^d$ done by adding the point at infinity. Since $\overline{R}^d \setminus B(a,r)$ is a topological closed ball, in view of Brouwer's fixed point theorem there exists a fixed point b of ϕ^{-1} in $\overline{R}^d \setminus B(a,r)$. Hence b is also a fixed point of ϕ and $b \neq a$. Then the map

$$\psi = i_{b,1} \circ \phi \circ i_{b,1}$$

($i_{b,1}$ equals identity if $b = \infty$) fixes ∞, which means that this map is affine, and $w = i_{b,1}(a)$ is an attracting fixed point of ψ. In addition $\psi(\tilde{M}) \subset \tilde{M}$, where $\tilde{M} = i_{b,1}(M)$, $w \in \tilde{M}$, and $\tilde{M}$ has a strong tangent at w. Let l be the line through w determined by the strongly tangent direction of $\tilde{M}$ at w. Since $\psi(w) = w$, since $\psi(l)$ is a straight line through w and since $\psi(\tilde{M}) \subset \tilde{M}$, we conclude that $\psi(l) = l$. Suppose now that $\tilde{M}$ is not contained in l. Consider $x \in \tilde{M} \setminus l$. Then for every $n \geq 0$

$$\psi^n(x) \in \psi(\tilde{M}) \setminus \psi(l) \subset \tilde{M} \setminus l$$

and since the map ψ is conformal and affine

$$\angle(\psi^n(x) - w, l) = \angle(\psi^n(x - w), \psi^n(l)) = \angle(x - w, l).$$

Since $\lim_{n\to\infty} \psi^n(x) = w$, we therefore conclude that l is not a strongly tangent line of $\tilde{M}$ at w. This contradiction shows that $\tilde{M} \subset l$. Since in addition $\tilde{M}$ is a continuum, it is a segment of l. □

And indeed to conclude the proof of Theorem 6.4.1 it suffices to pick an arbitrary index $i \in I$ (affine if exists) and to put $\phi = \phi_i$, $M = \overline{J}$ and $a = x_i$, the only attracting fixed point of ϕ_i belonging to J. □

6.5 Refined geometric rigidity

Throughout the section we assume that $d \geq 2$. By TD we will denote the *topological dimension* (we will only deal with subsets of $I\!R^d$ so all Hausdorff and topological dimensions are finite). The following is the main result of this section. It refines the main result of the previous section, cf. [MyU].

Theorem 6.5.1 *If $d \geq 3$, $S = \{\phi_i\}_{i\in I}$ is a CIFS and* $\mathrm{HD}(S(\infty)) < \mathrm{HD}(J)$*, then either*

(a) $\mathrm{HD}(J) > \mathrm{TD}(\overline{J})$ *or*

(b) $\overline{J}$ *is a proper compact subset of either a geometric sphere of dimension* $\mathrm{TD}(\overline{J})$ *or a* $\mathrm{TD}(\overline{J})$*-dimensional affine hyperspace, both contained in* $I\!R^d$.

In addition, if any one of the maps ϕ_i is a similarity map, then the latter case holds.

Since in the finite case the set $S(\infty)$ is empty, we get immediately the following.

Corollary 6.5.2 *If $d \geq 3$, $S = \{\phi_i\}_{i\in I}$ is a finite CIFS, then either*

(a) $\mathrm{HD}(J) > \mathrm{TD}(\overline{J})$ *or*

(b) $\overline{J}$ *is a proper compact subset of either a geometric sphere of dimension* $\mathrm{TD}(\overline{J})$ *or a* $\mathrm{TD}(\overline{J})$*-dimensional affine hyperspace, both contained in* $I\!R^d$.

In addition, if any one of the maps ϕ_i is a similarity map, then the latter case holds.

Remark 6.5.3 *Put $k = \mathrm{TD}(\overline{J})$. Since a compact subset of a k-dimensional sphere or hyperspace G has topological dimension k if and only if its interior in G is not empty, we see that in the second alternative of Theorem 6.5.1 and Corollary 6.5.2, $\overline{J}$ contains an open ball in the appropriate sphere or hyperspace and, for dynamical reasons, it turns out that there is an open subset Ω of that sphere or hyperspace such that $\overline{J} = \overline{\Omega}$.*

We first discuss some concepts and results from rectifialibity theory. A beautiful exposition of this can be found in [Ma1]. A set $Q \subset I\!R^d$ is called *k-rectifiable* if $\mathrm{H}^k(Q) > 0$ and there exist Lipschitz maps $g_i : I\!R^k \to I\!R^d$, $i = 1, 2, ...$, such that

$$\mathrm{H}^k\left(Q \setminus \bigcup_{i=1}^{\infty} g_i(I\!R^k)\right) = 0.$$

A set $T \subset I\!R^d$ is called *purely k-unrectifiable* if and only if $\mathrm{H}^k(Q \cap T) = 0$ for every k-rectifiable set Q.

It follows from Theorem 15.19 in [Ma1] that for H^k-a.e. point z in a k-rectifiable set $Q \subset I\!R^d$ there is a unique *approximate tangent k-plane* for Q at z. This tangent plane will be denoted in the sequel by T_zQ as a subset of $G(d,k)$. We recall that $G(d,k)$ is the *Grassmannian manifold* of all k-dimensional linear subspaces of $\mathbb{R}^d$ and that the existence of a tangent k-plane T_zQ for Q at z implies that, for every $0 < s < 1$,

$$\lim_{r\to 0}\frac{1}{r^k}\mathrm{H}^k\left(Q\cap B(z,r)\setminus Z(z,T_zQ,s)\right)=0,$$

where

$$Z(z,V,s)=\{x\in\mathbb{R}^d\ ;\ d(x-z,V)<s|x-z|\}.$$

The space $G(d,k)$ has a natural measure $\gamma_{d,k}$ (see Section 3.9 in [Ma1] for its definition and basic properties). Given $V\in G(d,k)$ we denote by $P_V:\mathbb{R}^d\to V$ the orthogonal projection from $\mathbb{R}^d$ onto V.

The following lemma is crucial since it gives rectifiability of the limit set provided the topological and Hausdorff dimension coincide.

Lemma 6.5.4 *If* $S=\{\phi_i\}_{i\in I}$ *is a CIFS and* $\mathrm{H}^{\mathrm{TD}(\overline{J})}(\overline{J_S})=\mathrm{H}^{\mathrm{TD}(\overline{J})}(J)>0$, $\mathrm{HD}(\overline{J})=\mathrm{TD}(\overline{J})$, *then the system* S *is regular,* $m=\frac{\mathrm{H}^k}{\mathrm{H}^k(J)}|_J$ *is the* k*-conformal measure on* J *and the closure* $\overline{J_S}$ *is* $\mathrm{TD}(\overline{J_S})$*-rectifiable.*

Proof. Put $k=\mathrm{TD}(\overline{J})$. Since $\mathrm{H}^k(J)>0$ and since $\mathrm{HD}(J)=k$, we conclude from Theorem 4.5.11 that the system S is regular and, using Theorem 4.5.1, we deduce that $\mathrm{H}^k(J)<\infty$ and $m=\frac{\mathrm{H}^k}{\mathrm{H}^k(J)}|_J$ is the k-conformal measure on J. It follows from Federer's theorem on p. 545 in [Fe] that the integralgeometric measure $\mathcal{I}_1^k(\overline{J})>0$. Since (see [Ma1], p. 86)

$$\mathcal{I}_1^k(\overline{J})=\int_{G(d,k)}\int_V \mathrm{H}^0(\overline{J}\cap \mathrm{P}_V^{-1}(a))d\mathrm{H}^k(a)d\gamma_{d,k}(V),$$

we therefore conclude that there exists a Borel set $G\subset G(d,k)$ with $\gamma_{d,k}(G)>0$ such that $\mathrm{H}^0(\overline{J}\cap\mathrm{P}_V^{-1}(a))>0$ for every $V\in G$ and all a in some set $W_V\subset V$ with $\mathrm{H}^k(W_V)>0$. In particular $\mathrm{P}_V(\overline{J})\supset W_V$ and therefore $\mathrm{H}^k(\mathrm{P}_V(\overline{J}))>0$ for all $V\in G$. Hence, it follows from Theorem 18.1(2) on p. 250 in [Ma1] that $\overline{J}$ is not purely k-unrectifiable. Therefore, combining Theorem 17.6 (notice that although this is not indicated in Mattilas's book, we need to know that $\mathrm{H}^k(\overline{J})>0$ for this theorem to make actually sense), Theorem 6.2(1) in [Ma1] and the fact that $\mathrm{H}^k(\overline{J})=\mathrm{H}^k(J)>0$, we conclude that $\Theta^k(\overline{J},x)=1$ for all x in some set $F\subset J$ with $\mathrm{H}^k(F)>0$, where the density functions Θ^k as well as Θ_*^k and Θ^{*k} were defined on p. 89 in [Ma1]. Fix now $x\in J$. It follows from the bounded distortion property (4f) that for all $i \in I$ and all $r > 0$ small enough

$$\begin{aligned} \mathrm{H}^k(\overline{J} \cap B(\phi_i(x), |\phi_i'(x)|r)) &\geq \mathrm{H}^k(\phi_i(J \cap B(x, K_r^{-1}r))) \\ &\geq K_r^{-k}|\phi_i'(x)|^h \mathrm{H}^k(J \cap B(x, K_r^{-1}r)), \end{aligned}$$

where K_r is such that $\lim_{r\to 0} K_r = 1$. Hence

$$\begin{aligned} \frac{\mathrm{H}^k(\overline{J} \cap B(\phi_i(x), |\phi_i'(x)|r))}{(2|\phi_i'(x)|r)^k} &\geq \frac{K_r^{-k}|\phi_i'(x)|^k}{(2|\phi_i'(x)|r)^k} \mathrm{H}^k(J \cap B(x, K_r^{-1}r)) \\ &= K_r^{-2k} \frac{\mathrm{H}^k(J \cap B(x, K_r^{-1}r))}{(2K_r^{-1}r)^k}. \end{aligned}$$

and letting $r \searrow 0$ we conclude that

$$\Theta_*^k(\overline{J}, \phi_i(x)) \geq \Theta_*^k(\overline{J}, x). \tag{6.24}$$

Let $\tilde{m}$ be the lift of the conformal measure m to the coding space I^∞ and let $\tilde{\mu}$ be its shift-invariant version produced in Theorem 2.2.4. Since by this theorem the dynamical system $(\sigma, \tilde{\mu})$ is ergodic, it therefore follows from Birkhoff's ergodic theorem and (6.24) that the function $\omega \mapsto \Theta_*^k(\overline{J}, \pi(\omega))$ is constant $\tilde{\mu}$-a.e. Since $\tilde{\mu}(\pi^{-1}(F)) > 0$, we therefore conclude that $\Theta_*^k(\overline{J}, \pi(\omega)) = 1$ for $\tilde{\mu}$-a.e. $\omega \in I^\infty$. Thus $\Theta_*^k(\overline{J}, x) = 1$ for H^k-a.e. $x \in \overline{J}$. Combining this with Theorem 6.2(1) in [Ma1] we see that $\Theta^k(\overline{J}, x)$ exists and is equal to 1 for H^k-a.e. $x \in \overline{J}$. Invoking now Theorem 17.6(1) in [Ma1] finishes the proof. □

We now pass directly to the proof of Theorem 6.5.1. Put $k = \mathrm{TD}(\overline{J_S})$ and suppose that $\mathrm{HD}(J) \leq k$. Since $\mathrm{HD}(S(\infty)) < \mathrm{HD}(J)$, using Proposition 1.0.1, we conclude that $\mathrm{HD}(\overline{J}) = \mathrm{HD}(J) \leq k$. Hence, $\mathrm{HD}(\overline{J}) = k$ and $\mathrm{H}^k(J) = \mathrm{H}^k(\overline{J}) > 0$. Thus the assumptions of Lemma 6.5.4 are satisfied. In view of this lemma the set $\overline{J}$ is k-rectifiable. By Theorem 15.19 in [Ma1] this equivalently means that for H^k-a.e. $z \in J$ there is a unique approximate tangent k-plane $T_z\overline{J}$ for $\overline{J}$ at z.

We fix now such a point, say $z_0 = \pi(\omega) \in J$, $\omega \in I^\infty$, and make the following renormalization. Set $\lambda_n = |\phi'_{\omega|_n}(\pi(\sigma^n(\omega)))|^{-1}$ and define then

$$\beta_n(z) = \lambda_n(z - z_0).$$

It follows from the bounded distortion property (4f) that each mapping $\beta_n \circ \phi_{\omega|_n} : X \to I\!\!R^d$ is locally Lipschitz continuous with Lipschitz constant K and from (4.20) that $\beta_n \circ \phi_{\omega|_n}(X) \subset B(0, DK)$. Therefore, the Ascoli-Arzelà theorem applies and there exists an unbounded increasing sequence of positive integers $\{n_j\}_{j=1}^{\infty}$ such that the sequence $\psi_j : X \to I\!R^d$ converges uniformly to a continuous function $\Psi : X \to I\!R^d$, where $\psi_j = \beta_{n_j} \circ \phi_{\omega|_{n_j}}$. The limit function $\Psi : X \to X$ is conformal. We shall prove the following.

Claim 6.5.5 $\Psi(\overline{J}) \subset T_{z_0}\overline{J}$.

Proof. Suppose on the contrary that the claim does not hold. Then there exists an open bounded set $\Omega \subset \Psi(\overline{J})$ such that

$$\eta = \mathrm{dist}(\Omega, I\!R^k) > 0\,. \tag{6.25}$$

Since Ω is an open subset of $\overline{J}$, we get $\mathrm{H}^k(\Omega) > 0$. Put $U = \Psi^{-1}(\Omega)$ and $U_j = f_j(U)$. Then

$$\begin{aligned} 0 < \mathrm{H}^k(\Omega) &= \int_U |\Psi'|^k d\mathrm{H}^k = \lim_{j\to\infty} \int_U |\psi_j'|^k d\mathrm{H}^k \lim_{j\to\infty} \mathrm{H}^k(\psi_j(U)) \\ &= \lim_{j\to\infty} \mathrm{H}^k(\beta_{n_j}(U_j)). \end{aligned}$$

Hence, there exists $\tau > 0$ and $j_0 \geq 1$ such that

$$0 < \tau \leq \mathrm{H}^k(\beta_{n_j}(U_j)) = \lambda_{n_j}^k \mathrm{H}^k(U_j) \tag{6.26}$$

for all $j \geq j_0$. Due to (6.25) and the boundedness of Ω we can choose $0 < s < 1$ such that $Z(0, I\!R^k, 2s) \cap \Omega = \emptyset$. Consider the cones $\mathcal{Z} = Z(0, T_{z_0}\overline{J}, s)$ and $\mathcal{Z}_j = Z(z_0, T_{z_0}\overline{J}, \lambda_{n_j}^{-1}s) = \beta_{n_j}^{-1}(\mathcal{Z})$. Fix also a ball $B = B(0, R)$ such that $\Omega \subset B(0, R/2)$ and set $B_j = \beta_{n_j}^{-1}(B) = B(z_0, R\lambda_{n_j}^{-1})$. Since $T_{z_0}\overline{J}$ is an approximate tangent k-plane of $\overline{J}$ at z_0, we have

$$\lim_{j\to\infty} \lambda_{n_j}^k \mathrm{H}^k(\overline{J} \cap B_j \setminus \mathcal{Z}_j) = \lim_{j\to\infty} \left(R\lambda_{n_j}^{-1}\right)^{-k} \mathrm{H}^k(\overline{J} \cap B_j \setminus \mathcal{Z}_j) = 0. \tag{6.27}$$

But, if j is sufficiently large, then $\beta_{n_j}(U_j) = \Psi_j \circ \Psi^{-1}(\Omega) \subset B \setminus \mathcal{Z}$ and therefore $U_j \subset \overline{J} \cap B_j \setminus \mathcal{Z}_j$. It then follows from (6.26) that

$$\lambda_{n_j}^k \mathrm{H}^k(\overline{J} \cap B_j \setminus \mathcal{Z}_j) \geq \lambda_{n_j}^k \mathrm{H}^k(U_j) \geq \tau > 0$$

and this contradicts (6.27). We thus proved the claim and therefore the "smooth or fractal" dichotomy announced in Theorem 6.5.1.

We are left to show that if one of the maps ϕ_i is a similarity map ($\lambda_i A_i + a_i$, $0 < \lambda_i < 1$), then $\overline{J}$ is contained in a k-dimensional hyperspace of $I\!R^d$. And indeed, suppose on the contrary that $\overline{J} \subset Q$, a geometric sphere in $\mathbb{R}^d$. Since $\phi_i(Q) = \lambda_i A_i(Q) + a_i$ is a geometric sphere of dimension k and the sphere $Q \cap \phi_i(Q)$ contains the k-dimensional set $\overline{J}$, this intersection is a k-dimensional sphere, and therefore equal to both Q and $\phi_i(Q)$. This contradicts the fact that $\phi_i : Q \to Q$ is a strict contraction with a Lipschitz constant equal to λ_i. $\square$

7

Dynamical Rigidity of CIFSs

In this chapter we deal with dynamical rigidity stemming from the work of Sullivan (see [Su3]) on conformal expanding repellers in the complex plane. We ask the fundamental question when two topologically conjugate infinite iterated function systems are conjugate in a smoother fashion. The answer is that such conjugacy extends to a conformal conjugacy on some neighborhoods of limit sets if and only if it is Lipschitz continuous. This turns out to equivalently mean that this conjugacy exchanges measure classes of appropriate conformal measures or that the multipliers of corresponding fixed points of all compositions of our generaters coincide.

7.1 General results

In this section we present the general rigidity results. We make no assumption about the space X and the dimension d.

We call two iterated function systems $F = \{f_i : X \to X, i \in I\}$ and $G = \{g_i : Y \to Y, i \in I\}$ topologically conjugate if and only if there exists a homeomorphism $h : J_F \to J_G$ such that

$$h \circ f_i = g_i \circ h$$

for all $i \in I$. Then by induction we easily get that $h \circ f_\omega = g_\omega \circ h$ for every finite word ω. The first result is the following.

Theorem 7.1.1 *Suppose that $F = \{f_i : X \to X\}_{i\in I}$ and $G = \{g_i : Y \to Y\}_{i\in I}$ are two topologically conjugate conformal iterated function systems. Then the following four conditions are equivalent.*

(1) $\exists C \geq 1\ \forall \omega \in I^*$

$$C^{-1} \leq \frac{\mathrm{diam}(g_\omega(Y))}{\mathrm{diam}(f_\omega(X))} \leq C.$$

(2) $|g'_\omega(y_\omega)| = |f'_\omega(x_\omega)|$ *for all* $\omega \in I^*$, *where* x_ω *and* y_ω *are the only fixed points of* $f_\omega : X \to X$ *and* $g_\omega : Y \to Y$ *respectively.*

(3) $\exists E \geq 1\ \forall \omega \in I^*$

$$E^{-1} \leq \frac{||g'_\omega||}{||f'_\omega||} \leq E.$$

(4) For every finite subset T *of* I, $\mathrm{HD}(J_{G,T}) = \mathrm{HD}(J_{F,T})$ *and the conformal measures* $m_{G,T}$ *and* $m_{F,T} \circ h^{-1}$ *are equivalent.*

Suppose additionally that both systems F *and* G *are regular. Then the following condition is also equivalent to the four conditions above.*

(5) $\mathrm{HD}(J_G) = \mathrm{HD}(J_F)$ *and the conformal measures* m_G *and* $m_F \circ h^{-1}$ *are equivalent.*

Proof. Let us first demonstrate that conditions (2) and (3) are equivalent. Indeed, suppose that (2) is satisfied and let K_F and K_G denote the distortion constants of the systems F and G respectively. Then for all $\omega \in I^*$, $||g'_\omega|| \leq K_G|g'_\omega(y_\omega)| = K_G|f'_\omega(x_\omega)| \leq K_G||f'_\omega||$ and similarly $||f'_\omega|| \leq K_F||g'_\omega||$. So suppose that (3) holds and (2) fails, that is that there exists $\omega \in I^*$ such that $|g'_\omega(y_\omega)| \neq |f'_\omega(x_\omega)|$. Without looss of generality we may assume that $|g'_\omega(y_\omega)| < |f'_\omega(x_\omega)|$. For every $n \geq 1$ let ω^n be the concatenation of n words ω. Then $g_{\omega^n}(y_\omega) = g^n_\omega(y_\omega) = y_\omega$ and similarly $f_{\omega^n}(x_\omega) = x_\omega$. So, $x_{\omega^n} = x_\omega = \pi_F(\omega^\infty)$ and $y_{\omega^n} = y_\omega = \pi_G(\omega^\infty)$. Moreover $|g'_{\omega^n}(y_\omega)| = |g'_\omega(y_\omega)|^n$ and $|f'_{\omega^n}(x_\omega)| = |f'_\omega(x_\omega)|^n$. Hence

$$\lim_{n\to\infty} \frac{|g'_{\omega^n}(y_\omega)|}{|f'_{\omega^n}(x_\omega)|} = 0.$$

On the other hand, by (3) and the bounded distortion property

$$\frac{|g'_{\omega^n}(y_\omega)|}{|f'_{\omega^n}(x_\omega)|} \geq \frac{K_G^{-1}||g'_{\omega^n}||}{||f'_{\omega^n}||} \geq E^{-1}K_G^{-1}$$

for all $n \geq 1$. This contradiction finishes the proof of equivalence of conditions (2) and (3). Since the equivalence of (1) and (3) follows immediately from the bounded distortion property, the proof of the equivalence of conditions (1)–(3) is finished. We shall now prove that (3) $\Rightarrow$ (5). Indeed, it follows from (3) that $E^{-1}\psi_{G,n}(t) \leq \psi_{F,n}(t) \leq E\psi_{G,n}(t)$ for all $t \geq 0$ and all $n \geq 1$. Hence $P_G(t) = P_F(t)$ and therefore by Theorem 4.2.13, $\mathrm{HD}(J_G) = \mathrm{HD}(J_F)$. Denote this common value by h.

Although we keep the same symbol for the homeomorphism establishing conjugacy between the systems F and G, it will never cause misunderstandings.
Suppose now that both systems are regular (in fact assuming (3) regularity of one of these systems implies regularity of the other). Then for every $\omega \in I^*$

$$(K_F E)^{-h} \leq \frac{K_F^{-h}||f'_\omega||^h}{||g'_\omega||^h} \leq \frac{m_F(f_\omega(J_F))}{m_G(g_\omega(J_G))} \leq \frac{||f'_\omega||^h}{K_G^{-h}||g'_\omega||^h} \leq (EK_G)^h.$$

So, the measures m_G and $m_F \circ h^{-1}$ are equivalent, and even more

$$(K_F E)^{-h} \leq \frac{dm_F \circ h^{-1}}{dm_G} \leq (EK_G)^h.$$

Let us show now that (5)⇒(3). Indeed, if (5) is satisfied then the measure $\mu_F \circ h^{-1}$ is equivalent to μ_G. Since additionally $\mu_F \circ h^{-1}$ and μ_G are both ergodic, they are equal. Hence, using the equality $\mathrm{HD}(J_F) = \mathrm{HD}(J_G) := h$, we get

$$\begin{aligned}||g'_\omega||^h &\asymp \int |g'_\omega|^h \, dm_G = m_G(g_\omega(J_G)) \asymp \mu_G(g_\omega(J_G)) \\ &= \mu_F \circ h^{-1}(g_\omega(J_G)) = \mu_F(f_\omega(J_F)) \asymp m_F(f_\omega(J_F)) \\ &= \int |f'_\omega|^h \, dm_F \asymp ||f'_\omega||^h\end{aligned}$$

and raising the first and the last term of this sequence of comparabilities to the power $1/h$, we finish the proof of the implication (5)⇒(3).

The equivalence of (4) and conditions (1)–(3) is now a relatively simple corollary. Indeed, to prove that (3) implies (4) fix a finite subset T of I. By (3) $E^{-1} \leq ||f'_\omega||/||g'_\omega|| \leq E$ for all $\omega \in T^*$, and as every finite system is regular, the equivalence of measures $m_{G,T}$ and $m_{F,T} \circ h^{-1}$ follows from the equivalence of conditions (3) and (5) applied to the systems $\{f_i : i \in T\}$ and $\{g_i : i \in T\}$. If in turn (4) holds and $\omega \in I^*$, then $\omega \in T^*$, where T is the (finite) set of letters making up the word ω, and the measures $m_{G,T}$ and $m_{F,T} \circ h^{-1}$ are equivalent. Hence, by the equivalence of (2) and (5) applied to the systems $\{f_i\}_{i\in T}$ and $\{g_i\}_{i\in T}$ we conclude that $|g'_\omega(y_\omega)| = |f'_\omega(x_\omega)|$. Thus (2) follows and the proof of Theorem 7.1.1 is finished. □

We say that a CIFS $\{\phi_i : X \to X : i \in I\}$ is of bounded geometry if and only if there exists $C \geq 1$ such that for all $i, j \in I$, $i \neq j$

$$\max\{\mathrm{diam}(\phi_i(X)), \mathrm{diam}(\phi_j(X))\} \leq C\mathrm{dist}\big(\phi_i(X), \phi_j(X)\big).$$

The next theorem provides a necessary and sufficient condition for two systems of bounded geometry to be *bi-Lipschitz* equivalent.

Theorem 7.1.2 *If both systems $F = \{f_i : X \to X\}_{i\in I}$ and $G = \{g_i : Y \to Y\}_{i\in I}$ are of bounded geometry, then the topological conjugacy $h : J_F \to J_G$ is bi-Lipschitz continuous if and only if the following two conditions are satisfied.*

$$Q^{-1} \leq \frac{\mathrm{diam}(f_\omega(X))}{\mathrm{diam}(g_\omega(Y))} \leq Q \tag{7.1}$$

for some $Q \geq 1$ and all $\omega \in I^*$.

$$D^{-1} \leq \frac{\mathrm{dist}\big(g_i(Y), g_j(Y)\big)}{\mathrm{dist}\big(f_i(X), f_j(X)\big)} \leq D \tag{7.2}$$

for some $D \geq 1$ and all $i, j \in I\!N$, $i \neq j$.

Proof. First notice that (7.1) and (7.2) remain true, with modified constants Q and D if necessary, if X is replaced by J_F and Y is replaced by J_G. Suppose now that $x \in f_i(J_F)$ and $y \in f_j(J_F)$ with $i \neq j$. Then

$$\begin{aligned}
|h(y) - h(x)| &\leq \mathrm{diam}(g_i(J_G)) + \mathrm{dist}(g_i(J_G), g_j(J_G)) + \mathrm{diam}(g_j(J_G)) \\
&\leq Q\mathrm{diam}(f_i(J_F)) + D\mathrm{dist}(f_i(J_F), f_j(J_F)) \\
&\quad + Q\mathrm{diam}(f_j(J_F)) \\
&\leq 2QC\mathrm{dist}(f_i(J_F), f_j(J_F)) + D\mathrm{dist}(f_i(J_F), f_j(J_F)) \\
&\leq (2QC + D)\mathrm{dist}(f_i(J_F), f_j(J_F)) \\
&\leq (2QC + D)|y - x|.
\end{aligned}$$

Suppose in turn that $x \neq y$ both belong to the same element $f_k(J_F)$. Then there exist $\omega \in I^*$ ($|\omega| \geq 1$) and $i \neq j \in I\!N$ such that $x, y \in f_\omega(J_F)$, $x \in f_{\omega i}(J_F)$ and $y \in f_{\omega j}(J_F)$. From what has been proved so far we know that $|g_\omega^{-1}(h(y)) - g_\omega^{-1}(h(x))| \leq (2QC+D)|f_\omega^{-1}(y) - f_\omega^{-1}(x)|$. Since $|y-x| \asymp ||f'_\omega|| \cdot |f_\omega^{-1}(y) - f_\omega^{-1}(x)|$ and $|h(y) - h(x)| \asymp ||g'_\omega|| \cdot |g_\omega^{-1}(h(y)) - g_\omega^{-1}(h(x))|$, we get

$$|h(y) - h(x)| \preceq \frac{||g'_\omega||}{||f'_\omega||}|y - x| \asymp |y - x|,$$

where the comparability sign we can write because of (7.1) and equivalence of conditions (1) and (3) of Theorem 7.1.1. In the same way we show that h^{-1} is Lipschitz continuous which completes the proof of the first part of our theorem.

So suppose now that h is bi-Lipschitz continuous. We shall show that conditions (7.1) and (7.2) are satisfied. Indeed, to prove (7.1) suppose that a and b in J_F are taken so that $|h(a) - h(b)| \geq \frac{1}{2}\mathrm{diam}(g_\omega(J_G))$. Then

$$\mathrm{diam}(g_\omega(J_G)) \leq 2|h(a) - h(b)| \leq 2L|a - b| \leq 2L\mathrm{diam}(f_\omega(J_F)),$$

where L is a Lipschitz constant of h and h^{-1}. In the same way it can be shown that $\mathrm{diam}(f_\omega(J_F)) \leq 2L\mathrm{diam}(g_\omega(J_G))$ which completes the proof of property (7.1). In order to prove the right-hand side of property (7.2) we proceed as follows. Fix $i, j \in I$, $i \neq j$ and $a \neq b \in J_F$. Then

$$\begin{aligned}
&\mathrm{dist}(g_i(Y), g_j(Y)) \\
&\quad\leq \mathrm{dist}(g_i(J_G), g_j(J_G)) \leq |g_i(h(a)) - g_j(h(b))| \leq L|f_i(a) - f_j(b)| \\
&\quad\leq L\big(\mathrm{diam}(f_i(X)) + \mathrm{dist}(f_i(X), f_j(X)) + \mathrm{diam}(f_j(X))\big) \\
&\quad\leq L(2C+1)\mathrm{dist}(f_i(X), f_j(X)),
\end{aligned}$$

where the last inequality follows from boundedness of geometry of the system $\{f_i\}_{i \in I}$. □

Remark 7.1.3 *Suppose now that $I = \mathbb{N}$ and that the maps $i \mapsto \phi_i(X)$ are monotone, that is suppose that for all i and j, $i < j$ implies $\phi_i(X) < \phi_j(X)$. We claim that then the bounded geometry of the system is equivalent with the weaker condition*

$$\max\{\mathrm{diam}(\phi_i(X)), \mathrm{diam}(\phi_{i+1}(X))\} \leq C\mathrm{dist}(\phi_i(X), \phi_{i+1}(X)).$$

Indeed, if $i < j$, then

$$\begin{aligned}
&\max\{\mathrm{diam}(\phi_i(X)), \mathrm{diam}(\phi_j(X))\} \\
&\quad\leq \max_{i \leq k \leq j-1} \{\max\{\mathrm{diam}(\phi_k(X)), \mathrm{diam}(\phi_{k+1}(X))\}\} \\
&\quad\leq \max_{i \leq k \leq j-1} \{C\mathrm{dist}(\phi_k(X), \phi_{k+1}(X))\} \\
&\quad\leq C\mathrm{dist}(\phi_i(X), \phi_j(X)),
\end{aligned}$$

where in writing the last inequality we used the monotonicity of the map $i \mapsto \phi_i(X)$. The opposite implication is obvious.

Remark 7.1.4 *If $I = \mathbb{N}$ and both maps $i \mapsto f_i(X)$ and $i \mapsto g_i(X)$ are monotone, then condition (7.2) from Theorem 7.1.2 can be replaced by* *the following.*

$$C^{-1} \leq \frac{\mathrm{dist}(g_k(Y), g_{k+1}(Y))}{\mathrm{dist}(f_k(X), f_{k+1}(X))} \leq C \tag{7.3}$$

for some constant $C \geq 1$ and all $k \in I\!N$.

Indeed, assuming (7.3) this follows from the following computation.

$$\begin{aligned}
\mathrm{dist}(g_i(Y), g_j(Y)) &= \sum_{k=i}^{j-1} \mathrm{dist}(g_k(Y), g_{k+1}(Y)) + \sum_{k=i+1}^{j-1} \mathrm{diam}(g_k(X)) \\
&\leq \sum_{k=i}^{j-1} C\mathrm{dist}(f_k(X), f_{k+1}(X)) + Q \sum_{k=i+1}^{j-1} \mathrm{diam}(f_k(X)) \\
&\leq \max\{C, Q\} \left(\sum_{k=i}^{j-1} \mathrm{dist}(f_k(X), f_{k+1}(X)) \right. \\
&\quad \left. + \sum_{k=i+1}^{j-1} \mathrm{diam}(f_k(X)) \right) \\
&= \max\{C, Q\}\mathrm{dist}(f_i(X), f_j(X)).
\end{aligned}$$

7.2 One-dimensional systems

We call a CIFS one-dimensional if $d = 1$. Our presentation of conjugacy of such systems follows [HU1]. We start with an adaptation of the concept of *scaling function* introduced by D. Sullivan (see [Su2]) in the context of hyperbolic *cookie-cutter Cantor sets.* In order to define our scaling functions we will need the following basic lemma.

Lemma 7.2.1 *If $\{\phi_n : X \to X : n \geq 1\}$ is a one-dimensional CIFS, then for every closed subinterval K of X and $\omega \in I\!N^\infty$ the following limit exists*

$$\lim_{n\to\infty} \frac{|\phi_{\omega_n \omega_{n-1} \dots \omega_0}(K)|}{|\phi_{\omega_n \omega_{n-1} \dots \omega_0}(X)|} := S(\omega, K)$$

and the convergence is uniform with respect to K, n and ω.

Proof. We shall show that the above sequence satisfies an appropriate

Cauchy condition. Fix $k < n$. We then have

$$\begin{aligned} &\frac{|\phi_{\omega_n \ldots \omega_k \ldots \omega_0}(K)|}{|\phi_{\omega_n \ldots \omega_k \ldots \omega_0}(X)|} \Big/ \frac{|\phi_{\omega_k \ldots \omega_0}(K)|}{|\phi_{\omega_k \ldots \omega_0}(X)|} \\ &\quad = \frac{|\phi_{\omega_n \ldots \omega_{k+1}}(\phi_{\omega_k \ldots \omega_0}(K))|}{|\phi_{\omega_k \ldots \omega_0}(K)|} \Big/ \frac{|\phi_{\omega_n \ldots \omega_{k+1}}(\phi_{\omega_k \ldots \omega_0}(X))|}{|\phi_{\omega_k \ldots \omega_0}(X)|} \\ &\quad = \frac{|\phi'_{\omega_n \ldots \omega_{k+1}}(x_n)|}{|\phi'_{\omega_n \ldots \omega_{k+1}}(y_n)|} \end{aligned} \tag{7.4}$$

for some $x_n \in \phi_{\omega_k \ldots \omega_0}(K)$ and $y_n \in \phi_{\omega_k \ldots \omega_0}(X)$, where the last equality sign follows from the mean value theorem. Denote now $|\phi_{\omega_j \ldots \omega_0}(K)|/|\phi_{\omega_j \ldots \omega_0}(X)|$ by a_j. In view of (7.4) and (4.18) we get

$$|\log a_n - \log a_k| \le \frac{L}{1-s^\alpha}|x_n - y_n|^\alpha \le \frac{L}{1-s^\alpha}|\phi_{\omega_k \ldots \omega_0}(X)|^\alpha \le \frac{L}{1-s^\alpha} s^{k\alpha}.$$

Thus the sequence $\{\log a_n\}_{n=1}^\infty$ is a Cauchy sequence, and consequently $\{a_n\}_{n=1}^\infty$ itself is also a Cauchy sequence. □

Let $\widetilde{\mathbb{N}}^\infty$ denote the set of infinite sequences of the form $\ldots \omega_n \omega_{n-1} \ldots \omega_1 \omega_0$ and let $\widetilde{\mathbb{N}}^*$ denote the set of all finite words of the form $\omega_n \omega_{n-1} \ldots \omega_1 \omega_0$. Lemma 7.2.1 allows us to introduce the scaling function (comp. also [Su2], [PT] and [U4]). We shall now explore this notion in greater detail. The *weaker scaling function* S^w is defined on the space $\widetilde{\mathbb{N}}^\infty \times \mathbb{N}$, takes values in $(0,1)$, and is given by the formula

$$S^w(\{\omega_n\}_{n=0}^\infty, i) = \lim_{n\to\infty} \frac{|\phi_{\omega_n \omega_{n-1} \ldots \omega_0}(\phi_i(X))|}{|\phi_{\omega_n \omega_{n-1} \ldots \omega_0}(X)|},$$

where the limit exists due to Lemma 7.2.1.

The *stronger scaling function* S^s is defined similarly but on the larger space $\widetilde{\mathbb{N}} \times (\mathbb{N} \cup \mathcal{C})$, where $\mathcal{C}$ denotes the set of all connected components of $X \setminus \bigcup_{i=1}^\infty \phi_i(X)$. Frequently, given $\omega \in \mathbb{N}^*$ we will consider the function $S^s(\omega) : (\mathbb{N} \cup \mathcal{C}) \to (\prime, \infty)$ given by the formula $S^s(\omega)(Z) = S^s(\omega, Z)$, and similarly we define the function $S^w(\omega)$. The following theorem is an immediate consequence of Lemma 7.2.1.

Theorem 7.2.2 *Both scaling functions $S^w : \widetilde{\mathbb{N}}^\infty \times \mathbb{N}$ and $S^s : \widetilde{\mathbb{N}} \times (\mathbb{N} \cup \mathcal{C})$ are continuous.*

We now go on to consider two systems $F = \{f_i : i \in I\!N\}$ and $G = \{g_i : i \in I\!N\}$. Our first theorem about them concerns the relations between possibilities of smooth extensuion of the topological conjugacy $h : J_F \to J_G$ and the corresponding scaling functions.

Theorem 7.2.3 *If the topological conjugacy $h : J_F \to J_G$ extends in a diffeomorphic fashion onto X, then J_F and J_G have the same strong scaling functions. Conversely, if two topologically conjugate one-dimensional IFSs F and G have the same weak scaling functions and condition (b) of Theorem 7.1.2 is satisfied, then the topological conjugacy is Lipschitz continuous.*

Proof. Let us first prove the first part of this theorem. Indeed, let us keep the same notation h for its diffeomorphic extension to X and let D be an arbitrary closed subinterval of X. For $\omega \in \widetilde{I\!N}^\infty$ we can write

$$\begin{aligned}\frac{S(\omega, D)}{S(\omega, h(D))} &= \lim_{n\to\infty} \frac{|f_{\omega_n\ldots\omega_0}(D)|}{|f_{\omega_n\ldots\omega_0}(X)|} \frac{|g_{\omega_n\ldots\omega_0}(h(D))|}{|g_{\omega_n\ldots\omega_0}(Y)|} \\ &= \lim_{n\to\infty} \frac{|f_{\omega_n\ldots\omega_0}(D)|}{|g_{\omega_n\ldots\omega_0}(h(D))|} \Big/ \frac{|f_{\omega_n\ldots\omega_0}(X)|}{|g_{\omega_n\ldots\omega_0}(Y)|}.\end{aligned}$$

Now, by the mean value theorem there exist a_n and b_n respectively in $f_{\omega_n\ldots\omega_0}(D)$ and in $f_{\omega_n\ldots\omega_0}(X)$ such that

$$\frac{S(\omega, D)}{S(\omega, h(D))} = \lim_{n\to\infty} \frac{|f_{\omega_n\ldots\omega_0}(D)|}{|h(f_{\omega_n\ldots\omega_0}(D))|} \Big/ \frac{|f_{\omega_n\ldots\omega_0}(X)|}{|h(f_{\omega_n\ldots\omega_0}(X))|} = \lim_{n\to\infty} \frac{h'(b_n)}{h'(a_n)}.$$

Since h' is uniformly continuous with no zeros and since $|b_n - a_n| \to 0$ the last limit is equal to 1 which finishes the proof of the first part of our theorem.

In order to prove the second part of this theorem it suffices to show that condition (7.1) of Theorem 7.1.2 is satisfied. So, let $\tau = \tau_0 \ldots \tau_{q-1}$ be an arbitrary word. Our aim is to show that $|(g_\tau)'(h(x_\tau))| = |(f_\tau)'(x_\tau)|$, where x_τ is the only fixed point of the map $f_\tau : X \to X$. First notice that for every n

$$\begin{aligned}&\frac{|g_{\tau^{n+1}\tau_0}(Y)|}{|g_{\tau^n\tau_0}(Y)|} \\ &= \frac{|g_{\tau^{n+1}\tau_0}(Y)|}{|g_{\tau^{n+1}}(Y)|} \cdot \frac{|g_{\tau^{n+1}}(Y)|}{|g_{\tau^n\tau_0\ldots\tau_{q-2}}(Y)|} \cdot \frac{|g_{\tau^n\tau_0\ldots\tau_{q-2}}(Y)|}{|g_{\tau^n\tau_0\ldots\tau_{q-3}}(Y)|} \cdot \ldots \cdot \frac{|g_{\tau^n\tau_0\tau_1}(Y)|}{|g_{\tau^n\tau_0}(Y)|}.\end{aligned}$$

Hence

$$\lim_{n\to\infty}\frac{|g_{\tau^{n+1}\tau_0}(Y)|}{|g_{\tau^n\tau_0}(Y)|} \tag{7.5}$$
$$= S^w_{\tau^\infty}(\tau_0)S^w_{\tau^\infty\tau_0\ldots\tau_{q-2}}(\tau_{q-1})S^w_{\tau^\infty\tau_0\ldots\tau_{q-3}}(\tau_{q-2})\ldots S^w_{\tau^\infty\tau_0}(\tau_1)$$

and similarly

$$\lim_{n\to\infty}\frac{|f_{\tau^{n+1}\tau_0}(X)|}{|f_{\tau^n\tau_0}(X)|} \tag{7.6}$$
$$= S^w_{\tau^\infty}(\tau_0)S^w_{\tau^\infty\tau_0\ldots\tau_{q-2}}(\tau_{q-1})S^w_{\tau^\infty\tau_0\ldots\tau_{q-3}}(\tau_{q-2})\ldots S^w_{\tau^\infty\tau_0}(\tau_1).$$

Since $g_{\tau^{n+1}\tau_0}(Y) = g_\tau(g_{\tau_n\tau_0}(Y)$ and since $f_{\tau^{n+1}\tau_0}(X) = f_\tau(f_{\tau_n\tau_0}(X))$, it follows from the mean value theorem that there exists $x_n \in f_{\tau_n\tau_0}(X)$ and $y_n \in g_{\tau_n\tau_0}(Y)$ such that $|g_{\tau^{n+1}\tau_0}(Y)| = |g'_\tau(y_n)| \cdot |g_{\tau_n\tau_0}(Y)|$ and $|f_{\tau^{n+1}\tau_0}(X)| = |f'_\tau(y_n)| \cdot |f_{\tau_n\tau_0}(X)|$. Thus in view of our assumptions and (7.5) and (7.6) we get

$$\lim_{n\to\infty}\frac{|g'_\tau(y_n)|}{|f'_\tau(x_n)|} = \lim_{n\to\infty}\frac{|g_{\tau^{n+1}\tau_0}(Y)|}{|g_{\tau^n\tau_0}(Y)|}\Big/\frac{|f_{\tau^{n+1}\tau_0}(X)|}{|f_{\tau^n\tau_0}(X)|} = 1.$$

Now, a straightforward computation shows that $y_n \to y_\tau$ and $x_n \to x_\tau$, where y_τ and x_τ are fixed points of g_τ and f_τ respectively. Hence $|g'_\tau(y_\tau)| = |f'_\tau(x_\tau)|$ and the equivalence of this condition with condition (1) of Theorem 7.1.1 finishes the proof. □

We end this section with the rigidity result concerning real-analytic non-essentially affine one-dimensional systems. We recall from the previous chapter that a one-dimensional system $S = \{\phi_i : X \to X\}_{i\in I}$ is said to be real-analytic if and only if there exists a topological disk D such that all the maps ϕ_i extend in a conformal (so 1-to-1) fashion to D. Suppose that S is regular. Let m be the conformal measure associated to the system S and let μ be the only probability S-invariant measure equivalent with m. We call the system S essentially affine (cf. [Su3] and [HU1]) if and only if all the Jacobians $D_{\phi_i} = \frac{d\mu\circ\phi_i}{d\mu}$, $i \in I$ are constant. More detailed discussion concerning the concept of essential affinity will be provided in the two following sections. We shall now prove the theorem, which is in a sense much stronger than both Theorem 7.1.1 and Theorem 7.1.2.

Theorem 7.2.4 *If both systems $\{f_i : X \to X\}_{i\in I}$ and $\{g_i : Y \to Y\}_{i\in I}$ are regular and real-analytic and neither is essentially affine, then the following conditions are equivalent.*

(a) The conjugacy between the systems $\{f_i : X \to X\}_{i\in I}$ *and* $\{g_i : Y \to Y\}_{i\in I}$ *is real-analytic.*

(b) The conjugacy between the systems $\{f_i : X \to X\}_{i\in I}$ *and* $\{g_i : Y \to Y\}_{i\in I}$ *is Lipschitz continuous.*

(c) $|g'_\omega(y_\omega)| = |f'_\omega(x_\omega)|$ *for all* $\omega \in \mathbb{N}^*$, *where* x_ω *and* y_ω *are the only fixed points of* $f_\omega : X \to X$ *and* $g_\omega : Y \to Y$ *respectively.*

(d) $\exists S \geq 1 \; \forall \omega \in \mathbb{N}^*$

$$S^{-1} \leq \frac{\mathrm{diam}(g_\omega(Y))}{\mathrm{diam}(f_\omega(X))} \leq S.$$

(e) $\exists E \geq 1 \; \forall \omega \in \mathbb{N}^*$

$$E^{-1} \leq \frac{||g'_\omega||}{||f'_\omega||} \leq E.$$

(f) $\mathrm{HD}(J_G) = \mathrm{HD}(J_F)$ *and the measures* m_G *and* $m_F \circ h^{-1}$ *are equivalent.*

(g) The measures m_G *and* $m_F \circ h^{-1}$ *are equivalent.*

Proof. The implication (a)⇒(b) is obvious.That (b)⇒(c) results from the fact that (b) implies condition (1) of Theorem 7.1.1, which in view of that theorem is equivalent with condition (2) of Theorem 7.1.1, which finally is the same as condition (c) of Theorem 7.2.4. The implications (c)⇒(d)⇒(e)⇒(f) have been proved in Theorem 7.1.1. The implication (f)⇒(g) is again obvious. We are left to prove that (g)⇒(a). And indeed, if (g) holds, then $\mu_F = \mu_G \circ h$ meaning that $J_h = \frac{d\mu_G \circ h}{d\mu_F} = 1$. Since $h \circ f_\omega = g_\omega \circ h$, the chain rule implies that $J_h \circ f_\omega \cdot J_{f_\omega} = J_{g_\omega} \circ h \cdot J_h$ and consequently

$$J_{f_\omega} = J_{g_\omega} \circ h.$$

Since the system G is not essentially affine, there exists a contraction $g_i \in S$ such that the Jacobian $D_{g_i} = \frac{d\mu \circ g_i}{d\mu}$ is not constant and which, due to Corollary 6.1.5, is real-analytic. This implies that D_{g_i} has only finitely many extremal points, since otherwise the equation $D'_{g_i} = 0$ would have an accumulation point in Y which in turn would imply that D_{g_i} would be constant on Y, contrary to the definition of g_i. Hence $D_{g_i}^{-1} \circ D_{f_i}$ is well defined and 1-to-1 on an open set $V \subset X$, and $h = D_{g_i}^{-1} \circ D_{f_i}$ on $V \cap F_F$. Consider now $\omega \in \mathbb{N}^*$ such that $f_\omega(X) \subset V$. Then the map $g_\omega^{-1} \circ (D_{g_i}^{-1} \circ D_{f_i}) \circ f_\omega : X \to X$ extends h and, due to Corollary 6.1.5 again, is real-analytic.

We shall now explore the case when $d \geq 2$. Because of the different techniques used we treat the cases $d = 2$ and $d \geq 3$ separately. Although the exposition of the case $d \geq 3$ is more elegant and clear, we decided to present the case $d = 2$ first since it is applied in the case $d \geq 3$. Our exposition follows [MPU] in the case when $d = 2$ and [U5] if $d \geq 3$.

7.3 Two-dimensional systems

At the very beginning of this section we would like to make use of the following two results concerning holomorphic univalent maps of the plane.

Theorem 7.3.1 *(see Ex. 5 on p. 224 in [Ne]) A necessary and sufficient condition for a holomorphic univalent function $\phi : \mathbb{D} \to \mathbb{C}$ with $\phi'(z) = 1$ to map the unit disk $\mathbb{D}$ onto a convex domain is that*

$$Re\left\{1 + \frac{z\phi''(z)}{\phi'(z)}\right\} \geq 0$$

for all $z \in \mathbb{D}$.

and

Theorem 7.3.2 *(see Theorem 1.5 on p. 3 in [CG]) If $\phi : \mathbb{D} \to \mathbb{C}$ is a holomorphic univalent function, then*

$$\left|\frac{z\phi''(z)}{\phi'(z)} - \frac{2|z|^2}{1 - |z|^2}\right| \leq \frac{4|z|}{1 - |z|^2}$$

for all $z \in \mathbb{D}$.

An easy argument combining the above two results leads to the following remarkable convexity result.

Theorem 7.3.3 *If $\phi : B(w, R) \to \mathbb{C}$ is a holomorphic univalent function, then the image $\phi(B(w, R/8))$ is convex.*

Generalizing the concept of one-dimensional systems from the previous section we now call the system $S = \{\phi_i\}_{i \in I}$ one-dimensional if there exists a set $D : \overline{J} \subset D \subset V$ composed of finitely many real-analytic curves with pairwise disjoint closures such that $\phi_i(D) \subset D$ for all $i \in I$ (see [MPU]). We prove the following result about one-dimensional systems.

Lemma 7.3.4 *If a non-empty open subset of $\overline{J}$ is contained in a one-dimensional real-analytic curve, then the system S is 1-dimensional.*

Proof. Since $\overline{J}$ is compact it suffices to show that each point in $\overline{J}$ has a neighborhood contained in a real-analytic curve. The assumptions of the lemma state that there exists a point $x \in \overline{J}$, an open ball $B(x)$ centered at x and M, a real-analytic curve, open-ended, containing $\overline{J} \cap B(x)$. Fix now an arbitrary point $z \in \overline{J}$. Since $x \in \overline{J}$ there exists $\omega \in I^*$ such that $\phi_\omega(z) \in \overline{J} \cap B(x)$; moreover $\phi_\omega(V) \subset B(x)$. Then the set $\phi_\omega(V) \cap M$ contains $\phi_\omega(V) \cap \overline{J}$, an open neighborhood of $\phi_\omega(z)$ in $\overline{J}$, and consists of countably many real-analytic curves. Let Γ be one of them, the connected component of $\phi_\omega(V) \cap M$ containing $\phi_\omega(z)$. It contains an open neighborhood of $\phi_\omega(z)$ in $\overline{J}$. Then $\phi^{-1}(\Gamma)$ contains an open neighborhood of z in $\overline{J}$. □

We now provide the following definition of essential affinity, generalizing that from the previous section.

Definition 7.3.5 *We say that the system S is essentially affine if S is conjugate by a conformal homeomorphism with a system consisting only of conformal affine contractions (i.e. of the form $az + b$).*

The first aim of this section is to establish in the following theorem several conditions equivalent with essential affinity.

Theorem 7.3.6 *Suppose that the system $S = \{\phi_i\}_{i \in I}$ is regular and denote the corresponding conformal measure by m. Then the following conditions are equivalent.*

(a) For each $i \in I$ the extended Jacobian $\tilde{D}_{\phi_i} : U \to \mathbb{R}$ is constant, where U is the neighborhood of X produced in Corollary 6.1.5.

(b) There exist a continuous function $u : X \to \mathbb{R}$ and constants $c_i \in \mathbb{R}$, $i \in I$, such that

$$\log |\phi_i'| = u - u \circ \phi_i + c_i$$

for all $i \in I$.

(c) There exist a continuous function $u : \overline{J} \to \mathbb{R}$ and constants $c_i \in \mathbb{R}$, $i \in I$, such that

$$\log |\phi_i'| = u - u \circ \phi_i + c_i$$

for all $i \in I$.

(d1) The conformal structure on $\overline{J}$ admits a Euclidean isometric refinement so that all maps ϕ_i, $i \in I$, become affine conformal; more precisely there exists an atlas $\{\psi_t : U_t \to \mathbb{C}\}$ with open

disks U_t, consisting of conformal injections such that $\bigcup_t U_t \supset \overline{J}$, all $U_t \cap U_s$ and $U_t \cap \phi_i(U_s)$ are connected and the compositions $\psi_t \circ \psi_s^{-1}$ and $\psi_t \circ \phi_i \circ \psi_s^{-1}$, respectively on $\psi_s(U_t \cap U_s)$ and $\psi_s \circ \phi_i^{-1}(U_t \cap \phi_i(U_s))$, are conformal affine with $|(\psi_t \circ \psi_s^{-1})'| \equiv 1$.

(d2) As (d1) *but no assumptions on $|(\psi_t \circ \psi_s^{-1})'|$ (i.e. the atlas is only conformal affine).*

(d3) The system S is essentially affine.

(eh) There exist a cover $\{B_\lambda\}_{\lambda \in \Lambda}$ of $\overline{J}$ consisting of open disks and a family of harmonic functions $\gamma_\lambda : B_\lambda \to I\!R$, $\lambda \in \Lambda$, such that for all $\lambda, \lambda' \in \Lambda$ and all $i \in I$

$$\gamma_\lambda - \gamma_{\lambda'} = \text{const} \tag{7.7}$$

on $B_\lambda \cap B_{\lambda'}$ and

$$arg_\lambda \phi_i' - \gamma_\lambda + \gamma_{\lambda'} \circ \phi_i = \text{const} \tag{7.8}$$

on $\phi_i^{-1}(B_\lambda' \cap \phi_i(B_\lambda))$, where $arg_\lambda \phi_i' : B_\lambda \to I\!R$ is a continuous branch of argument of ϕ_i' defined on the simply connected set B_λ. All the sets $B_\lambda \cap B_{\lambda'}$ and $\phi_i^{-1}(B_\lambda' \cap \phi_i(B_\lambda))$ are connected.

(er) As (eh) *but harmonic replaced by real-analytic.*

(ec) As (eh) *but harmonic replaced by continuous.*

(f) $\nabla \tilde{D}\phi_i(z) = 0$ for all $z \in \overline{J}$ and all $i \in I$ if S is one-dimensional. If S is not one-dimensional then

$$\det(\nabla \tilde{D}\phi_i \circ \phi_\omega(z), \nabla \tilde{D}\phi_i(z)) = 0$$

for all $z \in \overline{J}$ and all $i \in I, \omega \in I^*$.

Proof. We shall prove the following implications: (a)⇒(b)⇒(c)⇒(d1) ⇒(d2)⇒(a), (d3)⇒(d2)⇒(eh)⇒(er)⇒(ec)⇒(d2)⇒(d3), (a)⇒(f) and (f)⇒(er). Since the implication (d2)⇒ (d3) is rather easy and the opposite is obvious, we could simplify our considerations proving that (d3) implies (eh) instead of proving that (d3) implies (*d*2). We chose however this approach since it can be easily adapted to the case of conformal expanding repellers (see [PU1], cf. [Su3]). In the next section ($d \geq 3$), we follow the other one.

- (a)⇒(b). Since for every $i \in I$, $\tilde{D}_{\phi_i} = (\rho \circ \phi_i) \cdot |\phi_i'|^\delta \cdot \rho^{-1}$, we have

$$\log(|\tilde{D}_{\phi_i}|) = \log(|\rho| \circ \phi_i) + \delta \log |\phi_i'| - \log |\rho|.$$

Thus to finish the proof of the implication (a)⇒(b) it suffices to set $c_i = \frac{1}{\delta} \log(\tilde{D}_{\phi_i})$ and $u = \frac{1}{\delta} \log |\rho|$.

• The implication (b)⇒(c) is obvious.

(c)⇒(d1). Fix an element $v \in \overline{J}$ and an element $\tau \in I^\infty$. Given $n \geq 1$ and a word $\omega \in I^n$ we denote by $\overline{\omega}$ the flipped word $\omega_n\omega_{n-1}\dots\omega_1$. Our first aim is to show that the series

$$\sum_{n\geq 1}\left(\log|\phi'_{\tau_n}(\phi_{\overline{\tau|_{n-1}}}(z))| - \log|\phi'_{\tau_n}(\phi_{\overline{\tau|_{n-1}}}(v))|\right) \tag{7.9}$$

converges absolutely uniformly on V, where for $n = 1$ we set $\phi_{\overline{\tau|_{n-1}}} = \mathrm{Id}_V$. Indeed, it follows from the bounded distortion property (4f) and (4e) that

$$\begin{aligned}\left|\log|\phi'_{\tau_n}(\phi_{\overline{\tau|_{n-1}}}(z))| - \log|\phi'_{\tau_n}(\phi_{\overline{\tau|_{n-1}}}(v))|\right| &\leq KL\left|\phi_{\overline{\tau|_{n-1}}}(z) - \phi_{\overline{\tau|_{n-1}}}(v)\right| \\ &\leq KLs^{(n-1)}|z-v| \\ &\leq KL\mathrm{diam}(V)s^{(n-1)}.\end{aligned} \tag{7.10}$$

Since

$$\sum_{n\geq 1} KL\mathrm{diam}^\alpha(V)s^{(n-1)\alpha} \leq \frac{KL\mathrm{diam}^\alpha(V)}{1-s} < \infty,$$

the proof of the absolute uniform convergence of the series defined by (7.9) is complete. We now can define the function $u_v : V \to I\!R$ by setting

$$u_v(z) = u(v) + \sum_{n\geq 1}\left(\log|\phi'_{\tau_n}(\phi_{\overline{\tau|_{n-1}}}(z))| - \log|\phi'_{\tau_n}(\phi_{\overline{\tau|_{n-1}}}(v))|\right). \tag{7.11}$$

The function $u_v : V \to I\!R$, as the sum of an absolutely convergent series of harmonic functions, is harmonic. Iterating the formula appearing in Theorem 3.1(c), we obtain for every $n \geq 1$ and every $z \in \overline{J}$

$$\begin{aligned}u(z) - u(v) = &\sum_{k=1}^{n}\left(\log|\phi'_{\tau_k}(\phi_{\overline{\tau|_{k-1}}}(z))| - \log|\phi'_{\tau_k}(\phi_{\overline{\tau|_{k-1}}}(v))|\right) \\ &+ u(\phi_{\overline{\tau|_n}}(z)) - u(\phi_{\overline{\tau|_n}}(v)).\end{aligned}$$

Since, by the bounded distortion property, $|\phi_{\overline{\tau|_n}}(z) - \phi_{\overline{\tau|_n}}(v)| \leq s^n$ and since the function $u : \overline{J} \to I\!R$ as continuous on a compact set is uniformly continuous, it follows from the last display that $u_v(z) = u(z)$ for all $z \in \overline{J}$, i.e. u_v is a harmonic extension of u on V. From now on we will drop the subscript v and write simply $u : V \to I\!R$. Since all the functions $\log|\phi'_i|$ and $u - u\circ\phi_i + c_i$, $i \in I$, are harmonic on V, each set

$$Z_i = \{z \in V : \log|\phi'_i(z)| = u(z) - u\circ\phi_i(z) + c_i\},$$

$i \in I$, is either equal to V or is a real-analytic set.

Suppose first that $Z_i = V$ for all $i \in I$. For every $w \in \overline{J}$ consider a ball $B(w) \subset V$ centered at w. Let $l_w : B(w) \to \mathbb{R}$ be a harmonic conjugate function to the harmonic function $u : B(w) \to \mathbb{R}$ so that $u + il_w : B(w) \to \mathbb{C}$ is holomorphic. Write $G_w = \exp(u + il_w)$ and denote by $\psi_w : B(w) \to \mathbb{C}$ a primitive function of G_w. Since $\psi'_w(w) = G_w(w) \neq 0$, there exists a disk $U_w \subset B(w)$ centered at w and such that $\psi_w|_{U_w}$ is injective. Using Theorem 7.3.3 we may assume that in addition all the sets U_w are so small that all the images $\phi_i(U_w)$, $i \in I$, $w \in \overline{J}$, are convex. We claim that the family $\{\psi_w : U_w \to \mathbb{C}\}_{w \in \overline{J}}$ forms an atlas demanded in (d1). Indeed, fix $w, v \in \overline{J}$ and consider an arbitrary point $z \in U_w \cap U_v$. Then

$$\begin{aligned}(\psi_w \circ \psi_v^{-1})'(\psi_v(z)) &= \psi'_w(z) \cdot (\psi'_v(z))^{-1} = G_w(z) \cdot G_v^{-1}(z) \\ &= \exp\bigl(i(l_w(z) - l_v(z))\bigr)\end{aligned}$$

and therefore $(\psi_w \circ \psi_v^{-1})'$ is constant with absolute value 1 on $\psi_v(U_v \cap U_w)$, since h_w and h_v differ by an additive constant on the connected set $U_w \cap U_v$ as harmonic conjugates to the same harmonic function u.

To discuss $\left(\psi_v \circ \phi_i \circ \psi_w^{-1}\right)'$ fix again arbitrary $w, v \in \overline{J}$ and for every $i \in I$ consider the intersection $U_v \cap \phi_i(U_w)$. As the intersection of two convex sets, this set is convex, and consequently connected. Take now an arbitrary point $z \in \phi_i^{-1}(U_v \cap \phi_i(U_w))$. Since $Z_i = V$, we therefore have

$$\begin{aligned}&\left|\left(\psi_v \circ \phi_i \circ \psi_w^{-1}\right)'(\psi_w(z))\right| \\ &= |\psi'_{\phi_i(w)}(\phi_i(z)) \cdot \phi'_i(z) \cdot (\psi'_w(z))^{-1}| = |G_v(\phi_i(z)) \cdot \phi'_i(z) \cdot G_w^{-1}(z) \\ &= |\exp\bigl(u(\phi_i(z) + il_v(\phi_i(z)) - u(z) - il_w(z)| \cdot |\phi'_i(z)|\bigr) \\ &= \exp(u(\phi_i(z) - u(z))|\phi'_i(z)| \\ &= e^{c_i}.\end{aligned}$$

Hence the function $\left(\psi_v \circ \phi_i \circ \psi_w^{-1}\right)'$, as holomorphic and having constant absolute value, is constant on the connected set $\psi_w \circ \phi_i^{-1}(U_v \cap \phi_i(U_w))$.

Suppose in turn that $Z_i \neq V$ for some $i \in I$. Since equation (c) of Theorem 3.1 is satisfied on compact $\overline{J}$, then $\overline{J} \subset Z_i$. Since $\overline{J}$ is infinite its non-empty open part is contained in a real analytic curve, so the system is one-dimensional. Hence by Lemma 2.1 there are finitely many real-analytic pairwise disjoint curves whose union M contains $\overline{J}$. Since $\phi_i(\overline{J}) \subset \overline{J}$ for all $i \in I$, decreasing M if necessary, we may assume that $\phi_i(M) \subset M$ for all $i \in I$.

Change coordinates holomorphically on a neighborhood of M so that $M \subset I\!R$. (This uses the consequence of our assumptions that there is no closed curve among the components of M; with relaxed assumptions allowing the existence of such a curve we would change it to the unit circle and then use charts being branches of $z \mapsto \log(iz)$.) Since the function $u : M \to I\!R$ is real-analytic, it uniquely extends to a complex-analytic function $\tilde{u}$ on an open neighborhood of M in V. Now we proceed similarly as in the previous case; we define ψ_w, $w \in \overline{J}$, to be a primitive of $e^{\tilde{u}}$ on a sufficiently small neighborhood of $w \in V$ and we check that $(\psi - w \circ \psi_v^{-1})' = 1$ on $\psi_v(U_v \cap U_w)$. Now note that $\tilde{u} - \tilde{u} \circ \phi_i + c_i = \widetilde{\log|\phi_i'|}$, where the latter expression is a holomorphic extension of $\log|\phi_i'|$, which extends the equality (c). Note that $\widetilde{\log|\phi_i'|} = \log \pm \phi_i'$, where $\pm$ depends on whether ϕ_i' is positive or negative. We use the fact it is real. The equality extends the equality on $\overline{J}$ because the functions on both sides are holomorphic. We conclude with

$$|\big(\psi_{\phi_i(w)} \circ \phi_i \circ \psi_w^{-1}\big)'(\psi_w(z))| = e^{c_i}$$

for all $z \in \phi_i^{-1}(U_v \cap \phi_i(U_w))$. Hence $\big(\psi_{\phi_i(w)} \circ \phi_i \circ \psi_w^{-1}\big)'$ is constant on the connected set $\psi_w \circ \phi_i^{-1}(U_v \cap \phi_i^{(}U_w))$. The proof of the implication (c)$\Rightarrow$(d1) is complete.

Remark. As an intermediate step in the proof of the implication (c)$\Rightarrow$(d1) we proved (bh) (compare later (eh)), namely the property (b) with u harmonic on a neighborhood of $\overline{J}$, here V, provided the system S is not one-dimensional ($Z_i = V$ for all i). For S one-dimensional we can also prove (bh) but indirectly, via (d1). Indeed assuming (d1) and M in $I\!R$ we set the harmonic extension $u = \log|\psi_v'|$ independent of v.

• The implications (d1)$\Rightarrow$(d2) and (d3)$\Rightarrow$(d2) are obvious.

• (d2)$\Rightarrow$(a). Let $\{\psi_\lambda : U_\lambda \to \mathbb{C}\}_{\lambda\in\Lambda}$ be a finite conformal affine atlas for the system S. Fix $\beta \in \Lambda$, take a number $n_0 \geq 1$ so large that $\mathrm{diam}(V)s^{n_0}$ is less than a Lebesgue number of the cover $\{U_\lambda\}_{\lambda\in\Lambda}$ of $\overline{J}$, consider any number $n \geq n_0$ and for every $\omega \in I^n$ choose one element $\lambda(\omega) \in \Lambda$ such that $\phi_\omega(V) \subset U_{\lambda(\omega)}$. Next, given $n \geq n_0$ and $\omega \in I^n$ consider the map

$$(\psi_{\lambda(\omega)} \circ \phi_\omega \circ \psi_\beta^{-1})' \circ \psi_\beta$$

defined on U_β. Since our atlas is affine, this function is constant on every sufficiently small neighborhood of every point in $\overline{J} \cap U_\beta$ and therefore, as real-analytic, it is constant on U_β. Denote its value there by $c_{\beta,\omega}$. Since for every $z \in U_\beta$

$$\sum_{|\omega|=n} c_{\beta,\omega}^\delta |\psi_\beta'(z)|^\delta \cdot |\psi_{\lambda(\omega)}'(\phi_\omega(z))|^{-\delta} = \sum_{|\omega|=n} |\phi_\omega'(z)|^\delta = \mathcal{L}^n(1\!\!1), \quad (7.12)$$

since by Theorem 1.3

$$\lim_{n\to\infty} \mathcal{L}^n(1\!\!1)(z) = \rho(z) \quad (7.13)$$

and since the product $|\psi_\beta'(z)|^\delta \cdot |\psi_{\lambda(\omega)}'(\phi_\omega(z))|^{-\delta}$ is uniformly bounded away from zero and infinity, we conclude that there exists a constant $M \geq 1$ such that for all $z \in U_\beta$ and all $n \geq 1$

$$M^{-1} \leq \sum_{|\omega|=n} c_{\beta,\omega}^\delta \leq M. \quad (7.14)$$

Fix now an $\epsilon > 0$ and $n_1 \geq n_0$ so large that for all $n \geq n_1$ and all $\omega \in I^n$

$$\sup\{|\psi_{\lambda(\omega)}' \circ \phi_\omega|^{-\delta}\} - \inf\{|\psi_{\lambda(\omega)}' \circ \phi_\omega|^{-\delta}\} < \epsilon/M.$$

Then, using (7.12), we conclude that for all $n \geq n_1$ and all $z_1, z_2 \in U_\beta$

$$\left| \sum_{|\omega|=n} \left(c_{\beta,\omega}^\delta |\psi_{\lambda(\omega)}'(\phi_\omega(z_2))|^{-\delta} - c_{\beta,\omega}^\delta |\psi_{\lambda(\omega)}'(\phi_\omega(z_1))|^{-\delta} \right) \right| \leq \epsilon$$

and therefore

$$\lim_{n\to\infty} \left| \sum_{|\omega|=n} \left(c_{\beta,\omega}^\delta \psi_{\lambda(\omega)}'(\phi_\omega(z_2))|^{-\delta} - c_{\beta,\omega}^\delta \psi_{\lambda(\omega)}'(\phi_\omega(z_1))|^{-\delta} \right) \right| = 0.$$

Combining this, (7.12) and (7.13) we conclude that there exists a constant $c_\beta \geq 0$ such that for all $z \in U_\beta$

$$\lim_{n\to\infty} \sum_{|\omega|=n} c_{\beta,\omega}^\delta |\psi_{\lambda(\omega)}'(\phi_\omega(z))|^{-\delta} = c_\beta.$$

Combining in turn this, (7.12) and (7.13) we conclude that for all $z \in U_\beta$

$$\rho(z) = c_\beta |\psi_\beta'(z)|^\delta. \quad (7.15)$$

Fix now $i \in I$, $w \in U_\beta \cap \overline{J}$, and choose $\lambda \in \Lambda$ such that $\phi_i(w) \in U_\lambda$ and a connected neighborhood $V_w \subset U_\beta$ of w such that $\phi_i(V_w) \subset U_\lambda$. Then for every $z \in V_w$

$$\begin{aligned}\tilde{D}_{\phi_i}(z) &= \rho \circ \phi_i(z) |\phi_i'(z)|^\delta \rho(z)^{-1} = c_\lambda |\psi_\lambda'(\phi_i(z))|^\delta \cdot |\phi_i'(z)|^\delta \cdot c_\beta^{-1} |\psi_\beta'(z)|^{-\delta} \\ &= c_\lambda c_\beta^{-1} \left(|\psi_\lambda'(\phi_i(z))| \cdot |\phi_i'(z)| \cdot |\psi_\beta'(z)|^{-1} \right)^\delta\end{aligned}$$

and therefore, since our system S is affine, $\tilde{D}_{\phi_i}$ is constant on V_w. Since, by Theorem 2.2, $\tilde{D}_{\phi_i}$ is real-analytic on U, we thus conclude that $\tilde{D}_{\phi_i}$ is constant on U. The proof of the implication (d2)$\Rightarrow$(a) is finished.

• (d2)$\Rightarrow$(eh). We can assume the sets U_t appearing in condition (d2) are open balls. Since $\overline{J}$ is compact, we may choose from the family $\{U_t\}$ a finite subcover $\{B_\lambda\}_{\lambda\in\Lambda}$ of $\overline{J}$. Define then for every $\lambda \in \Lambda$ the map $\gamma_\lambda : B_\lambda \to I\!\!R$ to be a continuous branch of $\arg\psi'_\lambda$ and additionally for every $i \in I$, $\arg_\lambda \phi'_i : B_\lambda \to I\!\!R$ to be a continuous branch of argument of ϕ'_i. These branches exist since B_λ is simply connected and ψ'_λ and ϕ'_i nowhere vanish. Of course all the maps γ_λ, $\lambda \in \Lambda$, are harmonic. Consider now two indices $\lambda, \lambda' \in \Lambda$ such that $B_\lambda \cap B_{\lambda'} \neq \emptyset$. Since our atlas is affine, $\psi_\lambda(z) = \psi_\lambda \circ \psi_{\lambda'}^{-1}(\psi_{\lambda'}(z)) = a(\psi_{\lambda'}(z)) + b$ for all $z \in B_\lambda \cap B_{\lambda'}$ and some $a, b \in I\!\!\!C$. We conclude that $\gamma_\lambda - \gamma_{\lambda'}$ is on $B_\lambda \cap B_{\lambda'}$ equal to $\arg(a)$ up to an integer multiple of 2π. This means that (7.7) is satisfied. Since all the contractions $\{\phi_i\}_{i\in I}$ are affine in the atlas $\psi_\lambda : B_\lambda \to I\!\!\!C$, we conclude that given $\lambda, \lambda' \in \Lambda$, $i \in I$ there exist constants $d, c \in I\!\!\!C$ such that for every $z \in \phi_i^{-1}(B_{\lambda'} \cap \phi_i(B_\lambda))$

$$\psi_{\lambda'} \circ \phi_i(z) = \psi_{\lambda'} \circ \phi_i \circ \psi_\lambda^{-1}(\psi_\lambda(z)) = d\psi_\lambda(z) + c.$$

We conclude that $\arg_\lambda \phi'_i - \gamma_\lambda + \gamma_{\lambda'} \circ \phi_i$ is equal to $\arg(d)$ up to an integer multiple of 2π on the connected set $\phi_i^{-1}(B_{\lambda'} \cap \phi_i(B_\lambda))$. This means that (7.8) is satisfied. Thus the proof of the implication (d2)$\Rightarrow$(eh) is complete.

• The implications (eh)$\Rightarrow$(er)$\Rightarrow$(ec) are obvious.

• (ec)$\Rightarrow$(d2). The general idea is here the same as in the proof of the implication (c) $\Rightarrow$ (d1). Surprisingly, we do not get directly (c)$\Rightarrow$(d1). For this we need to go via (d2)$\Rightarrow$(a)$\Rightarrow$(d1).

Let $4\delta > 0$ be a Lebesgue number of the cover $\{B_\lambda\}_{\lambda\in\Lambda}$ of $\overline{J}$. By compactness of $\overline{J}$ there exists a finite set T and points $v_t \in \overline{J}$, $t \in T$, such that the family $\{B(v_t, \delta)\}_{t\in T}$ is a cover of $\overline{J}$. Since 4δ is a Lebesgue number of the cover $\{B_\lambda\}_{\lambda\in\Lambda}$, for every $t \in T$ there exists at least one element $\lambda(t) \in \Lambda$ such that $B(v_t, 2\delta) \subset B_{\lambda(t)}$. Fix now $t_0 \in T$, $\tau \in I^\infty$, that is similarly as in the implication (c)$\Rightarrow$(d1). Then for each integer $n \geq 1$ choose $t_n \in T$ such that $\phi_{\overline{\tau|_n}}(v_{t_0}) \in B(v_{t_n}, \delta)$. Since $\phi_{\overline{\tau|_n}}$ on $B(v_{t_0}, \delta)$ shrinks distances by a factor at least $s < 1$ for $n \geq 1$, we get $\phi_{\overline{\tau|_n}}(B(v_{t_0}, \delta)) \subset B(v_{t_n}, (1+s)\delta)$. Now, for every $i \in I$ and every $\lambda \in \Lambda$ let $\arg_\lambda \phi'_i : B_\lambda \to I\!\!R$ be a continuous branch of argument of ϕ'_i. It follows from (4.5) that

$$|\arg_\lambda(t)\phi_i'(y) - \arg_\lambda(t)\phi_i'(x)| \leq L|y-x|$$

for all $t \in T$, all $i \in I$ and all $x, y \in B(v_t, \delta)$, where $L = 6/d$. Hence for all $z \in B(v_{t_0}, \delta)$

$$\begin{aligned}\sum_{n\geq 1} |\arg_{\lambda(t_{n-1})}\phi_{\tau_n}'(\phi_{\overline{\tau|_{n-1}}}(z)) - \arg_{\lambda(t_{n-1})}\phi_{\tau_n}'(\phi_{\overline{\tau|_{n-1}}}(v_{t_0}))|& \\ \leq \sum_{n\geq 1} Ls^{\alpha(n-1)}|z - v_{t_0}|^\alpha& \qquad (7.16)\\ \leq L\mathrm{diam}^\alpha(V)\frac{1}{1-s^\alpha} < \infty.&\end{aligned}$$

Iterating formula (7.8) we obtain for every $n \geq 1$ and every $z \in B(v_{t_0}, \delta)$

$$\begin{aligned}&\gamma_{\lambda(t_0)}(z) - \gamma_{\lambda_(t_0)}(v_{t_0})\\ &= \sum_{k=1}^{n} \arg_{\lambda(t_{k-1})}(\phi_{\tau_k}'(\phi_{\overline{\tau|_{k-1}}}(z)) - \arg_{\lambda(t_{k-1})}\phi_{\tau_k}'(\phi_{\overline{\tau|_{k-1}}}(v_{t_0}))\\ &+ \gamma_{\lambda(t_n)}(\phi_{\overline{\tau|_n}}(z)) - \gamma_{\lambda(t_n)}(\phi_{\overline{\tau|_n}}(v_{t_0})).\end{aligned}$$

Since for all $t \in T$, $B(v_t, (1+s)\delta) \subset B(v_t, 2\delta) \subset B_{\lambda(t)}$, all the functions $\gamma_{\lambda(t)}|_{B(v_t,(1+s)\delta)}$ are uniformly continuous. Therefore, since the set T is finite, since $\phi_{\overline{\tau|_n}}(z), \phi_{\overline{\tau|_n}}(v_{t_0}) \in B(v_{t_n}, (1+s)\delta)$ and since $|\phi_{\overline{\tau|_n}}(z) - \phi_{\overline{\tau|_n}}(v_{t_0})| \leq \delta s^n$, applying (7.16) we conclude that for all $z \in B(v_{t_0}, \delta)$

$$\begin{aligned}&\gamma_{\lambda(t_0)}(z)\\ &\qquad = \gamma_{\lambda(t_0)}(v_{t_0}) + \sum_{k=1}^{\infty} \arg_{\lambda(t_k)}(\phi_{\tau_k}'(\phi_{\overline{\tau|_{k-1}}}(z))\\ &\qquad - \arg_{\lambda(t_k)}\phi_{\tau_k}'(\phi_{\overline{\tau|_{k-1}}}(v_{t_0})).\end{aligned}$$

Thus the function $\gamma_{\lambda(t_0)}|_{B(v_{t_0},\delta)}$ as the sum of an absolutely uniformly convergent series of harmonic functions is harmonic. So, all the functions $\gamma_{\lambda(t)} : B(v_t, \delta) \to I\!R$, $t \in T$, are harmonic.

Remark that in the case when S is not 1-dimensional the equation (ec) assumed only on $\overline{J}$ (analogously to (c)) would be sufficient for γ_λ extended by the formula above to satisfy (ec) on V; in particular (eh) would be proved.

However, if S is one-dimensional the existence of γ_λ satisfying (ec) on $\overline{J}$ is always true. Just take for γ an argument of the direction tangent to M, the union of a finite family of real-analytic curves containing $\overline{J}$.

Now, for every $t \in T$ by $l_t : B(v_t, \delta) \to \mathbb{R}$ denote the harmonic conjugate to $\gamma_{\lambda(t)}$. Thus the function $G_t = \exp(l_t + i\gamma_{\lambda(t)}) : B(v_t, \delta) \to \mathbb{C}$ is holomorphic. Denote by $\psi_t : B(v_t, \delta) \to \mathbb{C}$ a primitive of G_t. Fix $w \in \overline{J}$ and choose $t \in T$ such that $w \in B(v_t, \delta)$. Since $\psi_t'(w) = \exp(l_t(w) + i\gamma_{\lambda(t)}(w)) \neq 0$, there exists a disk $U_w \subset B(v_t, \delta)$ such that $\psi_t|_{U_w}$ is injective. Applying Theorem 7.3.3 as before, we may assume the disks U_w to be so small that all the sets $\phi_i(U_w)$ are convex. We claim that the family $\{\psi_w : U_w \to \mathbb{C}\}_{w \in \overline{J}}$ forms an affine atlas for the iterated function system S. Indeed, fix $w, v \in \overline{J}$ and consider $t, t' \in T$ such that $U_w \subset B(v_t, \delta) \subset B_{\lambda(t)}$ and $U_v \subset B(v_{t'}, \delta) \subset B_{\lambda(t')}$. Then for every $z \in U_w \cap U_v$ we get

$$\begin{aligned}(\psi_w \circ \psi_v^{-1})'(\psi_v(z)) &= \psi_w'(z)(\psi_v'(z))^{-1} = G_{\lambda(t)}(z) G_{\lambda(t')}^{-1}(z) \\ &= \exp\big(l_t(z) + i\gamma_{\lambda(t)}(z) - l_{t'}(z) - i\gamma_{\lambda(t')}(z)\big) \\ &= \exp\big(i(\gamma_{\lambda(t)}(z) - \gamma_{\lambda(t')}(z)\big) \exp\big(l_t(z) - l_{t'}(z)\big).\end{aligned}$$

Since by (7.7) $\gamma_{\lambda(t)} - \gamma_{\lambda(t')}$ is constant on $z \in U_w \cap U_v \subset U_{\lambda(t)} \cap U_{\lambda(t')}$ and since l_t and $l_{t'}$ differ on $U_{\lambda(t)} \cap U_{\lambda(t')}$ by an additive constant as harmonic conjugates to harmonic functions $\gamma_{\lambda(t)}$ and $\gamma_{\lambda(t')}$ respectively, we conclude that $(\psi_w \circ \psi_v^{-1})'$ is constant on $\psi_v(U_w \cap U_v)$.

Now fix $w, v \in \overline{J}$, $i \in I$, and write $C = \phi_i^{-1}(\phi_i(U_w) \cap U_v))$. Since $\phi_i(U_w) \cap U_v))$ is a convex set and therefore connected, its continuous image C is also connected. Then there are $t, t' \in T$ such that $U_w \subset B(v_t, \delta) \subset B_{\lambda(t)}$, $U_v \subset B(v_{t'}, \delta) \subset B_{\lambda(t')}$ and C is contained in a connected component of $B_{\lambda(t)} \cap \phi_i^{-1}(B_{\lambda'(t)})$. Using the chain rule we then get for all $z \in C$

$$\begin{aligned}(\psi_v \circ \phi_i \circ \psi_w^{-1})'(\psi_v(z)) &= \psi_v'(\phi_i(z))\phi_i'(z)(\psi_w'(z))^{-1} \\ &= G_{t'}(\phi_i(z))\phi_i'(z) G_t^{-1}(z) \\ &= \exp\big(i(\gamma_{\lambda(t')}(\phi_i(z))) + l_{t'}(\phi_i(z)) + \log|\phi_i'(z)| \\ &\quad + i\arg_{\lambda(t)}\phi_i'(z) - i\gamma_{\lambda(t)}(z) - l_t(z)\big) \\ &= \exp\big(l_{t'}(\phi_i(z)) + \log|\phi_i'(z)| - l_t(z)\big) \\ &\quad \times \exp\big(i(\arg_{\lambda(t)}\phi_i'(z) - \gamma_{\lambda(t)}(z) + \gamma_{\lambda(t')}(\phi_i(z))\big).\end{aligned}$$

Hence, using (7.8) we conclude that the derivative $(\psi_v \circ \phi_i \circ \psi_w^{-1})'$ has a constant argument on $\psi_v(C)$ and consequently $(\psi_v \circ \phi_i \circ \psi_w^{-1})'$ is constant on $\psi_v(C)$. The proof of the implication (ec)$\Rightarrow$(d2) is complete.

- The implication (a)$\Rightarrow$(f) is obvious.

• (f)⇒(er). Suppose first that the system S is 1-dimensional. Then the condition $\nabla\tilde{D}_{\phi_i} \equiv 0$ on $\overline{J}$ is similar (formally weaker) to $\tilde{D}_{\phi_i}$ constant in (a). We prove (er) similarly, via (c)⇒(d1)⇒(eh).

Assume now that S is not one-dimensional. Suppose that $\nabla\tilde{D}_{\phi_i} = 0$ on $\overline{J}$ for all $i \in I$. Since S is not one-dimensional, it implies that $\nabla D_{\phi_i} = 0$ on U for all $i \in I$. Thus $\tilde{D}_{\phi_i} = 0$ is constant on U for all $i \in I$, since U is connected. So, item (a) is proved in this case and therefore, in view of what we have already proved, so is (er).

So, we may assume that there exists $j \in I$ and $w \in \overline{J}$ such that $\nabla D_{\phi_j}(w) \neq 0$. By continuity of the function $\nabla\tilde{D}_{\phi_j}$ there thus exists a neighborhood $W \subset V$ of $w \in \mathbb{C}$ on which $\nabla\tilde{D}_{\phi_j}$ nowhere vanishes. Let us consider on W the *line field* l orthogonal to $\nabla\tilde{D}_{\phi_j}$. By the definition of the limit set J, for every $z \in \overline{J}$ there exists $\tau \in I^*$ such that $\phi_\tau(z) \in \overline{J} \cap W$. Then define

$$l(z) = (\phi_\tau^{-1})'_{\phi_\tau(z)}(l(\phi_\tau(z))), \tag{7.17}$$

where, temporarily changing notation, $(\phi_\tau^{-1})'_{\phi_\tau(z)}$ denotes the derivative of the map ϕ_τ^{-1} evaluated at the point $\phi_\tau(z)$ and the display above expresses its action on a line element. We want to show first that in this manner we define a line field on $\overline{J}$. So, we need to show that if $\phi_\tau(z), \phi_\eta(z) \in \overline{J} \cap W$, then

$$(\phi_\tau^{-1})'_{\phi_\tau(z)}(l(\phi_\tau(z))) = (\phi_\eta^{-1})'_{\phi_\eta(z)}(l(\phi_\eta(z))). \tag{7.18}$$

Suppose on the contrary that (7.18) fails with some z, τ, η as required above. Then there exists a point $x \in W \cap \overline{J}$ and $\gamma \in I^*$ (in fact for every $x \in W$ there exists γ) such that $\phi_\gamma(x)$ is so close to z that

$$(\phi_\tau^{-1})'_{\phi_\tau(\phi_\gamma(x))}(l(\phi_\tau(\phi_\gamma(x)))) \neq (\phi_\eta^{-1})'_{\phi_\eta(\phi_\gamma(x))}(l(\phi_\eta(\phi_\gamma(x)))).$$

Hence

$$(\phi_{\tau\gamma}^{-1})'_{\phi_{\tau\gamma}(x)} l(\phi_{\tau\gamma}(x)) \neq (\phi_{\eta\gamma}^{-1})'_{\phi_{\eta\gamma}(x)} l(\phi_{\eta\gamma}(x)).$$

So, either

$$(\phi_{\tau\gamma}^{-1})'_{\phi_{\tau\gamma}(x)} l(\phi_{\tau\gamma}(x)) \neq l(x)$$

or

$$(\phi_{\eta\gamma}^{-1})'_{\phi_{\eta\gamma}(x)} l(\phi_{\eta\gamma}(x)) \neq l(x).$$

Suppose for example the first incompatibility of l's holds. Then

$$\det(\nabla\tilde{D}_{\phi_j} \circ \phi_{\tau\gamma}(x), \nabla\tilde{D}_{\phi_j}(x)) \neq 0$$

contrary to our assumption. Thus the line field l is well defined on $\overline{J}$ and it immediately follows from the method this field is constructed that it is invariant with respect to all the contractions ϕ_i, $i \in I$. Notice that formula (7.17) defines an invariant line field on V. We can use any $\tau \in I^*$ such that $\phi_\tau(V) \subset W$. The resulting l does not depend on τ because for any other such η (7.18) holds for $z \in \overline{J}$, so it holds on the whole of V. Otherwise the system would be one-dimensional because l is real-analytic so the equation holds on a real-analytic set. The argument $\arg l$ is of course defined up to integer multiplicity of π. Using again Theorem 7.3.3, one can find $\{B_\lambda\}$, a finite cover of $\overline{J}$ by disks contained in V, small enough that all the images $\phi_i(B_\lambda)$, $i \in I$, are convex. Then all the intersections $B_\lambda \cap B_{\lambda'}$ and $B_\lambda \cap \phi_i(B_{\lambda'})$ are connected. Define γ_λ as an arbitrary branch of $\arg l$ on B_λ. Then (7.7) and (7.8) follow from the invariance of l under S, with constants $c(\lambda, \lambda')$ and $c(\lambda, \lambda', i)$ that are multiples of π. Thus (er) is proved.

• (d2)⇒(d3). Let $\{\psi_t : U_t \to \mathbb{C}\}_{t \in T}$ be the atlas produced by $(d2)$. Fix $x \in J_S$, choose $s \in T$ such that $x \in U_s$ and then $\rho \in I^*$ such that $x \in \phi_\rho(J) \subset \phi_\rho(W) \subset U_s$. Consider now the iterated function system

$$S_\rho = \{\psi_s \circ \phi_\rho \circ \phi_i \circ \phi_\rho^{-1} \circ \psi_s^{-1}\}_{i \in I},$$

where the role of X is played by $\psi_s(\phi_\rho(X))$ and the role of W is played by $\psi_s(\phi_\rho(W))$. It follows from (d2) that each map $\psi_s \circ \phi_\rho \circ \phi_i \circ \phi_\rho^{-1} \circ \psi_s^{-1}$, $i \in I$, is affine on each sufficiently small neighborhood of each point of $\psi_s(\phi_\rho(\overline{J}))$. Hence, as holomorphic, this map must be affine on the whole connected domain $\psi_s(\phi_\rho(W))$. □

Let us prove now a technical fact about 1-dimensional systems.

Proposition 7.3.7 *Suppose that $F = \{f_i : X \to X\}_{i \in I}$ and $G = \{g_i : Y \to Y\}_{i \in I}$ are two not essentially affine topologically conjugate systems. Suppose also that the measures m_G and $m_F \circ h^{-1}$ are equivalent. If one of these systems is 1-dimensional, then so is the other one.*

Proof. Suppose on the contrary that G is not 1-dimensional. Then it follows from Theorem 7.3.6 that there exist $y \in J_G$, $j \in I$, $\omega \in I^*$ and a neighborhood $W_2 \subset \mathbb{C}$ of y such that the map

$$\mathcal{G} = (\tilde{D}_{g_j} \circ \gamma_\omega, \tilde{D}_{g_j})$$

is invertible on W_2. Since the measures m_G and $m_F \circ h^{-1}$ are equivalent, after an appropriate normalization $\mu_F = \mu_G \circ h$ meaning that $D_h = \frac{d\mu_G \circ h}{d\mu_F} = 1$. Since $h \circ f_\tau = g_\tau \circ h$ for all $\tau \in I^*$ and since $D_h = 1$,

$$\mathcal{G} \circ h = \mathcal{F}$$

on J, where $\mathcal{F} = (\tilde{D}_{f_j} \circ \gamma_\omega, \tilde{D}_{f_j})$. Write $x = h^{-1}(y)$. Then $h = \mathcal{G}^{-1} \circ \mathcal{F}$ on $W_1 \cap J_F$ for some open neighborhood W_1 of x in $\mathbb{C}$ such that $\mathcal{F}(W_1) \subset \mathcal{G}(W_2)$. Since $\mathcal{F}, \mathcal{G}^{-1}$ are real-analytic, the image $\mathcal{G}^{-1} \circ \mathcal{F}(W_1 \cap M_F)$ for a small enough such W_1 is a real-analytic curve and $\mathcal{G}^{-1} \circ \mathcal{F}(W_1 \cap M_F) \cap J_G$ contains an open neighborhood of y in J_G. Using now Lemma 7.3.4 we conclude that G is 1-dimensional. □

The main result of this section is contained in the following.

Theorem 7.3.8 *If two conformal regular iterated function systems $F = \{f_i : X \to X\}_{i \in I}$ and $G = \{g_i : Y \to Y\}_{i \in I}$ satisfying the open set condition are not essentially affine and are conjugate by a homeomorphism $h : J_F \to J_G$, then the following conditions are equivalent.*

(a) The conjugacy between the systems F and G extends in a conformal fashion to an open neighborhood of $\overline{J_F}$.

(b) The conjugacy between the systems F and G extends in a real-analytic fashion to an open neighborhood of $\overline{J_F}$.

(c) The conjugacy between the systems F and G is bi-Lipschitz continuous.

(d) $|g'_\omega(y_\omega)| = |f'_\omega(x_\omega)|$ for all $\omega \in I^*$, where x_ω and y_ω are the only fixed points of $f_\omega : X \to X$ and $g_\omega : Y \to Y$ respectively.

(e) $\exists S \geq 1 \; \forall \omega \in I^*$

$$S^{-1} \leq \frac{\operatorname{diam}(g_\omega(Y))}{\operatorname{diam}(f_\omega(X))} \leq S.$$

(f) $\exists E \geq 1 \; \forall \omega \in I^*$

$$E^{-1} \leq \frac{||g'_\omega||}{||f'_\omega||} \leq E.$$

(g) $\mathrm{HD}(J_G) = \mathrm{HD}(J_F)$ and the measures m_G and $m_F \circ h^{-1}$ are equivalent.

(h) The measures m_G and $m_F \circ h^{-1}$ are equivalent.

Proof. The implications (a)⇒(b) and (b)⇒(c) are obvious. That (c)⇒(d) results from the fact that (c) implies condition (1) of Theorem 7.1.1, which in view of that theorem is equivalent with condition

(2) of Theorem 7.1.1, which finally is the same as condition (d) of Theorem 7.3.8. The implications (d)⇒(e)⇒(f)⇒(g) have been proved in Theorem 7.1.1. The implication (g)⇒(h) is again obvious. We are left to prove that (h)⇒(a). We shall first prove that (h)⇒(b). So, suppose that (h) holds. Then, after an appropriate normalization $\mu_F = \mu_G \circ h$ meaning that $D_h = \frac{d\mu_G \circ h}{d\mu_F} = 1$. If F is one-dimensional, then by Proposition 7.3.7, so is G and the implication (h)⇒(b) follows from Theorem 7.2.4 and the fact that each real-analytic map between real-analytic curves extends to a (complex) analytic map defined on some of their neighborhoods in $\mathbb{C}$. Hence, we may assume that neither system F or G is 1-dimensional. Therefore, since G is not essentially affine, there exist $y \in J_G$, $j \in I$, $\omega \in I^*$ and a neighborhood $W_2 \subset \mathbb{C}$ of y such that the map

$$\mathcal{G} = (\tilde{D}_{g_j} \circ g_\omega, \tilde{D}_{g_j})$$

is invertible on W_2. Since $h \circ f_\tau = g_\tau \circ h$ for all $\tau \in I^*$ and since $D_h = 1$,

$$\mathcal{G} \circ h = \mathcal{F}$$

on $W_1 \cap J_f$, where $\mathcal{F} = (\tilde{D}_{f_j} \circ g_\omega, \tilde{D}_{f_j})$ and W_1 is a neighborhood of $x = h^{-1}(y) \subset \mathbb{C}$. Since $\mathcal{G}$ is invertible on W_2, $\mathcal{G}(y) = \mathcal{F}(x)$ and $\mathcal{F}$ is continuous, we may assume that $\mathcal{F}(W_1) \subset \mathcal{G}(W_2)$. Hence $\mathcal{G}^{-1} \circ \mathcal{F}$ is well defined on W_1 and $\mathcal{G}^{-1} \circ \mathcal{F}|_{W_1 \cap J_F} = h$. Consider now $\omega \in I^*$ such that $f_\omega(J_F) \subset W_1$. Since

$$\mathcal{G}^{-1} \circ \mathcal{F}(f_\omega(J_F)) = h \circ f_\omega(J_F) = g_\omega \circ h(J_F) = g_\omega(J_G) \subset g_\omega(V_G),$$

since $g_\omega(W_2)$ is open, and since f_ω and $\mathcal{G}^{-1} \circ \mathcal{F}$ are continuous, there exists an open neighborhood $V_1 \subset V_F$ of $\overline{J_F}$ such that $f_\omega(V_1) \subset W_1$ and $\mathcal{G}^{-1} \circ \mathcal{F}(f_\omega(V_1)) \subset g_\omega(W_2)$. Hence, the map

$$g_\omega^{-1} \circ (\mathcal{G}^{-1} \circ \mathcal{F}) \circ f_\omega : V_1 \to \mathbb{C}$$

is well defined and by Corollary 6.1.5 is real-analytic, and $g_\omega^{-1} \circ (\mathcal{G}^{-1} \circ \mathcal{F}) \circ f_\omega|_{J_F} = h$. Thus, the property (b) is proved.

(b)⇒(a). Let H be this real-analytic extension of h on some neighborhood W_F of J_F in $\mathbb{C}$. If F is one-dimensional and D_F is the family of real-analytic curves coming from the one-dimensionality of F, then, in view of Proposition 7.3.7, G is also one-dimensional and the real-analytic map $H|_{M_F} : D_F \to D_G$ has a (complex) analytic extension to some neighborhood of D_F in $\mathbb{C}$. So, we may assume that F is not one-dimensional. We may also assume W_F to be so small that H' is a linear isomorphism at every point of W_F. Define the function $\psi : W_F \to I\!R$ by the formula

$$\psi(z) = \frac{||H'(z)||}{||(H'(z))^{-1}||}.$$

Suppose that $\psi(\xi) = 1$ for some point $\xi \in W_F$. Since for every $\omega \in I^*$

$$\psi(f_\omega(\xi)) = \frac{||H'(f_\omega(\xi))||}{||(H'(f_\omega(\xi)))^{-1}||} = \frac{||g'_\omega(H(\xi)) \cdot H'(\xi) \cdot (f'_\omega(\xi))^{-1}||}{||\big(g'_\omega(H(\xi)) \cdot H'(\xi) \cdot (f'_\omega(\xi))^{-1}\big)^{-1}||}$$
$$= \frac{||H'(\xi)||}{||(H'(\xi))^{-1}||} = \psi(\xi)$$

and since $\overline{\{f_\omega(\xi) : \omega \in I^*\}} \supset \overline{J}_F$, we conclude that $\psi = 1$ identically on $\overline{J}_F$. Since ψ is real-analytic and since F is not 1-dimensional, using Lemma 7.3.4, we conclude that $\psi = 1$ on an open neighborhood of $\overline{J}_F$. But this means that H is conformal. So, we may assume that $\psi(z) \neq 1$ for every $z \in W_F$. Define then the field $\{E_z\}_{z \in W_F}$ on W_F as follows.

$$E_z = \left\{ w \in C\!\!\!\!I : \frac{|H'(z)w|}{|w|} = |H'(z)| \right\} \cup \{0\}.$$

For every $z \in W_F$, the set E_z is a linear subspace of $C\!\!\!\!I$ of dimension ≥ 1. Its codimension is ≥ 1 since $\psi(z) \neq 1$. In conclusion $\dim(E_z) = 1$ for all $z \in W_F$. Obviously E_z depends continuously on z. Since the maps $f_i : C\!\!\!\!I \to C\!\!\!\!I$, $i \in$, are conformal, $f'_i(z)(E_z) = E_{f_i(z)}$ for all $i \in I$ and putting locally $\gamma_\lambda(z) = \arg(E_z)$, it therefore follows from Theorem 7.3.6(ec) that the system F is essentially affine. This contradiction finishes the proof. □

7.4 Rigidity in dimension $d \geq 3$

In this section we strengthen the rigidity results of the previous section, cf. [U5]. We begin with the following.

Lemma 7.4.1 *Suppose that $\phi : I\!R^d \to I\!R^d$, $d \geq 3$, is a conformal diffeomorphism that has an attracting fixed point a ($\phi(a) = a$, $|\phi'(a)| < 1$). If M is an open connected C^1-submanifold of $I\!R^d$ such that $\phi(M) \subset M$ and $a \in M$, then M is either a subset of a ϕ-invariant affine subspace of the same dimension as M, or a subset of a ϕ-invariant geometric sphere of the same dimension as M.*

Proof. Since a is an attracting fixed point of ϕ, there exists a radius $r > 0$ so small that $\phi^{-1}\big(\overline{R}^d \setminus B(a, r)\big) \subset \overline{R}^d \setminus B(a, r)$, where $\overline{R}^d$ is the Alexandrov compactification of $I\!R^d$ obtained by adding the point at infinity. Since $\overline{R}^d \setminus B(a,r)$ is a closed topological ball, in view of Brouwer's fixed point theorem there exists a fixed point b of ϕ^{-1} in $\overline{R}^d \setminus B(a,r)$. Hence b is also a fixed point of ϕ and $b \neq a$. Then the map

$$\psi = i_{b,1} \circ \phi \circ i_{b,1}^{-1}$$

($i_{b,1}$ equals identity if $b = \infty$) fixes ∞, which means that this map is affine, and $w = i_{b,1}(a)$ is an attracting fixed point of ψ. In addition $\psi(\tilde{M}) \subset \tilde{M}$, where $\tilde{M} = i_{b,1}(M)$, $w \in \tilde{M}$, and $\psi : I\!R^d \to I\!R^d$, as an affine map, can be written in the form $\lambda A + c$, where $\lambda > 0$ and A is an orthogonal matrix. Since $\psi(\tilde{M}) \subset \tilde{M}$, and since ψ is a diffeomorphism, $\psi'(z)(T_z\tilde{M}) = T_{\psi(z)}\tilde{M}$. In particular $\psi'(w)E = E$, where $E = T_w\tilde{M}$. Without looss of generality we may assume that $\tilde{M}$ is contained in the basin of immediate attraction to w. We shall show that

$$T_z\tilde{M} = E$$

for every $z \in \tilde{M}$. And indeed, take an arbitrary point point $z \in \tilde{M}$. Since $\psi'(x) = \lambda A$ for all $x \in I\!R^d$ and since λA is conformal, we get for all $n \geq 0$ that

$$\begin{aligned}\angle(T_z\tilde{M}, E) &= \angle(A^n(T_z\tilde{M}), A^n E) \\ &= \angle((\psi^n)'(z)T_z\tilde{M}), E) = \angle(T_{\psi^n(z)}\tilde{M}, E),\end{aligned}$$

where $\angle$ denotes the *angle between linear hyperspaces.* Since $\lim_{n\to\infty} T_{\psi^n(z)}\tilde{M} = T_w\tilde{M} = E$, we conclude that $\angle(T_z\tilde{M}, E) = 0$, or equivalently $T_z\tilde{M} = E$. Since the only integral manifolds of a constant field of linear subspaces are affine subspaces, we conclude that $\tilde{M}$ is contained in an affine subspace. Since $\tilde{M}$ is an open subset of it, this affine subspace is ϕ-invariant. Since $M = i_{b,1}(\tilde{M})$, we are done. □

We call the system $S = \{\phi_i\}_{i\in I}$ at most q-dimensional, $1 \leq q \leq d$, if there exists M_S, either a q-dimensional linear subspace of $I\!R^d$ or a q-dimensional geometric sphere contained in $I\!R^d$, such that $\overline{J} \subset M_S$ and $\phi_i(M_S) = M_S$ for all $i \in I$. We call the system $S = \{\phi_i\}_{i\in I}$ q-dimensional if q is the minimal number with this property.

Lemma 7.4.2 *If a non-empty open subset of $\overline{J}$ is contained in a q-dimensional real-analytic submanifold, then the system S is at most q-dimensional.*

Proof. The assumptions of the lemma state that there exists a point $x \in \overline{J}$, an open ball $B(x)$ centered at x and M, a p-dimensional open connected real-analytic submanifold M containing $\overline{J} \cap B(x)$, where $1 \leq p \leq q$ is the minimal integer with this property. Fix now an arbitrary auxiliary point $z \in \overline{J}$. Since $x \in \overline{J}$, there exists $\omega \in I^*$ such that $\phi_\omega(z) \in \overline{J} \cap B(x)$; moreover $\phi_\omega(V) \subset B(x)$. Then the set $\phi_\omega(V) \cap M$ contains $\phi_\omega(V) \cap \overline{J}$, an open neighborhood of $\phi_\omega(z)$ in $\overline{J}$, and consists of countably many connected p-dimensional real-analytic submanifolds. Taking the length of ω large enough we may assume that this countable family is a single manifold. Then $N = \phi_\omega^{-1}(\phi_\omega(V) \cap M)$ is a connected p-dimensional real-analytic submanifold (there are no branching points since ϕ_ω^{-1} is 1-to-1) containing $\overline{J}$ and contained in V. For the purpose of this proof it is not important whether N is in fact independent of ω or not. Fix an arbitrary $i \in I$. Let $x_i \in J$ be the only attracting fixed point of ϕ. Since the connected component $C_{i,n}$ of $\phi_i^n(N) \cap N$ containing x_i is a real-analytic manifold of dimension $\leq \dim(N) \leq \dim(M) = p$, it follows from the definition of p (its minimality) that $\dim(C_{i,n}) = p$. Hence, the two connected p-dimensional real-analytic open manifolds $\phi_i^n(N)$ and N are extensions of the same p-dimensional real-analytic manifold $C_{i,n}$. Therefore, since $\lim_{n\to\infty} \operatorname{diam}(\phi_i^n(N)) = 0$ and since $x_i \in \phi_i^n(N)$, we conclude that for all $n \geq 1$ so large that $\operatorname{diam}(\phi_i^n(N)) < \operatorname{dist}(x_i, \partial N)$, we have $\phi_i^n(N) \subset N$. Hence, in view of Lemma 7.4.2 applied with $\phi = \phi_i^n$, we gain that N is an open subset of a p-dimensional set M_S, either an affine subset or a geometric sphere contained in $I\!R^d$ invariant under ϕ_i^n. Now, for every $j \in I$, $\phi_j(M_S) \cap M_S \neq \emptyset$ since $J \subset M_S$. Since in addition $\phi_j(M_S) \cap M_S$ is either an affine subset or a geometric sphere contained in $I\!R^d$, we conclude from the minimality of p that $\phi_J(M_S) = M_S$. $\square$

We define essentially affine systems similarly as in the plane case.

Definition 7.4.3 *We say that the system S is essentially affine if S is conjugate by a conformal homeomorphism with a system consisting only of conformal affine contractions (i.e. of the form $\lambda A + b$).*

The first goal of this section is to prove the following characterizations of essential affinity.

Theorem 7.4.4 *Suppose that the system $S = \{\phi_i\}_{i\in I}$ is regular and denote the corresponding conformal measure by m. Then the following conditions are equivalent.*

(a) For each $i \in I$ the extended Jacobian $\tilde{D}_{\phi_i} : U \to I\!R$ is constant, where U is the neighborhood of X produced in Corollary 2.3.

(b) There exist a continuous function $u : X \to I\!R$ and constants $c_i \in I\!R$, $i \in I$, such that

$$\log|\phi_i'| = u - u\circ\phi_i + c_i$$

for all $i \in I$.

(c) There exist a continuous function $u : \overline{J} \to I\!R$ and constants $c_i \in I\!R$, $i \in I$, such that

$$\log|\phi_i'| = u - u\circ\phi_i + c_i$$

for all $i \in I$.

(d) The system S is essentially affine.

(er) There exist a real-analytic function $\gamma : V \to \mathrm{Lis}(d)$ such that

$$\gamma\circ\phi_i \cdot \frac{\phi_i'}{|\phi_i'|}\cdot\gamma^{-1} = k_i \in LC(d)$$

for every $i \in I$, where $\mathrm{Lis}(d)$ is the group of all linear isometries on $I\!R^d$ and $LC(d)$ is the group of all linear conformal (of the form λA) homeomorphisms of $I\!R^d$. The composition of linear maps we denote here and in the sequel either by $\cdot$ or we put no sign.

(ec) The same as (er) but γ is required to be continuous only.

(g) If S is not q-dimensional, then the vectors

$$(\nabla\tilde{D}\phi_i \circ \phi_{\omega^{(j)}(z)})_{j=1}^q$$

are linearly dependent for all $z \in \overline{J}$, all $i \in I$ and all sequnces $(\omega^{(j)})_{j=1}^q \in (I^*)^q$.

(f) If S is q-dimensional, $1 \le q \le d$, then either S is essentially affine or there exists a field of linear subspaces in TM_S of dimension and co–dimension greater than or equal to 1 defined on a neighborhood of $\overline{J}$ in M_S and invariant under the action of derivatives of all maps ϕ_i, $i \in I$.

Proof. We shall prove the following implications: (a)⇒(b)⇒(c)⇒(d)⇒(a), (d)⇒(er)⇒(ec)⇒(d) and (a)⇒(g)⇒(f)⇒(d).

- (a)⇒(b). Since for every $i \in I$, $\tilde{D}_{\phi_i} = (\rho\circ\phi_i)\cdot|\phi_i'|^\delta\cdot\rho^{-1}$, we have

$$\log(|\tilde{D}_{\phi_i}|) = \log(|\rho|\circ\phi_i) + \delta\log|\phi_i'| - \log|\rho|.$$

Thus to finish the proof of the implication (a)⇒(b) it suffices to set $c_i = \frac{1}{\delta}\log(\tilde{D}_{\phi_i})$ and $u = \frac{1}{\delta}\log|\rho|$.

• The implication (b)⇒(c) is obvious.

(c)⇒(d). If all the maps ϕ_i, $i \in I$, are affine , there is nothing to prove. So, we may assume that there exists $j \in I$ such that the map ϕ_j is not affine. For every $n \geq 1$ let $a^{(n)}$ be the inversion center of ϕ_j^n. We shall prove that the sequence $\{a^{(n)}\}_{n=1}^{\infty}$ does not converge to ∞. Indeed, suppose on the contrary that this sequence converges to ∞. Since $a^{(n)} = \phi_{j^n}^{-1}(\infty) = (\phi_j^{-1})^n(\infty)$, we therefore get

$$\begin{aligned}\infty = \lim_{n\to\infty} a^{(n-1)} &= \lim_{n\to\infty} \phi_j((\phi_j^{-1})^n) \\ &= \phi_j(\lim_{n\to\infty}(\phi_j^{-1})^n) = \phi_j(\lim_{n\to\infty} a^{(n)}) = \phi_j(\infty)\end{aligned}$$

which means that ϕ_j is affine. This contradiction shows that there exists a subsequence $\{k_n\}_{n=1}^{\infty}$ such that $a^{(k_n)} \to a$ for some $a \in I\!R^d$. Fix $v \in J$, the unique fixed point of $\phi_j : V \to V$. Iterating equation (c) n times, we get for every $z \in \overline{J}$ that

$$\begin{aligned}u(z) - u(v) &= \log|\phi'_{j^n}(z)| - \log|\phi'_{j^n}(v)| + u(\phi_{j^n}(z)) - u(\phi_{j^n}(v)) \\ &= -2\log||z - a^{(n)}|| + 2\log||v - a^{(n)}|| + u(\phi_{j^n}(z)) - u(v).\end{aligned}$$

Since $\phi_{j^n}(z)$ converges to v and since the function u is continuous, passing to the limit along the subsequence $\{k_n\}_{n=1}^{\infty}$, we get $u(z) = u(v) - 2\log||z-a|| + 2\log||v-a||$. Define the conformal map $G : I\!R^d \to I\!R^d$ by setting

$$G(z) = \mathrm{e}^{u(v)}||v-a||^2 i_{a,1}(z).$$

Then $\log|G'(z)| = u(v) + 2\log||v-a|| - 2\log||z-a|| = u(z)$. Therefore, using (c) again, we get for every $i \in I$ that $\log|\phi'_i| = \log|G'| - \log|G' \circ \phi_i| + c_i$, or equivalently that

$$|(G \circ \phi_i \circ G^{-1})'(z)| = \mathrm{e}^{c_i} \tag{7.19}$$

for all $z \in G(\overline{J})$. Suppose that $G \circ \phi_i \circ G^{-1}$ is not affine and let w and $\lambda > 0$ denote respectively its inversion center and the scalar coefficient. Then (7.19) takes the form $\lambda||z-w||^{-2} = \mathrm{e}^{c_i}$ on $G(\overline{J})$ or

$$||z-w||^2 = \lambda \mathrm{e}^{-c_i} \text{ on } G(\overline{J}).$$

So, $G(\overline{J})$ is contained in the sphere $S(w, \sqrt{\lambda \mathrm{e}^{-c_i}})$ centered at w and of radius $\sqrt{\lambda \mathrm{e}^{-c_i}}$. Since for every $n \geq 0$, $G \circ \phi_i^n \circ G^{-1}(G(\overline{J})) = G \circ \phi_i^n(\overline{J}) \subset G(\overline{J})$, we conclude that all the descending sets $G \circ \phi_i^n \circ G^{-1}(G(\overline{J}))$ are contained in the sphere $S(w, \sqrt{\lambda e^{-c_i}})$. Let $H \subset S(w, \sqrt{\lambda e^{-c_i}})$ be a minimal sphere (in the sense of inclusion) containing at least one of the sets $G \circ \phi_i^n \circ G^{-1}(G(\overline{J}))$, $n \geq 0$. Thus there exists $k \geq 0$ such that $G \circ \phi_i^k \circ G^{-1}(G(\overline{J})) \subset H$. Then

$$G \circ \phi_i^{k+1} \circ G^{-1}(G(\overline{J})) \subset H \cap (G \circ \phi_i \circ G^{-1}(H)). \tag{7.20}$$

Since $G \circ \phi_i \circ G^{-1}(H)$ is either a sphere or an affine subspace of $\mathbb{R}^d$ and since $G \circ \phi_i^{k+1} \circ G^{-1}(G(\overline{J}))$ contains at least three points (is uncountable in fact), the intersection $H \cap (G \circ \phi_i \circ G^{-1}(H))$ is a sphere (at least 1-dimensional) again. Therefore by the minimality of H and by (7.20) we conclude that $H \cap (G \circ \phi_i \circ G^{-1}(H)) \supset H$, which means that $G \circ \phi_i \circ G^{-1}(H) \supset H$. Therefore, since $\dim(G \circ \phi_i \circ G^{-1}(H)) = \dim(H)$, we conclude that

$$G \circ \phi_i \circ G^{-1}(H) = H. \tag{7.21}$$

Let x_i be the unique fixed point of the map $\phi_i : V \to V$. Since $G \circ \phi_i^n \circ G^{-1}(z) \to G(x_i)$ uniformly on $G(V) \subset G(\overline{J})$, it follows from (7.19) that $c_i > 0$. So, for every $z \in S(w, \sqrt{\lambda e^{-c_i}}) \supset H$,

$$|(G \circ \phi_i \circ G^{-1})'(z)| = \lambda_i ||z - w||^{-2} = \lambda_i \lambda_i^{-1} e^{c_i} = e^{c_i}.$$

This implies that $G \circ \phi_i \circ G^{-1}$ is a uniform contraction on H and therefore $G \circ \phi_i^n \circ G^{-1}(z) \to G(x_i)$ uniformly on H. This however contradicts (7.21) and finishes the proof of the implication (c)⇒(d).

• (d)⇒(a). Let $G : \mathbb{R}^d \to \mathbb{R}^d$ be a conformal homeomorphism providing conjugacy of S with a system consisting only of conformal affine contractions. Then for every $i \in I$

$$g_i = |(G \circ \phi_i \circ G^{-1})'(z)|$$

is a number independent of $z \in G(V)$. By the chain rule we have for every $z \in V$

$$\mathcal{L}^n(\mathbb{1})(z) = \sum_{|\omega|=n} |\phi_\omega(z)'|^h = \sum_{|\omega|=n} |G'(z)|^h |G'(\phi_\omega(z))|^{-\delta} \prod_{i=1}^n g_{\omega_i}^h.$$

Fix now $j \in I$. Then for every $n \geq 1$ and all $z \in V$ we get

$$\frac{\mathcal{L}^n(1\!\!1)(\phi_j(z))}{\mathcal{L}^n(1\!\!1)(z)}|\phi_j'(z)|^\delta = \frac{\sum_{|\omega|=n}|G'(\phi_j(z))|^\delta|G'(\phi_\omega(\phi_j(z)))|^{-\delta}\prod_{i=1}^n g_{\omega_i}^\delta}{\sum_{|\omega|=n}|G'(z)|^\delta|G'(\phi_\omega(z))|^{-\delta}\prod_{i=1}^n g_{\omega_i}^\delta}$$
$$\times\, g_j^\delta|G'(z)|^\delta|G'(\phi_j(z))|^{-\delta}$$
$$= \frac{\sum_{|\omega|=n}|G'(\phi_\omega(z))|^\delta}{\sum_{|\omega|=n}|G'(\phi_\omega(\phi_j(z))|^\delta}g_j^\delta.$$

Since $||\phi_\omega(x) - x_\omega|| \leq \text{const}\, s^{|\omega|}$, where x_ω is the only fixed point of $\phi_\omega : V \to V$, we conclude that

$$\frac{\mathcal{L}^n(1\!\!1)(\phi_j(z))}{\mathcal{L}^n(1\!\!1)(z)}|\phi_j'(z)|^h \to g_j^h$$

uniformly on V. Hence, applying Theorem 7.1.1, we conclude that

$$\tilde{D}_{\phi_j} = \rho(\phi_j(z))\rho(z)|\phi_j'(z)|^h = \lim_{n\to\infty}\frac{\mathcal{L}^n(1\!\!1)(\phi_j(z))}{\mathcal{L}^n(1\!\!1)(z)}|\phi_j'(z)|^h = g_j^h$$

on X. Since $\tilde{D}_{\phi_j}$ is real-analytic on U, we conclude that $\tilde{D}_{\phi_j} = g_j^h$ on U. The proof of the implication (d)⇒(a) is complete.

• (d)⇒(er). Define

$$\gamma = \frac{G'}{|G'|}.$$

Since for every $i \in I$, $G \circ \phi_i \circ G^{-1}$ is affine, we conclude that $G' \circ \phi_i \circ G^{-1} \cdot \phi_i' \circ G^{-1} \cdot (G')^{-1} \circ G^{-1} = k_i \in LC(d)$. Hence

$$\left(\left(\frac{G'}{|G'|}\right) \circ \phi_i \cdot \left(\frac{\phi_i'}{|\phi_i'|}\right) \cdot \left(\frac{G'}{|G'|}\right)^{-1}\right) \circ G^{-1} = \frac{k_i}{|k_i|}$$

and it suffices to take $\gamma = \frac{G'}{|G'|}$. Thus the proof of the implication (d)⇒(er) is complete.

• The implication (er)⇒(ec) is obvious.

• (ec)⇒(d). If all the maps ϕ_i, $i \in I$ are affine, there is nothing to prove. So, assume that there is $j \in I$ such that ϕ_j is not affine. Then no iterate ϕ_{j^n} is affine. Let $a^{(n)}$ denote the inversion center of ϕ_{j^n}. Fix $v \in J$. By (ec) the following holds for every $z \in V$ and every $n \geq 1$

$$\begin{aligned}
&(\gamma(v))^{-1}\gamma(z)\\
&= \left(\frac{\phi'_{j^n}(v)}{|\phi'_{j^n}(v)|}\right)^{-1} \cdot \big(\gamma \circ \phi_{j^n}(v)\big)^{-1} k_j^n k_j^{-n} \big(\gamma \circ \phi_{j^n}(z)\big) \left(\frac{\phi'_{j^n}(z)}{|\phi'_{j^n}(z)|}\right)\\
&= \left(\frac{\phi'_{j^n}(v)}{|\phi'_{j^n}(v)|}\right)^{-1} \cdot \big(\gamma \circ \phi_{j^n}(v)\big)^{-1} \big(\gamma \circ \phi_{j^n}(z)\big) \left(\frac{\phi'_{j^n}(z)}{|\phi'_{j^n}(z)|}\right)\\
&= (T_n(v))^{-1}\big(\gamma \circ \phi_{j^n}(v)\big)^{-1}\big(\gamma \circ \phi_{j^n}(z)\big)T_n(z),
\end{aligned}$$

where $T_n(w) = \mathrm{Id} - 2Q(w - a^{(n)})$ and in the canonical coordinates Q is given by the matrix

$$Q(x) = \frac{x_i x_j}{||x||^2}.$$

We shall now prove that the sequence $\{a^{(n)}\}_{n=1}^{\infty}$ does not converge to ∞. Indeed, suppose on the contrary that $\lim_{n\to\infty} a^{(n)} = \infty$. Since $a^{(n)} = \phi_{j^n}^{-1}(\infty) = (\phi_j^{-1})^n(\infty)$, we therefore get

$$\begin{aligned}
\infty = \lim_{n\to\infty} a^{(n-1)} &= \lim_{n\to\infty} \phi_j((\phi_j^{-1})^n) = \phi_j(\lim_{n\to\infty} (\phi_j^{-1})^n)\\
&= \phi_j(\lim_{n\to\infty} a^{(n)}) = \phi_j(\infty)
\end{aligned}$$

which means that ϕ_j is affine. This contradiction shows that there exists a subsequence $\{k_n\}_{n=1}^{\infty}$ such that $a^{(k_n)} \to a$ for some $a \in I\!\!R^d$. Then for every $n \geq 1$,

$$(\gamma(v))^{-1}\gamma(z) = (T_{k_n}(v))^{-1}\big(\gamma \circ \phi_{j^{n_k}}(v)\big)^{-1}\big(\gamma \circ \phi_{j^{n_k}}(z)\big)T_{k_n}(z),$$

and taking the limit when $n \to \infty$, we obtain

$$(\gamma(v))^{-1}\gamma(z) = \big(\mathrm{Id} - 2Q(v-a)\big)^{-1}\big(\mathrm{Id} - 2Q(z-a)\big)$$

or, equivalently,

$$\gamma(z) = \gamma(v)\big(\mathrm{Id} - 2Q(v-a)\big)^{-1}\big(\mathrm{Id} - 2Q(z-a)\big).$$

Define

$$G = \gamma(v)\big(\mathrm{Id} - 2Q(v-a)\big)^{-1} \circ i_{a,1}.$$

Then

$$G'(z) = \gamma(v)\big(\mathrm{Id} - 2Q(v-a)\big)^{-1} \frac{1}{||z-a||^2}\big(\mathrm{Id} - 2Q(z-a)\big).$$

Hence

$$\frac{G'(z)}{|G'(z)|} = \gamma(v)\big(\mathrm{Id} - 2Q(v-a)\big)^{-1}\big(\mathrm{Id} - 2Q(z-a)\big). = \gamma(z)$$

Therefore for every $i \in I$, (ec) takes the form

$$\frac{G' \circ \phi_i(z)}{|G' \circ \phi_i(z)|} \cdot \frac{\phi_i'(z)}{|\phi_i'(z)|} \cdot \left(\frac{G'(z)}{|G'(z)|}\right) = k_i.$$

Suppose that $G \circ \phi_i \circ G^{-1}$ is not affine. Then for y, the inversion center of $G \circ \phi_i \circ G^{-1}$, we get in canonical coordinates that

$$\delta_{mn} - 2\frac{(z_m - y_m)(z_n - y_n)}{||z-y||^2} = (k_i)_{mn}$$

for all $z \in V$ and all $m, n \in \{1, 2, \ldots, d\}$, where δ denotes the Kronecker symbol here. But this is impossible and we conclude that $G \circ \phi_i \circ G^{-1}$ is affine. The proof of the implication (ec)$\Rightarrow$(d) is complete.

- The implication (a)$\Rightarrow$(g) follows from the implication (a)$\Rightarrow$(d).

- (g)$\Rightarrow$(f). Conjugating the system S by a conformal diffeomorphism we may assume that $M_S = I\!\!R^q$. Given $i \in I$ and $(\omega^{(j)})_{j=1}^q \in (I^*)^q$ let

$$A = (i, \omega^{(1)}, \ldots, \omega^{(q)})$$

and let $H_A : I\!\!R^q \to I\!\!R^q$ be the map defined by the formula

$$H_A(z) = \big(\tilde{D}_{\phi_i} \circ \phi_{\omega^{(1)}}(z), \ldots, \tilde{D}_{\phi_i} \circ \phi_{\omega^{(q)}}(z)\big).$$

Suppose first that for every $i \in I$ there exists A such that $H_A' = 0$ on $\overline{J}$. Since S is not $(q-1)$-dimensional, this implies that $H_A' = 0$ on a neighborhood of $\overline{J}$ in $I\!\!R^q$. But then $\tilde{D}_{\phi_i}$ is constant on an open subset of $I\!\!R^q$ having a non-empty intersection with $\overline{J}$. Since by Corollary 6.1.5 $\tilde{D}_{\phi_i}$ is real-analytic, it is therefore constant on the appropriate set U_q produced in this corollary. Hence, in view of already proved implication (a)$\Rightarrow$(d), the system $\{\phi_i : I\!\!R^q \to I\!\!R^q\}$ is conjugate by a conformal diffeomorphism $\rho : R^q \to I\!\!R^q$ with an affine system. Since ρ extends to a conformal diffeomorphism from $I\!\!R^d$ to $I\!\!R^d$ and since an extension of an affine map in $I\!\!R^q$ to an affine map in $I\!\!R^d$ is also affine (if $q \leq 1$ we need to be certain that these extensions are of the form $\lambda A + b$), we are done in this case.

So, suppose that there exists $i \in I$ such that for every A with the first element equal to i there exists $x \in \overline{J}$ such that $H_A'(x) \neq 0$. Choose $w \in \overline{J}$ and $A = (i, \omega^{(1)}, \ldots, \omega^{(q)})$ such that $\dim \mathrm{Ker} H_A'(w)$ is minimal, say equal to $p \leq q-1$. By the assumptions of (g), $\dim \mathrm{Ker} H'_A(w) \geq 1$. So

$$1 \leq \dim \mathrm{Ker} H'_A = p \leq q-1$$

on W, a neighborhood of w in $I\!R^q$. By the definition of the limit set J, for every $z \in \overline{V}$ there exists $\tau \in I^*$ such that $\phi_\tau(z) \in W$. Then define

$$l(z) = (\phi_\tau^{-1})'_{\phi_\tau(z)}(\mathrm{Ker} H'_A(\phi_\tau(z))),$$

where, temporarily changing notation, $(\phi_\tau^{-1})'_{\phi_\tau(z)}$ denotes the derivative of the map ϕ_τ^{-1} evaluated at the point $\phi_\tau(z)$. We want to show first that we define in this manner a line field on $\overline{V}$. So, we need to show that if $\phi_\tau(z), \phi_\eta(z) \in W$, then

$$(\phi_\tau^{-1})'_{\phi_\tau(z)}(l(\phi_\tau(z))) = (\phi_\eta^{-1})'_{\phi_\eta(z)}(l(\phi_\eta(z))). \tag{7.22}$$

Suppose on the contrary that (7.22) fails with some z, τ, η as required above. Then there exists a point $x \in W$ and $\gamma \in I^*$ (in fact for every $x \in W$ there exists γ) such that $\phi_\gamma(x)$ is so close to z that

$$(\phi_\tau^{-1})'_{\phi_\tau(\phi_\gamma(x))}(l(\phi_\tau(\phi_\gamma(x)))) \neq (\phi_\eta^{-1})'_{\phi_\eta(\phi_\gamma(x))}(l(\phi_\eta(\phi_\gamma(x)))).$$

Hence

$$(\phi_{\tau\gamma}^{-1})'_{\phi_{\tau\gamma}(x)} l(\phi_{\tau\gamma}(x)) \neq (\phi_{\eta\gamma}^{-1})'_{\phi_{\eta\gamma}(x)} l(\phi_{\eta\gamma}(x)).$$

So, either

$$(\phi_{\tau\gamma}^{-1})'_{\phi_{\tau\gamma}(x)} l(\phi_{\tau\gamma}(x)) \neq \mathrm{Ker} H'_A(x)$$

or

$$(\phi_{\eta\gamma}^{-1})'_{\phi_{\eta\gamma}(x)} l(\phi_{\eta\gamma}(x)) \neq \mathrm{Ker} H'_A(x).$$

Without loss of generality we may assume that the first inequality holds. Since $(H_A \circ \phi_{\tau\gamma})'(x) = H'_A(\phi_{\tau\gamma}(x))\phi'_{\tau\gamma}(x)$, we get $\mathrm{Ker}\big((H_A \circ \phi_{\tau\gamma})'(x)\big) = \phi'_{\tau\gamma}(x)^{-1}(\mathrm{Ker} H'_A(\phi_{\tau\gamma}(x)))$ and therefore

$$\mathrm{Ker}\big((H_A \circ \phi_{\tau\gamma})'(x)\big) \neq \mathrm{Ker} H'_A(x). \tag{7.23}$$

If now $\phi_\gamma(x)$ is sufficiently close to z, then $\phi_{\tau\gamma}(x)$ is so close to $\phi_\tau(z)$ that $\phi_{\tau\gamma}(x) \in W$. Then

$$+\dim\big(\mathrm{Ker} H'_A(\phi_{\tau\gamma}(x))\big) = p = \dim(\mathrm{Ker} H'_A(x)). \tag{7.24}$$

Consider now linearly independent vectors $\big(\nabla \tilde{D}\phi_i \circ \phi_{\omega^{(k_1)}}(x), \ldots, \nabla \tilde{D}\phi_i \circ \phi_{\omega^{(k_t)}}(x)\big)$, $t = q - \dim(\mathrm{Ker} H'_A(x))$. If $v \in \mathrm{Ker} H'_A(x)$, then $< \nabla \tilde{D}\phi_i \circ \phi_{\omega^{(k_j)}}(x), v >= 0$ for all $j = 1, 2, \ldots, t$. Suppose that each vector $\nabla \tilde{D}\phi_i \circ \phi_{\omega^{(j)}\tau\gamma}(x)$, $j = 1, \ldots, q$, is a linear combination of the vectors

$\big(\nabla\tilde{D}\phi_i \circ \phi_{\omega^{(k_1)}}(x), \ldots, \nabla\tilde{D}\phi_i \circ \phi_{\omega^{(k_t)}}(x)\big)$, $t = q - \dim(\mathrm{Ker}H'_A(x))$. Then $< \nabla\tilde{D}\phi_i \circ \phi_{\omega^{(j)}\tau\gamma}(x), v >= 0$ for all $j = 1, \ldots, q$ and all $v \in \mathrm{Ker}H'_A(x)$. Hence $\mathrm{Ker}\big((H_A \circ \phi_{\tau\gamma})'\big)(x) \supset \mathrm{Ker}H'_A(x)$. Thus using (7.24) we conclude that $\mathrm{Ker}\big((H_A \circ \phi_{\tau\gamma})'\big)(x) = \mathrm{Ker}H'_A(x)$. This contradicts (7.23) and shows that there exists $1 \leq u \leq q$ such that the vectors $\big(\nabla\tilde{D}\phi_i \circ \phi_{\omega^{(k_j)}}(x)\big)_{j=1}^{t}$ together with the vector $\nabla\tilde{D}\phi_i \circ \phi_{\omega^{(u)}\tau\gamma}(x)$ form a linearly independent set. Hence, if $B = (i, \omega^{(u)}\tau\gamma, \omega^{(k_1)}, \ldots, \omega^{(k_t)}, i, \ldots, i)$ ($(q - (t+1))$ i's at the end), then the rank of $H'_B(x)$ is greater than or equal to $t+1$. Thus $\mathrm{Ker}H'_B(x) = q - \mathrm{rank}(H'_B(x)) \leq q - (t+1) = q - q + \dim(\mathrm{Ker}H'_A(x)) - 1 = p - 1$, which is a contradiction with the definition of p and finishes the proof of the implication (g)⇒(f).

• (f)⇒(d). In order to prove this implication suppose that there exists a field of linear subspaces E_x in TM_S of dimension and co–dimension greater than or equal to 1 defined on a neighborhood of $\overline{J}$ in M_S and invariant under the action of derivatives of all maps ϕ_i, $i \in I$. Conjugating our system by a conformal diffeomorphism, we may assume that $M_S = I\!R^q$. Fix an element $j \in I$. In the course of the proof of Lemma 7.4.1 we have shown that besides one attracting fixed point $x_j \in X$, the map ϕ_j has a different fixed point $y_j \in I\!R^d$. Conjugate the system S by the inversion $i_{y_j,1}$ (equal to the identity if $y_j = \infty$) and denote the resulting system by S_1. Put $\psi_i = i_{y_j,1} \circ \phi_i \circ i_{y_j,1}$ for all $i \in I$. The field $F_x = i'_{y_j,1}(E_x)$ is defined on a neighborhood W of J_{S_1} and it is S_1-invariant. Since $\psi_j : I\!R^q \to I\!R^q$ is linear, inspecting the appropriate part of the proof of Lemma 7.4.1 we see that the field $\{F_x\}_{x\in W}$ is constant, say equal to F. So, the field of affine subspaces $\{x + F\}_{x\in W}$, as the unique field of integral manifolds of the S_1-invariant field $\{F\}$ of linear subspaces, is S_1-invariant, which means that $\psi(x+F) = \psi_i(x)+F$ for every $i \in I$ and every $x \in W$. So, if ψ_i is not affine for some $i \in I$, then $x+F$ must contain $\psi_i^{-1}(\infty)$, the center of inversion of ψ_i, for every $x \in W$. Since W is open in $I\!R^q$ and since $\dim(F) \leq q - 1$, this is impossible and proves that ψ_i is affine. The implication (f)⇒(d) is thus proved. □

Our next result is the following rather unexpected fact.

Proposition 7.4.5 *Suppose that $F = \{f_i : X \to X\}_{i\in I}$ and $G = \{g_i : Y \to Y\}_{i\in I}$ are two not essentially affine topologically conjugate systems. If the measures m_G and $m_F \circ h^{-1}$ are equivalent, then the systems F and G are of the same dimension.*

Proof. Suppose on the contrary that the dimensions of F and G are not equal. Without loss of generality we may assume that $p = \dim F < q = \dim G$. Since G is not essentially affine, it follows from Theorem 7.4.4 that there exist $y \in J_G$, $i \in I$, a sequence $(\omega^{(j)})_{j=1}^q \in (I^*)^q$ and a neighborhood $W_G \subset M_G$ of y such that the map

$$\mathcal{G} = \left(\tilde{D}g_i \circ g_{\omega^{(1)}}, \ldots, \tilde{D}g_i \circ g_{\omega^{(q)}}\right)$$

is invertible on W_G. Since the measures m_G and $m_F \circ h^{-1}$ are equivalent, after an appropriate normalization $\mu_F = \mu_G \circ h$ which means that $D_h = \frac{d\mu_G \circ h}{d\mu_F} = 1$. Since $h \circ f_\tau = g_\tau \circ h$ for all $\tau \in I^*$ and since $D_h = 1$,

$$\mathcal{G} \circ h = \mathcal{F}$$

on J_F, where $\mathcal{F} = \left(\tilde{D}f_i \circ f_{\omega^{(1)}}, \ldots, \tilde{D}f_i \circ f_{\omega^{(q)}}\right)$. Write $x = h^{-1}(y)$. Then $h = \mathcal{G}^{-1} \circ \mathcal{F}$ on $W_F \cap J_F$ for some open neighborhood W_F of x in M_F such that $\mathcal{F}(W_F) \subset \mathcal{G}(W_G)$. Since by Corollary 6.1.5, the maps $\mathcal{F}$ and $\mathcal{G}^{-1}$ are real-analytic, the image $\mathcal{G}^{-1} \circ \mathcal{F}(W_F)$ for an adequate W_F small enough, is a real-analytic submanifold of dimension $\leq q$ and $\mathcal{G}^{-1} \circ \mathcal{F}(W_F) \cap J_G$ contains an open neighborhood of y in J_G. So, invoking Lemma 7.4.2, we conclude that G is at most p-dimensional. This contradiction finishes the proof. □

The main result of this section, concerning of course smooth conjugacies is contained in the following.

Theorem 7.4.6 *If two conformal regular iterated function systems $F = \{f_i : X \to X : i \in I\}$ and $G = \{g_i : Y \to Y : i \in I\}$ both satisfying the open set condition are not essentially affine and are conjugate by a homeomorphism $h : J_F \to J_G$, then the following conditions are equivalent.*

(a) The conjugacy between the systems F and G extends in a conformal fashion to an open neighborhood of X.

(b) The conjugacy between the systems F and G extends in a real-analytic fashion to an open neighborhood of X.

(c) The conjugacy between the systems F and G is bi-Lipschitz continuous.

(d) $|g'_\omega(y_\omega)| = |f'_\omega(x_\omega)|$ for all $\omega \in I^*$, where x_ω and y_ω are the only fixed points of $f_\omega : X \to X$ and $g_\omega : Y \to Y$ respectively.

(e) $\exists S \geq 1\ \forall \omega \in I^*$

$$S^{-1} \leq \frac{\operatorname{diam}(g_\omega(Y))}{\operatorname{diam}(f_\omega(X))} \leq S.$$

(f) $\exists E \geq 1\ \forall \omega \in I^*$

$$E^{-1} \leq \frac{||g'_\omega||}{||f'_\omega||} \leq E.$$

(g) $\mathrm{HD}(J_G) = \mathrm{HD}(J_F)$ *and the measures* m_G *and* $m_F \circ h^{-1}$ *are equivalent.*

(h) The measures m_G *and* $m_F \circ h^{-1}$ *are equivalent.*

Proof. The implications (a)⇒(b) and (b)⇒(c) are obvious. That (c)⇒(d) results from the fact that (c) implies condition (1) of Theorem 7.1.1, which in view of that theorem is equivalent with condition (2) of Theorem 7.1.1 which finally is the same as condition (d) of Theorem 7.4.6. The implications (d)⇒(e)⇒(f)⇒(g) have been proved in Theorem 7.1.1. The implication (g)⇒(h) is again obvious. We are left to prove that (h)⇒(a). We shall first prove that (h)⇒(b). So, suppose that (h) holds. Then, after an appropriate normalization $\mu_F = \mu_G \circ h$, which means that $D_h = \frac{d\mu_G \circ h}{d\mu_F} = 1$. Since G is not essentially affine, it follows from Theorem 7.4.4 that there exist $y \in J_G$, $i \in I$, a sequence $(\omega^{(j)})_{j=1}^q \in (I^*)^q$ and a neighborhood $W_G \subset M_G$ of y such that the map

$$\mathcal{G} = \left(\tilde{D}g_i \circ g_{\omega^{(1)}}, \ldots, \tilde{D}g_i \circ g_{\omega^{(q)}}\right)$$

is invertible on W_G. Since $h \circ f_\tau = g_\tau \circ h$ for all $\tau \in I^*$ and since $D_h = 1$, we have

$$\mathcal{G} \circ h = \mathcal{F}$$

on J_F, where $\mathcal{F} = \left(\tilde{D}f_i \circ f_{\omega^{(1)}}, \ldots, \tilde{D}f_i \circ f_{\omega^{(q)}}\right)$. Fix W_F, an open neighborhood of $x = h^{-1}(y)$ in M_F so small that $\mathcal{F}(W_1) \subset \mathcal{G}(W_2)$. Hence $\mathcal{G}^{-1} \circ \mathcal{F}$ is well defined on W_1 and $\mathcal{G}^{-1} \circ \mathcal{F}|_{W_1 \cap J_F} = h$. Consider now $\omega \in I^*$ such that $f_\omega(J_F) \subset W_1$. Since

$$\mathcal{G}^{-1} \circ \mathcal{F}(f_\omega(J_F)) = h \circ f_\omega(J_F) = g_\omega \circ h(J_F) = g_\omega(J_G) \subset g_\omega(V_G),$$

since $g_\omega(V_G)$ is open and since f_ω and $\mathcal{G}^{-1} \circ \mathcal{F}$ are continuous, there exists an open neighborhood $V'_F \subset V_F$ of X such that $f_\omega(V'_F) \subset W_1$ and $\mathcal{G}^{-1} \circ \mathcal{F}(f_\omega(V'_F)) \subset g_\omega(V_G)$. Hence, the map

$$g_\omega^{-1} \circ (\mathcal{G}^{-1} \circ \mathcal{F}) \circ f_\omega : V'_F \to \mathbb{C}$$

is well defined, by Corollary 6.1.5 is real-analytic, and $g_\omega^{-1} \circ (\mathcal{G}^{-1} \circ \mathcal{F}) \circ f_\omega|_{J_F} = h$. Thus, property (b) is proved.

The last step of the proof of Theorem 7.4.6, that is the implication (b)⇒(a), can be carried out using ideas from the proof of Lemma 7.2.7

in [Pr] as follows. Let H be this real-analytic extension of h on a neighborhood of W_F of J_F in M_F. We may assume W_F to be so small that H' is a linear isomorphism at every point of W_F. Define the function $\psi : W_F \to I\!R$ by the formula

$$\psi(z) = \frac{||H'(z)||}{||(H'(z))^{-1}||}.$$

Suppose that $\psi(\xi) = 1$ for some point $\xi \in W_F$. Since for every $\omega \in I^*$

$$\psi(f_\omega(\xi)) = \frac{||H'(f_\omega(\xi))||}{||(H'(f_\omega(\xi)))^{-1}||} = \frac{||g'_\omega(H(\xi)) \cdot H'(\xi) \cdot (f'_\omega(\xi))^{-1}||}{||\left(g'_\omega(H(\xi)) \cdot H'(\xi) \cdot (f'_\omega(\xi))^{-1}\right)^{-1}||}$$
$$= \frac{||H'(\xi)||}{||(H'(\xi))^{-1}||} = \psi(\xi)$$

and since $\overline{\{f_\omega(\xi) : \omega \in I^*\}} \supset \overline{J}_F$, we conclude that $\psi = 1$ identically on $\overline{J}_F$. Since ψ is real-analytic and since F is q-dimensional, using Lemma 7.4.2 we conclude that $\psi = 1$ on an open neighborhood of $\overline{J}_F$. But this means that H is conformal. So, we may assume that $\psi(z) \neq 1$ for every $z \in W_F$. Define the field $\{E_z\}_{z \in W_F}$ on W_F as follows.

$$E_z = \left\{ w \in I\!R^q : \frac{||H'(z)w||}{||w||} = ||H'(z)|| \right\} \cup \{0\}.$$

For every $z \in W_F$, the set E_z is a linear subspace of $I\!R^q$ of dimension ≥ 1. Its codimension is ≥ 1 since $\psi(z) \neq 1$. Obviously E_z depends continuously on z. Since the maps $f_i : I\!R^q \to I\!R^q$ are conformal, $f'_i(z)(E_z) = E_{f_i(z)}$ and it thereofore follows from Theorem 7.4.4 that the system F is essentially affine. This contradiction finishes the proof. $\square$

8

Parabolic Iterated Function Systems

In this chapter we develop the general theory of *conformal parabolic iterated function systems* (see [MU7]).

8.1 Preliminaries

Our setting is this. Let X be a compact connected subset of a Euclidean space $I\!R^d$. Suppose that we have countably many conformal maps $\phi_i : X \to X$, $i \in I$, where I has at least two elements satisfying the following conditions.

(1) (Open set condition) $\phi_i(\text{Int}(X)) \cap \phi_j(\text{Int}(X)) = \emptyset$ for all $i \neq j$.

(2) $|\phi_i'(x)| < 1$ everywhere except for finitely many pairs (i, x_i), $i \in I$, for which x_i is the unique fixed point of ϕ_i and $|\phi_i'(x_i)| = 1$. Such pairs and indices i will be called parabolic and the set of parabolic indices will be denoted by Ω. All other indices will be called hyperbolic.

(3) $\forall n \geq 1\ \forall \omega = (\omega_1, ..., \omega_n) \in I^n$ if ω_n is a hyperbolic index or $\omega_{n-1} \neq \omega_n$, then ϕ_ω extends conformally to an open connected set $W \subset I\!R^d$ and maps W into itself.

(4) If i is a parabolic index, then $\bigcap_{n\geq 0} \phi_{i^n}(X) = \{x_i\}$ and the diameters of the sets $\phi_{i^n}(X)$ converge to 0.

(5) (Bounded distortion property) $\exists K \geq 1\ \forall n \geq 1\ \forall \omega = (\omega_1, ..., \omega_n) \in I^n\ \forall x, y \in V$ if ω_n is a hyperbolic index or $\omega_{n-1} \neq \omega_n$, then

$$\frac{|\phi_\omega'(y)|}{|\phi_\omega'(x)|} \leq K.$$

(6) $\exists s < 1\ \forall n \geq 1\ \forall \omega \in I^n$ if ω_n is a hyperbolic index or $\omega_{n-1} \neq \omega_n$, then $||\phi_\omega'|| \leq s$.

(7) (Cone condition) There exist $\alpha, l > 0$ such that for every $x \in \partial X \subset \mathbb{R}^d$ there exists an open cone $\mathrm{Con}(x, \alpha, l) \subset \mathrm{Int}(X)$ with vertex x, central angle of Lebesgue measure α, and altitude l.

(8) There exists a constant $L \geq 1$ such that

$$\big||\phi_i'(y)| - |\phi_i'(x)|\big| \leq L||\phi_i'|| \cdot |y - x|$$

for every $i \in I$ and every pair of points $x, y \in V$.

(9) For every $i \in \tilde{\Omega}$ put

$$X_i = \bigcup_{j \in I \setminus \{i\}} \phi_j(X).$$

We will also need the following.

$$\sum_{n \geq 0} ||\phi_{i^n}'||_{X_i}^{\alpha} < \infty.$$

We call such a system of maps $S = \{\phi_i : i \in I\}$ a subparabolic iterated function system. Let us note that conditions (1),(3),(5)–(7) are modeled on similar conditions which were used to examine hyperbolic conformal systems. If $\Omega \neq \emptyset$, we call the system $\{\phi_i : i \in I\}$ parabolic. As declared in (2) the elements of the set $I \setminus \Omega$ are called hyperbolic. We extend this name to all the words appearing in (5) and (6). By I^* we denote the set of all finite words with alphabet I and by I^∞ all infinite sequences with terms in I. It follows from (3) that for every hyperbolic word ω, $\phi_\omega(W) \subset W$. Note that our conditions insure that $\phi_i'(x) \neq 0$, for all i and $x \in V$. We provide below without proofs all the geometrical consequences of the bounded distortion properties (5) and (8), derived in Section 4.1 which remain true in the parabolic case. We have for all hyperbolic words $\omega \in I^*$ and all convex subsets C of W

$$\mathrm{diam}(\phi_\omega(C)) \leq ||\phi_\omega'||\mathrm{diam}(C) \tag{8.1}$$

and

$$\mathrm{diam}(\phi_\omega(V)) \leq D||\phi_\omega'||, \tag{8.2}$$

where the norm $||\cdot||$ is the supremum norm taken over V and $D \geq 1$ is a universal constant. Moreover,

$$\mathrm{diam}(\phi_\omega(X)) \geq D^{-1}||\phi_\omega'|| \tag{8.3}$$

and

$$\phi_\omega(B(x, r)) \supset B(\phi_\omega(x), K^{-1}||\phi_\omega'||r), \tag{8.4}$$

for every $x \in X$, every $0 < r \le \mathrm{dist}(X, \partial V)$, and every hyperbolic word $\omega \in I^*$. Also, there exists $0 < \beta \le \alpha$ such that for all $x \in X$ and for all hyperbolic words $\omega \in I^*$

$$\begin{aligned}\phi_\omega(\mathrm{Int}(X)) &\supset \mathrm{Con}\big(\phi_\omega(x), \beta, D^{-1}||\phi'_\omega||\big) \\ &\supset \mathrm{Con}\big(\phi_\omega(x), \beta, D^{-2}\mathrm{diam}(\phi_\omega(V))\big),\end{aligned} \tag{8.5}$$

where $\mathrm{Con}\big(\phi_\omega(x), \beta, D^{-1}||\phi'_\omega||\big)$ and $\mathrm{Con}\big(\phi_\omega(x), \beta, D^{-2}\mathrm{diam}(\phi_\omega(V))\big)$ denote cones with vertices at $\phi_\omega(x)$, angles β, and altitudes $D^{-1}||\phi'_\omega||$ and $D^{-2}\mathrm{diam}(\phi_\omega(V))$ respectively. In addition, for every $\omega \in I^*$ (not necessarily hyperbolic) and every $x \in X$, there exists $l(\omega, x) > 0$ such that

$$\phi_\omega(\mathrm{Int}(X)) \supset \mathrm{Con}\big(\phi_\omega(x), \beta, l(\omega, x)\big). \tag{8.6}$$

The important point in (8.6) is that by conformality we can get a cone with vertex x and opening angle β lying in $\phi_\omega(X)$, but we cannot say anything about the height of this cone unless ω is a hyperbolic word, in which case we have (8.5). For each $\omega \in I^* \cup I^\infty$, we define the length of ω by the uniquely determined relation $\omega \in I^{|\omega|}$. If $\omega \in I^* \cup I^\infty$ and $n \le |\omega|$, then by $\omega|_n$ we denote the word $\omega_1\omega_2\ldots\omega_n$. Our first aim in this section is to prove the existence of the limit set. More precisely, we begin with the following lemma.

Lemma 8.1.1 *For all $\omega \in I^\infty$ the intersection $\bigcap_{n\ge0} \phi_{\omega|_n}(X)$ is a singleton.*

Proof. Since the sets $\phi_{\omega|_n}(X)$ form a nested sequence of compact sets, the intersection $\bigcap_{n\ge0} \phi_{\omega|_n}(X)$ is not empty. Moreover, it follows from (4) that if ω is of the form τi^∞, $\tau \in I^*, i \in \Omega$, then the diameters of the intersection $\bigcap_{n=0}^{k} \phi_{\omega|_n}(X)$ tend to 0 and, in the other case, the same conclusion follows immediately from (6). In any case, $\bigcap_{n\ge0} \phi_{\omega|_n}(X)$ is a singleton. □

Improving slightly the argument just given, we get the following.

Lemma 8.1.2 $\lim_{n\to\infty} \sup_{|\omega|=n}\{\mathrm{diam}(\phi_\omega(X))\} = 0.$

Proof. Let $g(n) = \max_{i\in\Omega}\{\mathrm{diam}(\phi_{i^n}(X))\}$. Since Ω is finite it follows from (4) that $\lim_{n\to\infty} g(n) = 0$. Let $\omega \in I^\infty$. Given $n \ge 0$ consider the word $\omega|_n$. Look at the longest block of the same parabolic element appearing in $\omega|_n$. If the length of this block exceeds $\sqrt{n}$ then, since due to (2) all the maps ϕ_j, $j \in I$, are Lipschitz continuous with a Lipschitz constant ≤ 1, we have $\mathrm{diam}(\phi_{\omega|n}(X)) \le g(\sqrt{n})$. Otherwise, we can find in $\omega|_n$ at least $\frac{n-\sqrt{n}}{\sqrt{n}} = \sqrt{n}-1$ distinct hyperbolic indices. It then follows from (6) (and Lipschitz continuity with a Lipschitz constant ≤ 1 of all the maps ϕ_i, $i \in I$) that $\mathrm{diam}(\phi_{\omega|n}(X)) \leq s^{\sqrt{n}-1}$. □

We introduce on I^∞ the standard metric $d(\omega, \tau) = \mathrm{e}^{-n}$, where n is the largest number such that $\omega|_n = \tau|_n$. The corollary below is now an immediate consequence of Lemma 2.2.

Corollary 8.1.3 *The map* $\pi : I^\infty \to X$, $\pi(\omega) = \bigcap_{n\geq 0} \phi_{\omega|_n}(X)$, *is uniformly continuous.*

The limit set $J = J_S$ of the system $S = \{\phi_i\}_{i\in I}$ satisfies

$$J = \pi(I^\infty) = \bigcup_{i\in I} \phi_i(J).$$

We recall that the set J is not compact if the index set I is infinite. This of course is one of the main technical issues to handle.

Lemma 8.1.4 *If X is a topological disk contained in $\mathbb{C}$, then every parabolic point lies on the boundary of X.*

Proof. Suppose on the contrary that a parabolic point $x_i \in \mathrm{Int}(X)$. Let $D^1 = \{z \in \mathbb{C} : |z| < 1\}$ and let $R : D^1 \to \mathrm{Int}(X)$ be the Riemann map (conformal homeomorphism) such that $R(0) = x_i$. Consider the composition $R^{-1} \circ \phi_i \circ R : D^1 \to D^1$. Then $|(R^{-1} \circ \phi_i \circ R)'(0)| = |R'(0)|^{-1}|R'(0)| = 1$. Thus by Schwarz's lemma $R^{-1}\circ\phi_i\circ R$ is a rotation. Since $\phi_i = R \circ (R^{-1} \circ \phi_i \circ R) \circ R^{-1}$, it follows that $\phi_i(X) = R \circ (R^{-1} \circ \phi_i \circ R) \circ R^{-1}(X) = X$. This contradiction finishes the proof. □

8.2 Topological pressure and associated parameters

Our first goal is to relate the pressure of the "volume potential" function ζ to the way pressure was defined in Section 3.1. We consider the function : $I^\infty \to \mathbb{R}$ given by the formula

$$\zeta(\omega) = -\log|\phi'_{\omega_1}(\pi(\sigma(\omega)))|.$$

Using heavily condition (8), we shall prove the following.

Proposition 8.2.1 *The function ζ defined above is acceptable.*

Proof. Fix $n \geq 1$ and $\omega, \tau \in I^\infty$ such that $\omega|_n = \tau|_n$. It then follows from (8) that

$$\begin{aligned} |g(\omega) - g(\tau)| &= \big|\log|\phi'_{\omega_1}(\pi(\sigma(\omega)))| - \log|\phi'_{\omega_1}(\pi(\sigma(\tau)))|\big| \\ &\leq \frac{\big||\phi'_{\omega_1}(\pi(\sigma(\omega)))| - |\phi'_{\omega_1}(\pi(\sigma(\tau)))|\big|}{\min\{|\phi'_{\omega_1}(\pi(\sigma(\omega)))|, |\phi'_{\omega_1}(\pi(\sigma(\omega)))|\}} \\ &\leq L\frac{||\phi'_{\omega_1}||}{\min\{|\phi'_{\omega_1}(\pi(\sigma(\omega)))|, |\phi'_{\omega_1}(\pi(\sigma(\omega)))|\}} \\ &\quad \times |\pi(\sigma(\omega)) - \pi(\sigma(\tau))|. \end{aligned}$$

If ω_1 is a hyperbolic index, then using the bounded distortion property, we get

$$|g(\omega) - g(\tau)| \leq LK|\pi(\sigma(\omega)) - \pi(\sigma(\tau))|^\alpha.$$

On the other hand, since there are only finitely many parabolic indices, there is a positive constant M such that if ω_1 is parabolic, then

$$|g(\omega) - g(\tau)| \leq LM|\pi(\sigma(\omega)) - \pi(\sigma(\tau))|^\alpha.$$

Let $L' = L\max\{K, M\}$. Since X being compact is bounded, taking $n = 1$, it follows from the last inequalities that $\max_{i\in I}\{\sup(g|_{[i]}) - \inf(g|_{[i]})\} \leq L'\mathrm{diam}^\alpha(X) < \infty$. The uniform continuity of g follows from inequality $|g(\omega) - g(\tau)| \leq L'|\pi(\sigma(\omega)) - \pi(\sigma(\tau))|^\alpha$ and Corollary 8.1.3. □

We define for each $W \subset X$

$$\mathrm{P}_W(t) = \lim_{n\to\infty}\frac{1}{n}\sum_{|\omega|=n}||\phi'_\omega||_W^t,$$

where $||\phi'_\omega||_W = \sup\{|\phi'_\omega(x)| : x \in W\}$. Let us note that

$$\mathrm{P}_W(t) = \inf\{s : \sum_{n\geq 1}\sum_{|\omega|=n}||\phi'_\omega||_W^t \mathrm{e}^{-sn} < \infty\}.$$

We now introduce some notation. For each $i \in \Omega$, let $I^p_{gi} = \{\omega \in I^p : \omega_p \neq i\}$.

Lemma 8.2.2 $\mathrm{P}(\sigma, -t\zeta) = \mathrm{P}(t)$.

Proof. First, we show $\mathrm{P}(t) = \mathrm{P}_J(t)$. Clearly, $\mathrm{P}_J(t) \le \mathrm{P}(t)$. To prove the converse inequality, suppose $\mathrm{P}_J(t) < s$. Then using (5)

$$\sum_{n\ge 1}\sum_{|\omega|=n} ||\phi'_\omega||^t e^{-sn}$$

$$= \sum_{n\ge 1}\sum_{|\omega|=n,\omega_n\notin\Omega} ||\phi'_\omega||^t e^{-sn} + \sum_{n\ge 1}\sum_{i\in\Omega}\sum_{k=1}^{n}\sum_{\omega\in I_{gi}^{n-k}} ||\phi'_{\omega i^k}||^t e^{-sn}$$

$$\le K^t \sum_{n\ge 1}\sum_{|\omega|=n,\omega_n\notin\Omega} ||\phi'_\omega||_J^t e^{-sn} + K^t \sum_{n\ge 1}\sum_{i\in\Omega}\sum_{k=1}^{n}\sum_{\omega\in I_{gi}^{n-k}} ||\phi'_{\omega i}||_J^t e^{-sn}$$

$$\le K^t \sum_{n\ge 1}\sum_{|\omega|=n,\omega_n\notin\Omega} ||\phi'_\omega||_J^t e^{-sn} + K^t \sum_{n\ge 1}\sum_{i\in\Omega}\sum_{k=1}^{n}\sum_{\omega\in I_{gi}^{n-k}}$$

$$\times ||\phi'_{\omega i}||_J^t e^{-s(n-k+1)}$$

$$\le K^t \sum_{n\ge 1}\sum_{|\omega|=n} ||\phi'_\omega||_J^t e^{-sn} < \infty.$$

So, $\mathrm{P}(t) \le s$ and consequently $\mathrm{P}(t) \le \mathrm{P}_J(t)$. Next, we compute

$$\begin{aligned}
\mathrm{P}(\sigma, -t\zeta) &= \lim_{n\to\infty} \frac{1}{n} \log \sum_{|\omega|=n} \exp\big(\sup_{\tau\in[\omega]} \sum_{j=0}^{n-1} tg(\sigma^j(\tau))\big) \\
&= \lim_{n\to\infty} \frac{1}{n} \log \sum_{|\omega|=n} \exp\big(\sup_{\tau\in[\omega]} \sum_{j=1}^{n} t\log|\phi'_{\omega_j}(\pi(\sigma^j(\tau)))|\big) \\
&= \lim_{n\to\infty} \frac{1}{n} \log \sum_{|\omega|=n} \sup_{\tau\in[\omega]} |\phi'_\omega(\pi(\sigma^n\tau))|^t \\
&= \lim_{n\to\infty} \frac{1}{n} \log \sum_{|\omega|=n} ||\phi'_\omega||_J = \mathrm{P}_J(t) = \mathrm{P}(t).
\end{aligned}$$

□

Our next goal is to prove that for parabolic systems the image of all shift-invariant measures also satisfies a measure theoretic open set condition known for hyperbolic systems (see Theorem 4.4.1). At this point we make essential use of the cone condition.

Theorem 8.2.3 *If μ is a shift-invariant Borel probability measure on I^∞, then*

$$\mu\circ\pi^{-1}(\phi_\tau(X)\cap\phi_\rho(X)) = 0$$

for all incomparable words $\tau, \rho \in I^*$.

Proof. We begin by describing a situation which we show to be impossible. Suppose that there exist a point $x \in X$, an integer $q \geq 1$ and an increasing sequence $\{n_k\}_{k\geq 1}$ of positive integers along with pairwise different words $\rho^{(k)}, \tau^{(k)} \in E^{n_k}$ such that

$$x \in \phi_{\tau^{(k)}}(X) \cap \phi_{\rho^{(k)}}(X)$$

and $\rho^{(k)}|_{n_k-q} = \tau^{(k)}|_{n_k-q}$. Passing to a subsequence we may assume that $n_{k+1} - n_k > q$ for every $k \geq 1$. We shall construct by induction with respect to $k \geq 1$ a sequence $\{C_k\}_{k\geq 1}$ such that for every k, C_k consists of at least $k+1$ incomparable words from $\{\rho^{(j)}, \tau^{(j)} : j \leq k\}$.

Indeed, set $C_1 = \{\rho^{(1)}, \tau^{(1)}\}$. Suppose now that C_k has been defined. If $\rho^{(k+1)}$ does not extend any word in C_k, then we form C_{k+1} by adding $\rho^{(k+1)}$ to C_k. We can do a similar thing in case $\tau^{(k+1)}$ does not extend any word in C_k. If, on the other hand, $\rho^{(k+1)}$ extends some word κ in C_k and $\tau^{(k+1)}$ extends a word η in C_k, then κ and η are both extended by $\rho^{(k+1)}$ since for $j \leq k$, $|\rho^{(j)}|, |\tau^{(j)}| \leq n_j \leq n_k$, $n_{k+1} > n_k + q$, and $\rho^{(k+1)}|_{n_{k+1}-q} = \tau^{(k+1)}|_{n_{k+1}-q}$. Since the words in C_k are incomparable, $\kappa = \eta$ and this is the only word in C_k which is extended by both $\rho^{(k+1)}$ and $\tau^{(k+1)}$. In this case we form C_{k+1} by taking away κ and adding both $\rho^{(k+1)}$ and $\tau^{(k+1)}$. Now the sets $\{\phi_\kappa(X) : \kappa \in C_k\}$ are nonoverlapping since the words are incomparable. By (8.6) we get $k+1$ pairwise disjoint open cones each with vertex x and opening angle β. This is clearly impossible if k is large enough. So it is impossible to have such a point x.

Let μ be a shift-invariant probability measure and suppose that $\mu \circ \pi^{-1}(\phi_\tau(X) \cap \phi_\rho(X)) > 0$ for some incomparable words $\tau, \rho \in I^*$. Without loss of generality we may assume that $|\tau| = |\rho|$. Put $E = \phi_\tau(X) \cap \phi_\rho(X)$ and set

$$E_\infty = \bigcap_{k=0}^{\infty} \bigcup_{n=k}^{\infty} \bigcup_{|\omega|=n} \phi_\omega(E).$$

In view of what we have just proved $E_\infty = \emptyset$. On the other hand, by (8.6), for every $n \geq 0$, each element of X belongs to at most $1/\beta$ elements of $\phi_\omega(X)$, $\omega \in I^n$ (we assume that $\lambda_{d-1}(S^{d-1}) = 1$). Using this and the σ-invariance of measure μ, we get for every $n \geq 0$

$$\begin{aligned}\mu \circ \pi^{-1}\left(\bigcup_{|\omega|=n} \phi_\omega(E)\right) &\geq \beta^{-1} \sum_{|\omega|=n} \mu \circ \pi^{-1}(\phi_\omega(E)) \\ &\geq \beta^{-1} \sum_{|\omega|=n} \mu(\omega\pi^{-1}(E)) \\ &= \beta^{-1}\mu \circ \pi^{-1}(E),\end{aligned}$$

where $\omega A = \{\omega\kappa : \kappa \in A\}$ for every set $A \subset I^\infty$. Hence, for every $k \geq 0$,

$$\mu \circ \pi^{-1}\left(\bigcup_{n=k}^{\infty} \bigcup_{|\omega|=n} \phi_\omega(E)\right) \geq \beta^{-1}\mu \circ \pi^{-1}(E),$$

and therefore $\mu \circ \pi^{-1}(E_\infty) \geq \beta^{-1}\mu \circ \pi^{-1}(E) > 0$. This contradiction finishes the proof. □

Let

$$\theta = \theta(S) = \inf\{t \geq 0 : \mathrm{P}(t) < \infty\}.$$

Following the previous chapter we call θ the finiteness parameter of the system S. Recall that $\alpha = \{[i] : i \in I\}$ is the partition of I^∞ into initial cylinders of length 1. The following result has been proved in the hyperbolic context as Theorem 4.4.2.

Theorem 8.2.4 *If μ is a shift-invariant ergodic Borel probability measure on I^∞ such that $\mathrm{H}_\mu(\alpha) < \infty$, $\chi_\mu(\sigma) = \int \zeta d\mu < \infty$ and either $\chi_\mu(\sigma) > 0$ or $\mathrm{h}_\mu(\sigma) > 0$ ($\mathrm{h}_\mu(\sigma) > 0$ implies $\chi_\mu(\sigma) > 0$), then*

$$\mathrm{HD}(\mu \circ \pi^{-1}) = \frac{\mathrm{h}_\mu(\sigma)}{\chi_\mu(\sigma)}.$$

The same proof goes through. Let

$$\beta = \beta(S) = \sup\{\mathrm{HD}(\mu \circ \pi^{-1})\},$$

where the supremum is taken over all ergodic finitely supported (so shift-invariant) measures of positive entropy. Of course there are many such measures. We shall prove the following.

Proposition 8.2.5 *The pressure function $\mathrm{P}(t)$ has the following properties:*

(1) $\mathrm{P}(t) \geq 0$ *for all* $t \geq 0$

(2) $\mathrm{P}(t) > 0$ *for all* $0 \le t < \beta$.

(3) $\mathrm{P}(t) = 0$ *for all* $t \ge \beta$.

(4) $\mathrm{P}(t)$ *is non-increasing.*

(5) $\mathrm{P}(t)$ *is strictly decreasing on* $[\theta, \beta]$.

(6) $\mathrm{P}(t)$ *is continuous and convex on* (θ, ∞).

Proof. (1) Let i be a parabolic index and let x_i be the corresponding parabolic point. Then $\pi(i^\infty) = x_i$. Let μ be the Dirac measure supported on i^∞. Of course, μ is ergodic, finitely supported, and $\int tg\,d\mu = t\log|\phi_i'(x_i)| = 0$. Hence, by Theorem 2.1.6 and Proposition 8.2.1, $P(\sigma, tg) \ge \mathrm{h}_\mu(\sigma) + \int tgd\mu = 0$ and (1) is proved.

(2) Suppose that $0 \le t < \beta$. Then there exists an ergodic, σ-invariant, and finitely supported measure μ of positive entropy such that $\mathrm{HD}(\mu \circ \pi^{-1}) > t$. So, Theorem 8.2.4 applies to give $t < \mathrm{HD}(\mu \circ \pi^{-1}) \le \mathrm{h}_\mu(\sigma)/\chi_\mu(\sigma)$ which due to Theorem 2.1.6 and Proposition 8.2.1 implies that $\mathrm{P}(\sigma, tg) \ge \mathrm{h}_\mu(\sigma) + \int tg\,d\mu > 0$.

(3) Suppose that $\mathrm{P}(t) > 0$ for some $t \ge 0$. Then in view of Theorem 2.1.6 and Proposition 8.2.1 there exists an ergodic σ-invariant finitely supported measure μ such that $\mathrm{h}_\mu(\sigma) - t\chi_\mu(\sigma) > 0$. Therefore $\mathrm{h}_\mu(\sigma) > 0$ and hence, by Theorem 8.2.3, $t < \frac{\mathrm{h}_\mu(\sigma)}{\chi_\mu(\sigma)} = \mathrm{HD}(\mu \circ \pi^{-1}) \le \beta$. We are done.

(4) Suppose that $t_1 < t_2$. It is clear from the definition of pressure that $\mathrm{P}(t_2) = \infty$ implies $\mathrm{P}(t_1) = \infty$. So, we may assume $\theta \le t_1 < t_2$. Fix $\epsilon > 0$. By Theorem 2.1.6 and Proposition 8.2.1 there exists an ergodic finitely supported measure μ_2 such that $\mathrm{h}_{\mu_2}(\sigma) + \int t_2gd\mu_2 \ge \mathrm{P}(\sigma, t_2g) - \epsilon$. Then by Theorem 2.1.6 and Proposition 8.2.1, $\mathrm{P}(\sigma, t_1g) \ge \mathrm{h}_{\mu_2}(\sigma) + \int t_1gd\mu_2 = \mathrm{h}_{\mu_2}(\sigma) + \int t_2gd\mu_2 + \int(t_1 - t_2)gd\mu_2 \ge \mathrm{h}_{\mu_2}(\sigma) + \int t_2gd\mu_2 \ge \mathrm{P}(\sigma, t_2g) - \epsilon$. Letting $\epsilon \searrow 0$, we are done.

(5) Suppose $\theta \le t_1 < t_2 < \beta$. Since $\mathrm{P}(\sigma, t_2g) > 0$, in view of Theorem 2.1.6 and Proposition 8.2.1 there exists an ergodic σ-invariant and finitely supported measure μ_2 such that

$$\mathrm{h}_{\mu_2}(\sigma) + \int t_2gd\mu_2 \ge \max\left\{\frac{1}{2}, 1 - \frac{t_2 - t_1}{4\beta}\right\}\mathrm{P}(\sigma, t_2g). \tag{8.7}$$

Then $\mathrm{h}_{\mu_2}(\sigma) \ge \mathrm{P}(\sigma, t_2g)/2 > 0$ and therefore by Theorem 8.2.4, $\frac{\mathrm{h}_{\mu_2}(\sigma)}{\chi_{\mu_2}(\sigma)} = \mathrm{HD}(\mu_2 \circ \pi^{-1}) \le \beta$. Hence $\int -gd\mu_2 \ge \mathrm{h}_{\mu_2}(\sigma)/\beta \ge \mathrm{P}(\sigma, t_2g)/2\beta$. Thus, using (8.7), Theorem 2.1.6 and Proposition 8.2.1, we get

$$\begin{aligned}
\mathrm{P}(\sigma, t_1 g) &\geq \mathrm{h}_{\mu_2}(\sigma) + \int t_1 g d\mu_2 = \mathrm{h}_{\mu_2}(\sigma) + \int t_2 g d\mu_2 + \int (t_1 - t_2) g d\mu_2 \\
&\geq \mathrm{P}(\sigma, t_2 g) - \mathrm{P}(\sigma, t_2 g)\frac{t_2 - t_1}{4\beta} + \mathrm{P}(\sigma, t_2 g)\frac{t_2 - t_1}{2\beta} \\
&= \mathrm{P}(\sigma, t_2 g) + \mathrm{P}(\sigma, t_2 g)\frac{t_2 - t_1}{4\beta} > \mathrm{P}(\sigma, t_2 g).
\end{aligned}$$

An application of Hölder's inequality shows that each function

$$t \mapsto \sum_{|\omega|=n} \exp\Big(\sup_{\tau \in [\omega]} \sum_{j=0}^{n-1} g(\sigma^j(\tau))\Big)$$

is log convex. Therefore the map $t \mapsto \mathrm{P}(t)$, $t \in (\theta, \infty)$, is convex and consequently continuous. □

Let us remark that it is possible that $\beta = \theta$. We will call such systems "strange" and deal with them in more detail in sections 8.4 and 8.5. Also, although it can happen that $\theta_S = 0$, we always have $P(0) \geq \log 2$ and therefore $h > 0$.

8.3 Perron–Frobenius operator, semiconformal measures and Hausdorff dimension

It follows from Proposition 8.2.5 that β is the first zero of the pressure function. We shall provide below more characterizations of this number. Given $t > \theta(S)$ we define in a familiar fashion the associated Perron–Frobenius operator acting on $C(X)$ as follows

$$\mathcal{L}_t(f)(x) = \sum_{i \in I} |\phi_i'(x)|^t f(\phi_i(x)).$$

Notice that the nth composition of $\mathcal{L}$ satisfies

$$\mathcal{L}_t^n(f)(x) = \sum_{|\omega|=n} |\phi_\omega'(x)|^t f(\phi_\omega(x)).$$

Consider the dual operator $\mathcal{L}_t^*$ acting on the space of finite Borel measures on X as follows

$$\mathcal{L}_t^*(\nu)(f) = \nu(\mathcal{L}_t(f)).$$

Notice that the map $\nu \mapsto \mathcal{L}_t^*(\nu)/\mathcal{L}_t^*(\nu)(1)$ sending the space of Borel probability measures into itself is continuous and by the Schauder-

Tikhonov theorem it has a fixed point. In other words $\mathcal{L}_t^*(\nu) = \lambda\nu$, for some probability measure ν, where $\lambda = \mathcal{L}_t^*(\nu)(1) > 0$. A probability measure m is said to be (λ, t)-semiconformal provided that $\mathcal{L}_t^*(m) = \lambda m$. If $\lambda = 1$ we simply speak about t-semiconformal measures. Repeating a short argument from the proof of Theorem 3.6 of [MU1] we shall first prove the following.

Lemma 8.3.1 *If m is a (λ, t)-semiconformal measure for the system S with $\lambda > 0$, then $m(J) = 1$.*

Proof. For each $n \geq 1$ let $X_n = \bigcup_{|\omega|=n} \phi_\omega(X)$. The sets X_n form a descending family and $\bigcap_{n\geq 1} X_n = J$. Notice that $1\!\!1_{X_{|\omega|}} \circ \phi_\omega = 1\!\!1_X$ for all $\omega \in I^*$ and therefore, using the (λ, t)-semiconformality of m, we obtain for every $n \geq 1$

$$\begin{aligned}
\lambda^n m(X_n) &= \int 1\!\!1_{X_n} d\mathcal{L}_t^{*n}(m) = \int \mathcal{L}_t^n(1\!\!1_{X_n}) dm \\
&= \int \sum_{|\omega|=n} |\phi'_\omega|^t (1\!\!1_{X_n} \circ \phi_\omega) dm \\
&= \int \sum_{|\omega|=n} |\phi'_\omega|^t dm = \int 1\!\!1_X d\mathcal{L}_t^{*n}(m) \\
&= \int \lambda^n 1\!\!1_X dm = \lambda^n.
\end{aligned}$$

Thus, $m(X_n) = 1$ and therefore $m(J) = m(\bigcap_{n\geq 1} X_n) = 1$. □

We set

$$\psi_n(t) = \sum_{|\omega|=n} ||\phi'_\omega||^t.$$

We note that $\theta(S) = \inf\{t : \psi(t) = \psi_1(t) < \infty\}$. In order to demonstrate the existence of $(e^{P(t)}, t)$-semiconformal measures we shall prove the following.

Lemma 8.3.2 *If $t > \theta(S)$ and $\mathcal{L}_t^*(m) = \lambda m$ for some measure m on X, then $\lambda = e^{P(t)}$.*

Proof. We first show the easier part that $\lambda \le e^{P(t)}$. Indeed, for all $n \ge 1$

$$\lambda^n = \int \mathcal{L}_t^n(1\!\!1_X)dm = \int \sum_{|\omega|=n} |\phi'_\omega(x)|^t dm(x)$$
$$\le \int \sum_{|\omega|=n} ||\phi'_\omega||^t dm = \sum_{|\omega|=n} ||\phi'_\omega||^t$$

and therefore

$$\log\lambda \le \lim_{n\to\infty} \frac{1}{n} \log \sum_{|\omega|=n} ||\phi'_\omega||^t = P(t). \tag{8.8}$$

In order to prove the opposite inequality, for each $p \ge 1$, let $T_p = \sum_{\omega\in I_g^p} ||\phi'_\omega||^t$, where I_g^p is the set of those words $\omega \in I^p$ such that ω_{p-1}, ω_p are not the same parabolic element. For each n,

$$\psi_n(t) = \sum_{|\omega|=n} ||\phi'_\omega||^t \le \sum_{\omega\in I_g^n} ||\phi'_\omega||^t + \sum_{i\in\Omega} \sum_{\omega\in I_g^{n-1}} ||\phi'_\omega||^t ||\phi'_i||^t$$
$$+ \sum_{i\in\Omega} \sum_{\omega\in I_g^{n-2}} ||\phi'_\omega||^t ||\phi'_{ii}||^t + \cdots + \sum_{i\in\Omega} ||\phi'_{i^n}||^t \le \sum_{k=0}^{n} \#\Omega T_k,$$

where $T_0 = 1$. Take $0 \le q(n) \le n$ that maximizes T_k. Then $\psi_n \le (n+1)\#\Omega T_{q(n)}$ and therefore

$$\begin{aligned}
P(t) &= \lim_{n\to\infty} \frac{1}{n} \log\psi_n \\
&\le \liminf_{n\to\infty} \left(\frac{\log(n+1)}{n} + \frac{q(n)}{n} \cdot \frac{1}{q(n)} \log T_{q(n)} + \frac{1}{n} \log \#\Omega \right) \\
&\le \max\left\{ 0, \limsup_{n\to\infty} \frac{1}{n} \log T_n \right\}.
\end{aligned} \tag{8.9}$$

Let

$$\tilde{\mathcal{L}}_t^n(1) = \sum_{\omega\in I_g^n} |\phi'_\omega|^t.$$

It follows from condition (5) of a subparabolic iterated function system that for all $n \ge 1$, $\omega \in I_g^n$ and all $x \in X$

$$||\phi'_\omega||^t \le K^t |\phi'_\omega(x)|^t.$$

Summing we have $T_n \le K^t \tilde{\mathcal{L}}_t^n(1)(x)$ and integrating this inequality with respect to the measure m, we get

$$T_n \le K^t \int \tilde{\mathcal{L}}_t^n(1)(x)dm(x) \le K^t\lambda^n.$$

Thus, by (8.9)

$$\mathrm{P}(t) \leq \max\{0, \limsup_{n\to\infty} \frac{1}{n} \log T_n\} \leq \max\{0, \log \lambda\}.$$

If now $t < \beta(S)$, then by Proposition 8.2.5(2), $\mathrm{P}(t) > 0$, and we therefore get $\mathrm{P}(t) \leq \log \lambda$. Thus, we are done in this case. So, suppose that $t \geq \beta(S)$. Then by Proposition 8.2.5(3), $\mathrm{P}(t) = 0$ and in view of (8.8) we are left to show that $\lambda \geq 1$. In order to do it fix an arbitrary $0 < \eta < 1$. It follows from conditions (4) and (2) that for all n large enough, say $n \geq n_0$, $|\phi'_{i^n}(x)| \geq \eta^n$ for all $i \in \Omega$ and all $x \in X$. Fix $j \in \Omega$. We then have for all $n \geq n_0$

$$\lambda^n = \int 1\!\!1 d\mathcal{L}^{*n}(m) = \int \sum_{|\omega|=n} |\phi'_\omega|^t dm \geq \int |\phi'_{j^n}|^t dm \geq \int \eta^{tn} dm = \eta^{tn}.$$

Thus $\lambda \geq \eta^t$ and letting $\eta \nearrow 1$ we get $\lambda \geq 1$. □

Lemma 8.3.3 *For every* $t > \theta(S)$ *a* $(\mathrm{P}(t), t)$*-semiconformal measure exists.*

Proof. In view of Lemma 8.3.2 it suffices to prove the existence of an eigenmeasure of the conjugate operator $\mathcal{L}_t^*$. But this has been done in the paragraph preceding Lemma 8.3.1. □

Let $e = e(S)$ be the infimum of the exponents for which a t-semiconformal measure exists. We shall shortly see this infimum is a minimum. Also, let $h = h_S$ be the Hausdorff dimension of the limit set J. As an immediate consequence of Proposition 8.2.5(3) and Lemma 8.3.3 we get the following.

Lemma 8.3.4 $e(S) \leq \beta(S)$.

Now, suppose that m is t-semiconformal, or equivalently

$$\int \sum_{\omega \in I^n} |\phi'_\omega|^t (f \circ \phi_\omega)\, dm = \int f\, dm, \tag{8.10}$$

for every continuous function $f : X \to I\!R$. Since this equality extends to all bounded measurable functions f, we get

$$m(\phi_\omega(A)) = \sum_{\tau \in I^n} \int |\phi'_\tau|^t (1_{\phi_\omega(A)} \circ \phi_\tau)\, dm \geq \int_A |\phi'_\omega|^t\, dm \tag{8.11}$$

for all $n \geq 1$, $\omega \in I^n$ and all Borel subsets A of X.

Our next task in this section is to note that $h \leq e$. But this follows immediately from the following lemma whose proof, using (8.10), is the same as the proof of Theorem t4.4.1.

Lemma 8.3.5 *If m is a t-semiconformal measure, then $\mathcal{H}^t|_J \ll m$ and the Radon-Nikodym derivative $\frac{d\mathcal{H}^t}{dm}$ is uniformly bounded from above.*

Since obviously $\beta \leq h$, we have thus proved the following characterization of the Hausdorff dimension of the limit set; cf. Theorem 4.2.13, which is true in the hyperbolic context, and the discussion preceding it.

Theorem 8.3.6 *$e = \beta = h =$ the minimal zero of the pressure function.*

As an immediate consequence of Lemma 8.3.5, Lemma 8.3.3, Proposition 8.2.5(3) and Theorem 8.3.6 we get the following.

Corollary 8.3.7 *The h-dimensional Hausdorff measure of the limit set J is finite.*

8.4 The associated hyperbolic system. Conformal and invariant measures

In this section we describe how to associate to our parabolic system a new system which is hyperbolic and whose properties we applyto study the original system, in particular to prove the existence of h-conformal measures. However, we begin this section with a result describing the structure of t-semiconformal measures with exponents $t > h$. Let

$$\Omega_* = \{\phi_\omega(x_i) : i \in \Omega,\, \omega \in I^*\}.$$

So, Ω_* is the set of orbits of parabolic points. The following theorem allows us to conclude that a t-semiconformal measure is conformal provided the parabolic orbits do not mix.

Theorem 8.4.1 *If $t > h$ and m_t is a t-semiconformal measure, then m_t is supported on Ω_*, that is $m_t(\Omega_*) = 1$. If for every $\omega \in I^*$ and every $i \in \Omega$, $\pi^{-1}(\phi_\omega(x_i)) = \omega i^\infty$, then each t-semiconformal measure ($t > h$) is t-conformal.*

Proof. For every $r > h$ let m_r be an r-semiconformal measure. Note that the existence of at least one such measure (for every $r > h$) has been proved in Lemma 8.3.3, cf. also Proposition 8.2.5(3) and Theorem 8.3.6. It is then not difficult to see that for every $r > h$ there exists a Borel probability measure $\tilde{m}_r$ on I^∞ such that $\tilde{m}_r \circ \pi^{-1} = m_r$ and $\tilde{m}_r([\omega]) = \int |\phi'_\omega|^r dm_r$, for all $\omega \in I^*$. Now, fix $t > h$ and $h < s < t$. Let $\tilde{\Omega}_* = \{\omega i^\infty : i \in \Omega,\ \omega \in I^*\}$. If $\omega \notin \tilde{\Omega}_*$, then there exists an increasing infinite sequence $\{n_k\}_{k=1}^\infty$ such that either $\omega_{n_k} \notin \Omega$ or $\omega_{n_k-1} \neq \omega_{n_k}$. In either case, using condition (5) we get

$$\begin{aligned} m_t([\omega|_{n_k}]) &= \int |\phi'_{\omega_{n_k}}|^t dm_t \le ||\phi'_{\omega_{n_k}}||^t = ||\phi'_{\omega_{n_k}}||^{t-s} ||\phi'_{\omega_{n_k}}||^s \\ &\le ||\phi'_{\omega_{n_k}}||^{t-s} K^s \int |\phi'_{\omega_{n_k}}|^s dm_s \\ &= K^{-s} ||\phi'_{\omega_{n_k}}||^{t-s} m_s([\omega|_{n_k}]). \end{aligned} \tag{8.12}$$

It immediately follows from conditions (6) and (2) that $\lim_{k\to\infty} ||\phi'_{\omega_{n_k}}|| = 0$. Combining this and (8.12) we conclude that $\tilde{m}_t(I^\infty \setminus \tilde{\Omega}_*) = 0$ or equivalently $m_t(\tilde{\Omega}_*) = 1$. Since $\pi^{-1}(\Omega_*) \supset \tilde{\Omega}_*$, we get $m_t(\Omega_*) = \tilde{m}_t \circ \pi^{-1}(\Omega_*) \ge \tilde{m}_t(\tilde{\Omega}_*) = 1$. The proof of the first part of Theorem 8.4.1 is complete. The second part is an immediate consequence of (8.10) applied to the indicator functions of the sets of the form $\phi_\omega(A)$, where $\omega \in I^*$ and A is a Borel subset of X. □

Consider now the system S^* generated by I_*, where

$$I_* = \{\phi_{i^n j} : n \ge 1,\ i \in \Omega,\ i \neq j\} \cup \{\phi_k :\ k \in I \setminus \Omega\}.$$

It immediately follows from our assumptions that the following is true.

Theorem 8.4.2 *The system S^* is a hyperbolic conformal iterated function system.*

Proof. The conditions (4a)–(4d) immediately follow from our assumptions and the definition of the system S^*. We only need to prove condition (4e). So, fix $i \in \tilde{\Omega}$ and $j \in I \setminus \{i\}$. Consider arbitrary $n \ge 1$ and

$x, y \in X$. Write $t = \min\{|\phi_i'(x)| : i \in \tilde{\Omega}, x \in X\} > 0$. We then get

$$\begin{aligned}
&\Big||\phi_{i^n j}'(y)| - |\phi_{i^n j}'(x)|\Big| \\
&\quad = |\phi_{i^n j}'(x)| \left|1 - \frac{|\phi_{i^n j}'(y)|}{|\phi_{i^n j}'(x)|}\right| \leq ||\phi_{i^n j}'|| \left|\log \frac{|\phi_{i^n j}'(y)|}{|\phi_{i^n j}'(x)|}\right| \\
&\quad \leq \Big|\log|\phi_j'(y)| - \log|\phi_j'(x)|\Big| + \sum_{k=0}^{n-1} \Big|\log|\phi_i'(\phi_{i^k j}(y))| - \log|\phi_i'(\phi_{i^k j}(x))|\Big| \\
&\quad \leq \left(\frac{K}{||\phi_j'||}\Big||\phi_j'(y)| - |\phi_j'(x)|\Big| + \sum_{k=0}^{n-1} \frac{1}{t}\,\Big||\phi_i'(\phi_{i^k j}(y))| - |\phi_i'(\phi_{i^k j}(x))|\Big|\right) \\
&\quad \leq \left(KL|y-x| + \frac{1}{t}\sum_{k=0}^{n-1} L|\phi_{i^k j}(y) - \phi_{i^k j}(x)|\right) \\
&\quad \leq \left(KL|y-x| + \frac{L}{t}\sum_{k=0}^{n-1} ||\phi_{i^k}'||_{X_i} |\phi_j(y) - \phi_j(x)|\right) \\
&\quad \leq \left(KL + \frac{L}{t}\sum_{k=0}^{\infty} ||\phi_{i^k}'||_{X_i}\right)|y-x| \\
&\quad \leq L\left(K + \frac{T}{t}\right)|y-x|.
\end{aligned}$$

□

In particular all the families $t\text{Log} = \{t \log|\phi_b'|\}_{b \in I_*}$ are Hölder.

The limit set generated by the system S^* is denoted by .

The next lemma shows that as long as we are interested in the fractal geometry of the limit set J_S, we can replace this set by J_S^*.

Lemma 8.4.3 *The limit sets J and J^* of the systems S and S^* respectively differ only by a countable set: $J^* \subset J$ and $J \setminus J^*$ is countable.*

Proof. Indeed, it is obvious that $J^* \subset J$. On the other hand, the only infinite words generated by S but not generated by S^* are of the form ωi^∞, where ω is a finite word and i is a parabolic element of S. □

Definition 8.4.4 *If S is an iterated function system with limit set J, then a measure ν supported on J is said to be invariant for the system*

S provided

$$\nu(E) = \nu\left(\bigcup_{i\in I}\phi_i(E)\right)$$

and ν *is said to be ergodic for the system* S *provided* $\nu(E) = 0$ *or* $\nu(J \setminus E) = 0$ *whenever* $\nu(E \triangle \bigcup_{i\in I}\phi_i(E)) = 0$.

Let us set up some notation. Let $J_0 \subset J$ consist of all points with a unique code under S. For each $x = \pi(\omega) \in J_0$ express $\omega = i^n\tau$, where i is a parabolic element, $n \geq 0$, $\tau_1 \neq i$ and define $n(x) = n$. For each $k \geq 0$, put

$$B_k = \{x \in J_0 : n(x) = k\} \text{ and } D_k = \{x \in J_0 : n(x) \geq k\}.$$

The next theorem shows how to obtain invariant (σ-*finite*) measures for the parabolic system S provided that a probability S^*-invariant measure is given.

Theorem 8.4.5 *Suppose that* μ^* *on* J^* *is a probability measure invariant under* S^* *and* $\mu^*(J_0) = 1$. *Define the measure* μ *by setting for each Borel set* $E \subset J_0$,

$$\mu(E) = \sum_{k=0}\sum_{|\omega|=k}\mu^*(\phi_\omega(E)\cap D_k) = \sum_{k\geq 1}\sum_{i\in\Omega}\mu^*(\phi_{i^k}(E)) + \mu^*(E). \quad (8.13)$$

Then μ *is a* σ-*finite invariant measure for the system* S *and* μ^* *is absolutely continuous with respect to* μ. *If, for each* $i \in I$, *the measure* $\mu^* \circ \phi_i$ *is absolutely continuous with respect to the measure* μ^*, *then* μ *and* μ^* *are equivalent; and if* μ^* *is ergodic for the system* S^*, *then* μ *is ergodic for the system* S. *Moreover, in this last case* μ *is unique up to a multiplicative constant.*

Proof. Let us check first that μ is S-invariant. Indeed,

$$\begin{aligned}\mu\left(\bigcup_{j\in I}\phi_j(E)\right) &= \sum_{k\geq 1}^{\infty}\sum_{i\in\Omega}\mu^*\left(\phi_{i^k}(\bigcup_{j\in I}\phi_j(E))\right) + \sum_{j\in I}\mu^*(\phi_j(E)) \\ &= \sum_{k=1}^{\infty}\sum_{i\in\Omega}\sum_{j\in I}\mu^*(\phi_{i^k j}(E)) + \sum_{j\in I}\mu^*(\phi_j(E))\end{aligned}$$

$$= \sum_{k=1}^{\infty}\sum_{i\in\Omega}\sum_{j\in I}\mu^*(\phi_{i^kj}(E)\cap D_{k+1}) + \sum_{k=1}^{\infty}\sum_{i\in\Omega}\sum_{j\in I}\mu^*(\phi_{i^kj}(E)\cap B_k)$$
$$+ \sum_{j\in I}\mu^*(\phi_j(E))$$
$$= \sum_{k=2}^{\infty}\sum_{i\in\Omega}\mu^*(\phi_{i^k}(E)) + \sum_{k=1}^{\infty}\sum_{i\in\Omega}\sum_{j\neq i}\mu^*(\phi_{i^kj}(E)) + \sum_{j\in I}\mu^*(\phi_j(E))$$
$$= \sum_{k=2}^{\infty}\sum_{i\in\Omega}\mu^*(\phi_{i^k}(E)) + \mu^*(E) + \sum_{j\in\Omega}\mu^*(\phi_j(E)) = \mu(E),$$

where the last equality holds due to the invariance of μ^* under S^*. The invariance of μ has been proved. Since $J_0 = \bigcup_{n\geq 0} B_n$, in order to show that μ is σ-finite it suffices to demonstrate that $\mu(B_n) < \infty$ for every $n \geq 0$. And indeed, given $n \geq 0$ we have

$$\mu(B_n) = \sum_{k\geq 1}\sum_{i\in\Omega}\mu^*(\phi_{i^k}(B_n)) + \mu^*(B_n). \tag{8.14}$$

Now, for every $i \in \Omega$,

$$\phi_{i^k}(B_n) \subset B_k \cup B_{n+k}.$$

Hence, $\mu(B_n) \leq 2\#\Omega \sum_{k=0}^{\infty}\mu^*(B_k) = 2\#\Omega\mu^*\left(\bigcup_{k=0}^{\infty} B_k\right) = 2\#\Omega\mu^*(X) = 2\#\Omega$. Thus, μ is σ-finite. It follows in turn from (8.13) that $\mu(E) = 0$ implies $\mu^*(E) = \mu^*(E\cap D_0) = 0$. So, μ^* is absolutely continuous with respect to μ.

Now suppose that for each $i \in I$, the measure $\mu^*\circ\phi_i$ is absolutely continuous with respect to the measure μ^*. If $\mu^*(E) = 0$, then $\mu^*(\phi_\omega(E)) = 0$ for all $\omega \in I^*$. Thus, it follows from (8.13) that $\mu(E) = 0$ and the equivalence of μ and μ^* is shown. Suppose now that E is S-invariant, implying that $\bigcup_{i\in I}\phi_i(E) \subset E$. Then $\bigcup_{\omega\in I_*}\phi_\omega(E) \subset E$ and since μ^* is ergodic, either $\mu^*(E) = 0$ or $\mu^*(E^c) = 0$. Since μ is absolutely continuous with respect to μ^*, this implies that either $\mu(E) = 0$ or $\mu(E^c) = 0$. Hence μ is ergodic and the proof is complete. □

We can provide the following necessary and sufficient condition for the S-invariant measure produced in this theorem to be finite.

Theorem 8.4.6 *If the assumptions of Theorem 8.4.5 are satisfied, then the σ-finite measure μ produced there is finite if and only if*

$$\sum_{n\geq 1} n\mu^*(B_n) < \infty.$$

Proof. Let us set $B_n^i = \{x \in J_0 : x = \pi(j^n\tau), j \in \Omega\backslash\{i\}, \tau \in I^\infty, \tau_1 \neq j\}$ and $D^i = \bigcup_{n\geq 0} B_n^i$. By (8.14), we can write

$$\begin{aligned}
\mu(J) &= \sum_{n\geq 0} \mu(B_n) = \sum_{n\geq 0}\sum_{k\geq 0}\sum_{i\in\Omega} \mu^*(\phi_{i^k}(B_n)) \\
&= \sum_{k\geq 0}\sum_{n\geq 0} \mu^*(B_{k+n}) + \sum_{k\geq 0}\sum_{n\geq 0}\sum_{i\in I} \mu^*(\phi_{i^k}(B_n^i)) \\
&= \sum_{k\geq 0}\sum_{n\geq 0} \mu^*(B_{k+n}) + \sum_{k\geq 0}\sum_{i\in\Omega} \mu^*(\phi_{i^k}(D^i)) \\
&= \sum_{n\geq 0}(n+1)\mu^*(B_n) + \sum_{n\geq 0} \mu^*(B_n) = \sum_{n\geq 0}(n+2)\mu^*(B_n).
\end{aligned}$$

□

The main result of this section, relating conformal measures of the systems S and S^*, is provided by the following.

Theorem 8.4.7 *Suppose that S is a parabolic conformal iterated function system and the associated hyperbolic system S^* is regular. Then m, the h-conformal measure for S^*, is also h-conformal for S and m is the only h-semiconformal measure for S.*

Proof. We will first show that m is h-conformal for the system S over the limit set J. We will then associate with S one more hyperbolic system S^{**} and use some properties of this system to verify that m is h-conformal for S. Since $m(J^*) = 1$, the probability measure m clearly satisfies the first condition for conformality: $m(J) = 1$. Next, we will show that m satisfies equation (4.30) for all Borel subsets A of J. Since $J \setminus J^*$ is countable and m is atomless, it suffices to show that (4.30f) holds for Borel subsets of J^*. Also, since (4.30) holds whenever i is a hyperbolic index even for Borel subsets of X, we only need to verify (4.30) for parabolic indices. Let

$$\mathcal{G} = \{A : A \text{ is a Borel subset of } J^* \text{ and (4.30) holds } \forall i \in \Omega\}.$$

Since $\mathcal{G}$ is closed under monotone limits, it suffices to show that (4.30) holds for every subset U of J^* which is relatively open. Let

$$\Gamma = \{\omega \in I_*^\infty : \omega = (a_1b_1),(a_2b_2),(a_3b_3),\ldots;\forall n\ a_n,b_n \in \Omega,\ b_n \neq a_n, a_{n+1}\}.$$

Let $W = \pi(\Gamma)$. Using Theorem 2.2.4, Theorem 3.2.3, Proposition 4.2.5 and Birkhoff's ergodic theorem, we see that $m(W) = 0$ and $m(\phi_i(W)) = 0, \forall i \in \Omega$. Let us demonstrate that if $i \in \Omega$ and $\omega = (\omega_1,\omega_2,\omega_3,\ldots) \in I_*^\infty \setminus \Gamma$, then there is some l such that for every $k \geq l, (\omega_1,\ldots,\omega_k) \in I_*^*$ and the concatenation $i*\omega_1*\cdots*\omega_k$ can be parsed (or regrouped) so that it represents an element of I_*^*. To see this, first suppose that $\omega_1 \in I\setminus\Omega$. Then $l = 1$ since $i*\omega_1*\cdots*\omega_k$ can be parsed as $i\omega_1,\omega_2,\ldots,\omega_k$ which is an element of I_*^*. Now, suppose $\omega_1 = p^nq$ where $p \in \Omega$ and $p \neq q$. If $p = i$, then again $l = 1$, since $i*\omega_1*\cdots*\omega_k$ can be parsed as $i^{n+1}q,\omega_2,\ldots,\omega_k$ which is an element of I_*^*. If $i \neq p$ and $n > 1$, then $i*\omega_1*\cdots*\omega_k$ can be parsed as $(ip, p^{n-1}q,\omega_2,\omega_3,\ldots,\omega_k) \in I_*^*$ and also in this case $l = 1$. If, on the other hand, $n = 1$ and $p = i$, then $\omega_1 = a_1b_1$, where $a_1 \in \Omega$ and $b_1 \neq a_1$. If $b_1 \in I\setminus\Omega$, then $i*\omega_1*\cdots*\omega_k$ can be parsed as $(ia_1,b_1,\omega_2,\omega_3,\ldots,\omega_k) \in I_*^*$ and $l = 1$. So, suppose that $b_1 \in \Omega$. Now, consider ω_2. If $\omega_2 \in I\setminus\Omega$, then the concatenation $i*\omega_1*\cdots*\omega_k$ can be parsed as $(ia_1,b_1\omega_2,\omega_3,\ldots,\omega_k) \in I_*^*$ and $l = 2$. Otherwise $\omega_2 = p^nq$, where $p \in \Omega$, $q \neq p$ and $n \geq 1$. If $p = b_1$, then $i*\omega_1*\cdots*\omega_k$ can be parsed as $(ia_1, b_1^{n+1}q,\omega_3,\ldots,\omega_k) \in I_*^*$ and $l = 2$. If $p \neq b_1$ and $n > 1$, then $i*\omega_1*\cdots*\omega_k$ can be parsed as $(ia_1,b_1p,p^{n-1}q,\omega_3,\ldots,\omega_k) \in I_*^*$ and $l = 2$. If, on the other hand, $n = 1$, then $\omega_2 = a_2b_2$, where $a_2 \neq b_1, b_2$. If $b_2 \in I\setminus\Omega$, then $i*\omega_1*\cdots*\omega_k$ can be parsed as $(ia_1,b_1a_2,b_2,\omega_3,\ldots,\omega_k) \in I_*^*$ and $l = 2$. So, we may assume that $b_2 \in \Omega$. Now, excluding inductively in this manner the cases when $i*\omega_1*\cdots*\omega_k$ can be parsed in such a fashion that it would belong to I_*^*, we would end up with the conclusion that $\omega \in \Gamma$, contrary to our assumption. Now, let $U \subset J^*$ be relatively open. Then there is a set $M \subset I_*^*$, consisting of incomparable words and such that $U\setminus W \subset \bigcup_{\tau\in M}\phi_\tau(J^*) \subset U$, and if $\tau \in M$ then $i*\tau \in I_*^*$. Thus,

$$m(\phi_i(U)) = m(\phi_i(\bigcup\phi_\tau(J^*))) \cup (U\setminus\bigcup\phi_\tau(J^*)) = \sum_\tau m(\phi_i(\phi_\tau(J^*)))$$

$$= \sum_\tau \int_{J^*} |(\phi_i\circ\phi_\tau)'|^h dm = \sum_\tau \int_{\phi_\tau(J^*)} |\phi_i'|^h dm = \int_U |\phi_i'|^h dm,$$

where the third equality follows since m is h-conformal for the system S^* and in the fourth equality we additionally employed the change of variables formula. Now, we want to show

$$m(\phi_i(J) \cap \phi_j(J)) = 0$$

whenever $i \neq j$. Again, it suffices to verify this when J is replaced by J^* and at least one of the indices i and j is parabolic. As before there is a set $M_i \subset I_*^*$ of incomparable words and such that $J^* \setminus W \subset \bigcup_{\tau \in M_i} \phi_\tau(J^*) \subset J^*$, and if $\tau \in M_i$ then $i * \tau \in I_*^*$. Also, let $M_j \subset I_*^*$ have similar properties with respect to the index j. Then

$$\begin{aligned} m(\phi_i(J) \cap \phi_j(J)) &= m(\bigcup_{\tau,\rho \in M_i \times M_j} \phi_{i\tau}(J^*) \cap \phi_{j\rho}(J^*)) \\ &\leq \sum_{M_i \times M_j} m(\phi_{i\tau}(J^*) \cap \phi_{j\rho}(J^*)) = 0. \end{aligned}$$

Finally, to show that m is conformal, we must demonstrate that (4.30) and (4.31) hold whenever A is a Borel subset of X. Note that it suffices to show that $m(A) = 0$ implies $m(\phi_i(A)) = 0$, for all Borel subsets A of X and all parabolic indices i. In order to prove this, we introduce a new hyperbolic system. The index set for this system is $I_{**} = I^3 \setminus \{(i,i,i) : i \in \Omega\} \cup \{p^n q : p \in \Omega,\ q \neq p,\ n \geq 2\}$. Let us prove that the system S^{**} satisfies the bounded distortion property. To see this, read a word $\omega \in I_{**}^*$ as a word in $I^* : \omega = (\omega_1, \omega_2, \ldots, \omega_n)$. If $\omega_n \in I \setminus \Omega$, then we have bounded distortion by property (5) of the system S. If $\omega_n \in \Omega$ and $\omega_{n-1} \neq \omega_n$, then again by property (5) we have bounded distortion with constant K. If $\omega_{n-1} = \omega_n$, then $\omega_{n-2} \neq \omega_{n-1}$, by the definition of I_{**}^*. Then the word $\omega|_{n-1}$ satisfies the hypothesis of condition (5) and so

$$\begin{aligned} \frac{|\phi'_\omega(y)|}{|\phi'_\omega(x)|} &= \frac{|\phi'_{\omega|_{n-1}}(\phi_{\omega_n}(y))||\phi'_{\omega_n}(y)|}{|\phi'_{\omega|_{n-1}}(\phi_{\omega_n}(y))||\phi'_{\omega_n}(y)|} \\ &\leq K \max\left\{ \frac{||\phi'_i||}{\min\{\phi'_i(x) : x \in X\}} : i \in \Omega \right\}, \end{aligned}$$

where the last number is finite since Ω is. To see that S^{**} satisfies the open set condition, notice that $\phi_{ijk}(\text{Int}(X)) \cap \phi_{pqr}(\text{Int}(X)) = \emptyset$ for all $ijk \neq pqr$. Next consider $\phi_{i^n j}(\text{Int}(X)) \cap \phi_{p^m q}(\text{Int}(X))$, where $n, m \geq 2$. If $i \neq p$, this intersection is empty. Also if $i = p$ and $n \neq m$, the intersection is empty. Otherwise, $q \neq j$ and the intersection is empty. Finally, consider $\phi_{i^n j}(\text{Int}(X)) \cap \phi_{pqr}(\text{Int}(X))$, where $n \geq 2$. If $i \neq p$ or if $i = p$ and $q \neq i$, the intersection is empty. Otherwise, $i = p = q$ and in this case $r \neq i$ since the word (i,i,i) is not allowed in I_{**}. Finally, the hyperbolicity of the system S^{**} is an immediate consequence of property

(6). So, S^{**} is a hyperbolic conformal iterated function system. Also, since each element of I_*^∞ can be parsed into an element of I_{**}^∞, we have $J^{**} = J^* = J \setminus \{\text{eventually parabolic points}\}$. Also notice that if the system S^* is regular, then the system S^{**} is regular. To see this note that we have already shown that if m is conformal for S^*, then m is conformal for S over J. Thus m is conformal for S^{**} over J. So, for each n, $1 = \int_J dm = \int_J \sum_{\omega \in I_{**}^n} |\phi'_\omega(x)| dm$. But, for each $x \in J$, we have

$$\sum_{\omega \in I_{**}^n} ||\phi'_\omega||^h \geq \sum_{\omega \in I_{**}^n} |\phi'_\omega(x)|^h \geq (K^{**})^{-h} \sum_{\omega \in I_{**}^n} ||\phi'_\omega||^h,$$

where K^{**} is the distortion constant for the system S^{**} over X. Integrating this formula against the measure m we get

$$\sum_{\omega \in I_{**}^n} ||\phi'_\omega||^h \geq 1 \geq (K^{**})^{-h} \sum_{\omega \in I_{**}^n} ||\phi'_\omega||^h.$$

From this it immediately follows that $\mathrm{P}^{**}(h) = 0$. But, this is equivalent to saying that there is an h-conformal measure m^{**} for the system S^{**} over X. We only need to prove that $m^{**} = m$. Let G be open relative to J^*. Let W be a collection of incomparable words in I_{**} such that $G = \bigcup_{\omega \in W} \phi_\omega(J^*)$. Since m is conformal for S^{**} over J,

$$\begin{aligned} m(G) &= \sum_{\omega \in W} \int_J |\phi'_\omega| dm \leq \sum_{\omega \in W} K^h ||\phi'_\omega|| \leq \sum_{\omega \in W} K^h K^{**h} \int_J |\phi'_\omega| dm^{**} \\ &= K^h K^{**h} m^{**}(G). \end{aligned}$$

Interchanging m and m^{**} in the above estimate, we get

$$(K^h K^{**h})^{-1} m^{**}(G) \leq m(G) \leq K^h K^{**h} m^{**}(G).$$

From this it follows that m and m^{**} are equivalent. To show that $m = m^{**}$, let A be a Borel subset of X. Then $m(\phi_\omega(A)) = m(\phi_\omega(A \cap J)) + m(\phi_\omega(A \setminus J))$. But, since m^{**} is conformal over X, $m^{**}(A \setminus J) = \int_{A \setminus J} |\phi'_\omega|^h dm^{**} = 0$. So, since m is conformal for S over J, we have $m(\phi_\omega(A)) = \int_{A \cap J} |\phi'_\omega|^h dm = \int_A |\phi'_\omega|^h dm$. Also, one can show that (4.30) holds using the same procedure. Thus, m is conformal for S^{**} over X. Finally, to see that m is conformal for the entire system S over X, let $i \in \Omega$ and choose an arbitrary $q \neq i$, $q \in I$. Then $iq \in I_*$ and $iqi \in I_{**}$. Thus,

$$\int_{\phi_i(A)} |\phi'_{iq}|^h dm = m(\phi_{iq}(\phi_i(A)) = m(\phi_{iqi}(A))) = \int_A |\phi'_{iqi}|^h dm.$$

So, if $m(A) = 0$, then since $|\phi'_{iq}|^h$ is positive on $\phi_i(A)$, we have $m(\phi_i(A)) = 0$.

In order to prove the second part of our theorem suppose that ν is an arbitrary measure supported on J and satisfying

$$\nu(\phi_\omega(A)) \geq \int_A |\phi'_\omega|^h d\nu \tag{8.15}$$

for all Borel sets $A \subset X$ and all $\omega \in I^*$. We show that m is absolutely continuous with respect to ν. Indeed, for every $\omega \in I^*_*$ we have

$$\begin{aligned}\nu(\phi_\omega(X)) &\geq \int_X |\phi'_\omega|^h d\nu \geq K^{-h}||\phi'_\omega||^h \\ &\geq K^{-h}\int_X |\phi'_\omega|^h dm = K^{-h}m(\phi_\omega(X)).\end{aligned}$$

Next, consider an arbitrary Borel set $A \subset X$ such that $\nu(A) = 0$. Fix $\epsilon > 0$. Since ν is regular there exists an open subset G of X such that $A \cap J^* \subset G$ and $\nu(G) \leq \epsilon$. There now exists a family $\mathcal{F} \subset I^*_*$ of mutually incomparable words such that $A \cap J^* \subset \bigcup_{\omega\in\mathcal{F}} \phi_\omega(X) \subset G$. It easily follows from Lemma 4.2.4 that there exists a universal upper bound M on the multiplicity of the family $\{\phi_\omega(X) : \omega \in \mathcal{F}\}$. Hence, using the fact that m is supported on J^*, we obtain

$$\begin{aligned}m(A) = m(A \cap J^*) &\leq m\left(\bigcup_{\omega\in\mathcal{F}} \phi_\omega(X)\right) \leq \sum_{\omega\in\mathcal{F}} m(\phi_\omega(X)) \\ &\leq K^h \sum_{\omega\in\mathcal{F}} \nu(\phi_\omega(X)) \leq K^h M\nu\left(\bigcup_{\omega\in\mathcal{F}} \phi_\omega(X)\right) \leq K^h M\nu(G) \\ &\leq K^h M\epsilon.\end{aligned}$$

Thus, letting $\epsilon \searrow 0$, we get $m(A) = 0$, which finishes the proof of the absolute continuity of m with respect to ν. Our next aim is to show that $\nu(J \setminus J^*) = 0$. Suppose on the contrary that $\nu(J \setminus J^*) > 0$. Set $P = \{\phi_\omega(x_i) : i \in \Omega,\ \omega \in I^*\}$. Since $J \setminus J^* \subset P$, $\nu(P) > 0$. Write $\nu = \nu_0 + \nu_1$, where $\nu_0|_{X\setminus P} = 0$ and $\nu_1|_P = 0$. Thus $\nu_0(P) = \nu(P) > 0$. Since $\phi_\omega(P) \subset P$ for all $\omega \in I^*$, we get for every Borel set $A \subset X$ and every $\omega \in I^*$

$$\begin{aligned}\nu_0(\phi_\omega(A)) &\geq \nu_0(\phi_\omega(A \cap P)) = \nu(\phi_\omega(A \cap P)) \\ &\geq \int_{A\cap P} |\phi'_\omega|^h d\nu = \int_A |\phi'_\omega|^h d\nu_0.\end{aligned}$$

Hence, multiplying ν_0 by $1/\nu_0(X)$, we conclude from what has been proved that m is absolutely continuous with respect to ν_0. Since $\nu_0(X \setminus P) = 0$, this implies that $m(X \setminus P) = 0$, and consequently $m(P) = 1$. This is a contradiction since any conformal measure of a hyperbolic system is continuous. Thus $\nu(J^*) = 1$. Suppose now in addition that ν is h-conformal for S. Then, as we have just proved, ν is h-conformal for S^* (this statement includes that fact that $\nu(J^*) = 1$). The equality $\nu = m$ follows now from Theorem 4.2.9.

Finally, notice that using the argument from the proof of Lemma 3.10 from [MU1] and proceeding as in the proof of Theorem 5.7 from [MU7] one could prove that m is the unique h-semiconformal measure for S. □

Following the case of hyperbolic systems we call a parabolic system regular if there exists an h-conformal measure for S supported on J^*. Since such a measure is h-conformal for S^*, as an immediate consequence of Theorem 8.4.7 we get the following.

Corollary 8.4.8 *The parabolic system is regular if and only if the associated system S^* is regular.*

In trying to say something about parabolic systems which are not regular, we are led to introduce the class of "strange" systems, which by definition are those systems for which there is no t with $0 < \mathrm{P}(t) < \infty$. In the hyperbolic case the strange systems coincide with systems which are not strongly regular or equivalently with those with $\theta = h$. This last characterization continues to be true also for parabolic systems and this class may also be characterized by the requirement of the existence of a number α (which then turns out to be $\theta = h$) such that $\mathrm{P}(t) = \infty$ for all $t < \alpha$ and $\mathrm{P}(t) = 0$ for all $t \geq \alpha$. Let us remark that we do not want to call the strange systems "irregular" since the irregular hyperbolic systems are precisely those for which no conformal measure exists whereas for a strange parabolic system the following questions remain open.

Questions. Can there exist a strange parabolic system such that the associated hyperbolic system is regular? Can there exist a strange parabolic system with a purely atomic h-conformal measure?

Let

$$\psi^*(t) := \psi_{S^*}(t).$$

We shall prove the following.

Proposition 8.4.9 *If the system S is strange, then so is S^*.*

Proof. Since $h_{S^*} = h_S$, $\mathrm{P}^*(t) \le 0$ for all $t \ge h_S$. So, we are only left to show that $\mathrm{P}^*(t) = \infty$ for all $t < h_S$. And indeed, fix $t < h_S$. Since S is strange, $\mathrm{P}(t) = \infty$ and therefore $\psi(t) = \infty$. Since Ω is finite, this implies that $\sum_{i \in I \setminus \Omega} ||\phi_i'||^t = \infty$. But then $\psi^*(t) \ge \sum_{i \in I \setminus \Omega} ||\phi_i'||^t = \infty$. Hence $\mathrm{P}^*(t) = \infty$. □

Let us briefly touch on the packing measure of J. Since J^* is dense in J, as an immediate consequence of Theorem 4.5.2 and Theorem 8.4.7 we get the following.

Corollary 8.4.10 *Suppose that S is a parabolic iterated function system and the associated hyperbolic system S^* is regular. If $J \cap \mathrm{Int}(X) \neq \emptyset$ (that is, if the strong open set condition is satisfied), then the h-dimensional packing measure of J is positive.*

Let us remark here that in Corollary 8.3.7 we have proved that the h-dimensional Hausdorff measure of J is finite.

Finally, let us give some results about equivalent ergodic invariant measures for regular systems. As a consequence of Theorem 8.4.7 we have the following.

Corollary 8.4.11 *Suppose that S is a parabolic iterated function system and the associated hyperbolic system S^* is regular. Let m be the corresponding h-conformal measure for S^*. Then there exists a unique probability measure μ^* equivalent with m with Radon-Nikodym derivatives bounded away from 0 and infinity. The measure μ^* is ergodic and invariant under S^* and, up to a multiplicative constant, there exists a unique σ-finite measure μ equivalent with m and ergodic invariant under S.*

Proof. The first part of this corollary is an immediate consequence of Theorem 2.2.4. That m is h-conformal for S follows from Theorem 8.4.7. The last part is a consequence of this conformality (the measures $\mu^* \circ \phi_i$ are therefore absolutely continuous with respect to μ^*) and Theorem 8.4.5. □

Corollary 8.4.12 *If the assumptions of Corollary 8.4.11 are satisfied, then the σ-invariant measure μ produced there is finite if and only if*

$$\sum_{i\in\Omega}\sum_{n=1}^{\infty} n\int_{X_i} |\phi'_{i^n}|^h dm < \infty,$$

where, let us recall, $X_i = \bigcup_{j\neq i}\phi_j(X)$.

Proof. Since by Corollary 8.4.11 the measures m and μ^* are equivalent with Radon-Nikodym derivatives bounded away from 0 and infinity, it therefore follows from Theorem 8.4.6 that μ is finite if and only if the series $\sum_{n\geq 1} nm(B_n)$ converges. Since $m(B_n) = \sum_{i\in\Omega}\int_{X_i}|\phi'_{i^n}|^h dm$, the proof is complete. □

Corollary 8.4.13 *If for every $i\in\Omega$ there exists some β_i and a constant $C_i \geq 1$ such that for all $n\geq 1$ and for all $z\in X_i$*

$$C_i^{-1} n^{-\frac{\beta_i+1}{\beta_i}} \leq |\phi'_{i^n}(z)| \leq C_i n^{-\frac{\beta_i+1}{\beta_i}},$$

then the σ-finite invariant measure μ produced in Corollary 8.4.11 is finite if and only if

$$h > 2\max\left\{\frac{\beta_i}{\beta_i+1} : i\in\Omega\right\}.$$

Proof. The proof is an immediate consequence of Corollary 8.4.12. □

8.5 Examples

This section contains examples illustrating some of the ideas developed in this chapter.

Example 8.5.1 *Apollonian packing.*

Consider on the complex plane the three points $z_j = e^{2\pi i j/3}$, $j = 0, 1, 2$ and the following additional three points: $a_0 = \sqrt{3} - 2$, $a_1 = (2-\sqrt{3})e^{\pi i j/6}$ and $a_2 = (2-\sqrt{3})e^{-\pi i j/6}$. Let f_0, f_1, and f_2 be the Möbius transformations determined by the following requirements: $f_0(z_0) = z_0$, $f_0(z_1) = a_2$, $f_0(z_2) = a_1$, $f_1(z_0) = a_2$, $f_1(z_1) = z_1$, $f_1(z_2) = a_0$, $f_2(z_0) = a_1$, $f_2(z_1) = a_0$, and $f_2(z_2) = z_2$. Set $X = \overline{B}(0,1)$, the closed ball centered at the origin of radius 1. It is straightforward that the images

$f_0(X)$, $f_1(X)$ and $f_2(X)$ are mutually tangent (at the points a_0, a_1 and a_2, respectively) disks whose boundaries pass through the triples (z_0, a_1, a_2), (z_1, a_0, a_2) and (z_2, a_0, a_1) respectively. Of course all the three disks $f_0(X)$, $f_1(X)$ and $f_2(X)$ are contained in X and are tangent to X at the points z_0, z_1 and z_2 respectively. Let $S = \{f_0, f_1, f_2\}$ be the iterated function system on X generated by f_0, f_1 and f_2. Notice that all the maps f_0, f_1 and f_2 are parabolic with parabolic fixed points z_0, z_1 and z_2 respectively. It is not difficult to check that all the requirements of a parabolic system are satisfied. Observe that the limit set J of the parabolic system S coincides with the residual set of the Apollonian packing generated by the curvilinear triangle with vertices z_0, z_1, z_2. In [MU8], using a slightly different iterated function system, we have dealt with geometrical properties of J, proving that $1 < h = \mathrm{HD}(J) < 2$, $0 < \mathcal{H}^h < \infty$ and $\mathcal{P}^h(J) = \infty$. In this paper we want to study its dynamical properties. Let us first notice that the system S^* is regular. Indeed, changing bi-holomorphically the system of coordinates so that 1 is sent to ∞ it is not difficult to show (see [MU8] for details, cf. the next chapter) that

$$f_0^n(z) = \frac{(\sqrt{3} - n)z + n}{-nz + n + \sqrt{3}}$$

and

$$(f_0^n)'(z) = \frac{3}{(-nz + n + \sqrt{3})^2}.$$

By the symmetry of the situation this implies that

$$|(f_i^n \circ f_j)'(z)| \asymp \frac{1}{n^2}$$

for all $i \neq j$. Hence $\psi^*(t) \asymp \sum_{n \geq 1} \frac{1}{n^{2t}}$. Thus $\theta(S^*) = 1/2$ and $\psi^*(1/2) = \infty$. This means the system is hereditarily regular. Thus, the assumptions of Theorem 8.4.7 and Corollary 8.4.11 are satisfied in our case. Let m be the h-conformal measure for S and let μ be an S-invariant σ-finite measure equivalent with m. We shall prove the following central dynamical property of our system.

Theorem 8.5.2 *The invariant measure μ of the Apollonian system $\{f_0, f_1, f_2\}$ is finite.*

Proof. In the proof of regularity of S^*, we have observed that $|(f_i^n)'| \asymp$

$1/n^2$ on X_i, $i = 0, 1, 2$. Since $h > 1 = 2\frac{1}{1+1}$, it therefore follows from Corollary 8.4.13 that μ is finite. □

Example 8.5.3

A large class of examples appears already in the case when X is a compact subinterval of the real line $\mathbb{R}$. We call such systems one-dimensional. If the parabolic elements ϕ_i of a one-dimensional system S have, around parabolic fixed points x_i, a representation of the form

$$\phi_i(x) = x + a(x - x_i)^{1+\beta_i} + o((x - x_i)^{1+\beta_i}) \tag{8.16}$$

then (see e.g., [U4])

$$|\phi'_{i^n}(x)| \asymp n^{-\frac{\beta_i+1}{\beta_i}} \tag{8.17}$$

outside every open neighborhood of x_i. Hence the following theorem is a consequence of Theorem 8.4.7 and Corollary 8.4.13.

Theorem 8.5.4 *If S is a one-dimensional parabolic system with finite alphabet and satisfying (8.16), then S is regular and any S-invariant measure μ equivalent with the h_S-conformal measure is finite if and only if $h > 2\max\{\frac{\beta_i}{\beta_i+1} : i \in \Omega\}$.*

Proof. The regularity of S^* is checked in exactly the same way as in Example 8.5.1. So, the system S is regular by Corollary 8.4.8. Since the other assumptions of Corollary 8.4.13 are satisfied by (8.17), the proof of this theorem is an immediate consequence of Corollary 8.4.13. □

Corollary 8.5.5 *If S is a one-dimensional parabolic system with finite alphabet, and if for all $i \in \Omega$, $\beta_i \geq 1$ (or equivalently if all ϕ_i's are twice differentiable at x_i), then S is regular and the corresponding invariant measure μ equivalent with the h_S-conformal measure is infinite.*

Proof. The proof is an immediate consequence of Theorem 8.5.4 and the fact that $h \leq 1$. □

We would like to close this section with some examples of strange systems.

Example 8.5.6

Our aim here is to describe a class of one-dimensional systems which are strange. Towards this end consider an arbitrary hyperbolic system

$S = \{\phi_i : i \in I\}$ on the interval $X = [0,1]$ such that $\psi(\theta(S)) < \infty$ or equivalently $\mathrm{P}(\theta(S)) < \infty$ (examples of such systems may be found in the section Examples of [MU1]); we may assume that there is an interval $G = [0,\gamma)$ with $G \subset X \setminus \bigcup_{i\in I} \phi_i(X)$. Consider also a parabolic map $\phi : X \to G$ such that 0 is its parabolic point and ϕ has the following representation around 0:

$$\phi(x) = x - ax^{\beta+1} + o(x^{\beta+1}),$$

where $\theta(S)\frac{\beta+1}{\beta} > 1$ and $a > 0$. We shall prove the following.

Theorem 8.5.7 *If $F \subset I$ is a sufficiently large finite set, then the system $S_F = \{\phi\} \cup \{\phi_i : i \in I \setminus F\}$ is strange.*

Proof. In view of (6.2) and the relation between $\theta(S)$ and β there exists a constant $C \geq 1$ such that for each $i \in I$, $\sum_{n\geq 1} ||(\phi^n \circ \phi_i)'||^{\theta(S)} \leq C||\phi_i'||^{\theta(S)}$. Since $\psi_S(\theta(S)) < \infty$, for every sufficiently large finite set $F \subset I$ we have $(C+1)\sum_{i\in I\setminus F} ||\phi_i'||^{\theta(S)} < 1$. Hence

$$\begin{aligned}
\psi^*_{S_F}(\theta(S)) &= \sum_{i\in I\setminus F} ||\phi_i'||^{\theta(S)} + \sum_{i\in I\setminus F}\sum_{n\geq 1} ||(\phi^n \circ \phi_i)'||^{\theta(S)} \\
&\leq \sum_{i\in I\setminus F} ||\phi_i'||^{\theta(S)} + C \sum_{i\in I\setminus F} ||\phi_i'||^{\theta(S)} \\
&= (C+1) \sum_{i\in I\setminus F} ||\phi_i'||^{\theta(S)} < 1.
\end{aligned}$$

Hence $\mathrm{P}^*_{S_F}(\theta(S)) < 0$ and therefore, as $h_{S_F^*} = h_{S_F}$, $\mathrm{P}_{S_F}(t) = 0$ for all $t \geq \theta(S)$. On the other hand, since for every $t < \theta(S)$, $\psi_S(t) = \infty$ and since F is finite, $\psi_{S_F}(t) = ||\phi'||^t + \sum_{i\in I\setminus F} ||\phi_i'||^t = \infty$. Hence $\mathrm{P}_{S_F}(t) = \infty$. □

9
Parabolic Systems: Hausdorff and Packing Measures

In this chapter we continue our studies of a parabolic iterated function system $S = \{\phi_i\}_{i \in I}$, following closely [MU9]. We keep all the notation introduced in the previous chapter. Our main goal is to characterize conformal measures of finite parabolic systems in terms of Hausdorff and packing measures. This simultaneously provides the answer to the question about necessary and sufficient conditions for these two geometric measures to be finite and positive.

9.1 Preliminaries

For every integer $q \geq 1$ we denote

$$S^q = \{\phi_\omega : \omega \in I^q\}.$$

Of course $J_{S^q} = J_S$ and sometimes in the sequel it will be more convenient to consider an appropriate iterate S^q of S rather than S itself. The following proposition is an immediate consequence of condition (4) from the previous chapter.

Proposition 9.1.1 *If the alphabet I is finite, then*

$$S^*(\infty) = \{x_i : i \in \Omega\}, \text{ the set of parabolic fixed points.}$$

In Section 4 we shall prove the following.

Theorem 9.1.2 *If S is a finite parabolic IFS, then the system S^* is hereditarily regular and consequently an h-conformal measure for S^* exists.*

From now, unless otherwise stated, we will assume that the alphabet I is finite and m will denote the h-conformal measure produced in Theorem 9.1.2. Combining this theorem along with Corollary 8.3.7 and Corollary 8.4.10 (due to the existence of parabolic elements the strong open set condition is satisfied) we get the following.

Theorem 9.1.3 *If S is a finite parabolic IFS satisfying the strong open set condition, then $\mathrm{H}^t(J) < \infty$ and $\mathrm{P}^h(J) > 0$.*

The following main theorem of this chapter contains a complete description of the h-dimensional Hausdorff and packing measures of the limit set of a finite parabolic IFS.

Theorem 9.1.4 *Suppose that S is a finite parabolic IFS satisfying the strong open set condition. Then*

(a) If $h < 1$, then $0 < \mathrm{P}^h(J) < \infty$ and $\mathrm{H}^h(J) = 0$.
(b) If $h = 1$, then $0 < \mathrm{H}^h(J), \mathrm{P}^h(J) < \infty$.
(c) If $h > 1$, then $0 < \mathrm{H}^h(J) < \infty$ and $\mathrm{P}^h(J) = \infty$.,

This sort of theorem has appeared in the context of Kleinian groups in [Su4], in the context of parabolic rational functions in [DU3], for rational functions with no recurrent critical points in the Julia set in [U6] and for parabolic Cantor sets (which comprise 1-dimensional parabolic IFS) in [U4]. The idea of our proofs is different. It is based on the necessary and sufficient conditions for the Hausdorff and packing measures to be positive and finite, provided by Theorem 4.5.3 and Theorem 4.5.5. The inducing procedures proposed in [UZd] indicate that in the case of parabolic rational functions and perhaps even in the case of maps with no recurrent critical points in the Julia set, one can demonstrate appropriate versions of Theorem 9.1.4 as a corollary of Theorem 9.1.4 proved here. We shall also prove in Section 4 the following.

Theorem 9.1.5 *If S is a finite parabolic IFS, then*

$$\overline{\mathrm{BD}(J)} = \mathrm{HD}(J),$$

where $\overline{\mathrm{BD}(J)}$ denotes the upper ball-counting dimension called also box-counting dimension, Minkowski dimension or capacity.

The dynamical properties of the parabolic IFS proved in Sections 2 and 3 and needed for the proofs of Theorem 9.1.4 and Theorem 9.1.5 are provided in the beginning of Section 4 in a unified fashion. Therefore the reader interested in Theorem 9.1.4 and Theorem 9.1.5 only may actually read Section 4 independently of Section 2 and Section 3.

Section 2 mainly deals with dynamical properties of a single parabolic conformal diffeomorphism and can also be viewed as an introduction to the technically more complicated Section 3, which deals with dynamical properties of a single simple parabolic holomorphic map. Both sections provide a compact systematic description of the quantitative behaviour of parabolic maps needed for the proofs in Section 4. The qualitative behavior of a single parabolic holomorphic map considered in Section 3 is known as Fatou's flower theorem (see [Al] for additional historical information). Some quantitative results can be also found in these papers. At the ends of both Sections 2 and 3 some facts about parabolic iterated function systems are proved. We want to end this section with a short terminological convention. Given two sets $A, B \subset I\!R^d$ we denote

$$\mathrm{dist}(A,B) = \inf\{||a-b|| : (a,b) \in A \times B\} \quad \text{and}$$
$$\mathrm{Dist}(A,B) = \sup\{||a-b|| : (a,b) \in A \times B\}.$$

9.2 The Case $d \geq 3$

Definition 9.2.1 *We call a conformal map $A : I\!R^d \to I\!R^d$ parabolic if it has a fixed point $\omega \in I\!R^d$ and a point $\xi \in I\!R^d \setminus \{\omega\}$ such that $|A'(\omega)| = 1$ and $\lim_{n\to\infty} A^n \xi = \omega$.*

Let

$$\tilde{A} = i_{\omega,1}^{-1} \circ A \circ i_{\omega,1} = i_{\omega,1} \circ A \circ i_{\omega,1^{-1}}.$$

Then $\tilde{A}(\infty) = \infty$ and therefore

$$\tilde{A} = \lambda D + c,$$

where $\lambda > 0$, D is an orthogonal matrix, and $c \in I\!R^d$. From now on, without loss of generality, we will assume that $\omega = 0$, i.e., ω is the origin, and we will write i for $i_{0,1}$.

Lemma 9.2.2 *If $A : I\!R^d \to I\!R^d$ is a parabolic conformal map and if λ is the scalar involved in the formula for $\tilde{A}$, then $\lambda = 1$.*

Proof. If $\lambda < 1$, then $\tilde{A} : I\!R^d \to I\!R^d$ is a strict contraction and due to Banach's contraction principle, it has a fixed point $b \in I\!R^d$ such that $\lim_{n\to\infty} \tilde{A}^n(z) = b$ for every $z \in I\!R^d$. However, this is a contradiction, since $\lim_{n\to\infty} \tilde{A}^n(i(\xi)) = \infty$. Thus, $\lambda \geq 1$. Assume $\lambda > 1$. Then for every $z \in I\!\!R^d \setminus \{0\}$,

$$\begin{aligned}||A'(z)|| &= ||i'(\tilde{A}(i(z))\tilde{A}'(i(z))i'(z)|| = \lambda||\tilde{A}(i(z))||^{-2}||z||^{-2} \\ &= \lambda||z||^{-2}||\lambda D(||z||^{-2}(z)) + c||^{-2} \\ &= \lambda||(\lambda||z||^{-1}D(z) + c||z||)||^{-2} \\ &= \lambda^{-1}[||D(z/||z||) + (||z||/\lambda)c||]^{-2}.\end{aligned}$$

Since $\lim_{z\to 0} ||z|| = 0$ and since $||D(z)|| = ||z||$, we deduce that $||A'(0)|| = \lim_{z\to 0} ||A'(z)|| = \lambda^{-1} < 1$. This contradiction shows that $\lambda \leq 1$, and consequently $\lambda = 1$. □

Next, we want to estimate the rate at which $\tilde{A}^n(z)$ goes to $+\infty$.

Lemma 9.2.3 *If $A : I\!\!R^d \to I\!\!R^d$ is a parabolic conformal map, then there exists a non-zero vector $b \in I\!\!R^d$ and a positive constant κ such that for every $z \in I\!\!R^d$ and every positive integer n*

$$||\tilde{A}^n z - nb|| \leq ||z|| + \kappa.$$

Proof. By a straightforward induction, we get

$$\tilde{A}^n z = D^n z + \sum_{j=0}^{n-1} D^j(c).$$

Write $c = b + a$, where b is a fixed point (*a priori* perhaps 0) of D and a belongs to W, the orthogonal complement of the vector space of the fixed points of D. Since $\lim_{n\to\infty} \tilde{A}^n(i(\xi)) = \infty$, W is not the trivial subspace of $I\!\!R^d$. In addition, $D(W) = W$ and $D - \mathrm{Id} : W \to W$ is invertible. Since

$$(D - \mathrm{Id})\left(\sum_{j=0}^{n-1} D^j(a)\right) = D^n a - a$$

and since $||D^n a - a|| \leq 2||a||$, we therefore conclude that for every $n \geq 1$

$$\left\|\sum_{j=0}^{n-1} D^j(a)\right\| \leq 2||a|| \cdot \left\|(D - \mathrm{Id})|_W^{-1}\right\|.$$

Hence,

$$||\tilde{A}^n z - nb|| = \left\|D^n z + \sum_{j=0}^{n-1} D^j(a)\right\| \leq ||z|| + 2\left\|(D - \mathrm{Id})|_W^{-1}\right\| \cdot ||a||.$$

Again, since $\lim_{n\to\infty} \tilde{A}^n(i(\xi)) = \infty$, we finally conclude that $b \neq 0$. □

As an immediate consequence of this lemma we get the following.

Corollary 9.2.4 *Let $A : I\!R^d \to I\!R^d$ be a parabolic conformal map. For every compactum $F \subset I\!R^d$, there exists a constant $B_F \geq 1$ and integer $M_F \in I\!N$ such that for every $n \geq M_F$ and every $z \in F$*

$$B_F^{-1} n \leq ||\tilde{A}^n z|| \leq B_F n.$$

Lemma 9.2.5 *Let $A : I\!R^d \to I\!R^d$ be a parabolic conformal map. For every compactum $L \subset I\!R^d \setminus \{0\}$, there exist a constant $C_{L,1} \geq 1$ and integer $N_L \in I\!N$ such that for every $n \geq N_L$ and every $z \in L$*

$$C_{L,1}^{-1} n^{-2} \leq ||(A^n)'(z)|| \leq C_{L,1} n^{-2} \text{ and } \operatorname{diam}(A^n(L)) \leq C_{L,1} n^{-2}.$$

Proof. By the chain rule, we find for every $z \in I\!R^d \setminus \{0\}$

$$\begin{aligned} ||(A^n)'(z)|| &= ||i'(\tilde{A}^n(i(z)))|| \cdot ||(\tilde{A}^n)'(i(z))|| \cdot ||i'(z)|| \\ &= ||\tilde{A}^n(i(z))||^{-2} ||z||^{-2}. \end{aligned}$$

For every $z \in L$, $\operatorname{Dist}^{-2}(0, L) \leq ||z||^{-2} \leq \operatorname{dist}^{-2}(0, L)$, and in view of Corollary 9.2.4, if $n \geq M_{i(L)}$, then $B_{i(L)}^{-1} n \leq ||\tilde{A}^n z|| \leq B_{i(L)} n$. Consequently, if $z \in L$ and $n \geq M_{i(L)}$, we have

$$\left(B_{i(L)} \operatorname{Dist}(0, L)\right)^{-2} n^{-2} \leq ||(A^n)'(z)|| \leq B_{i(L)}^2 \operatorname{dist}^{-2}(0, L) n^{-2}.$$

□

Lemma 9.2.6 *Let $A : I\!R^d \to I\!R^d$ be a parabolic conformal map. For every compactum $L \subset I\!R^d \setminus \{0\}$, there exists a constant $C_{L,2} \geq 1$ such that for all integers k, n with $n \geq k \geq 1$,*

$$\operatorname{Dist}(A^k(L), A^n(L)) \leq C_{L,2} \left|k^{-1} - (n+1)^{-1}\right|$$

and

$$\operatorname{Dist}(A^n(L), 0) \leq C_{L,2} n^{-1}.$$

Proof. Let us start with the second inequality. If $n \geq M_{i(L)}$ and $z \in L$, then, by Corollary 9.2.4, we get $||A^n z|| = ||\tilde{A}^n(i(z))||^{-1} \leq B_{i(L)} n^{-1}$ and the second inequality follows provided $C_{L,2}$ is sufficiently large.

Towards obtaining the first inequality, for every set $Y \subset I\!R^d$, let $\operatorname{conv}(Q)$ denote the convex hull of Y. Obviously, $\operatorname{conv}(Y) \subset$

$B(Y, \mathrm{diam}(Y))$ and $\mathrm{diam}(\mathrm{conv}(Y)) = \mathrm{diam}(Y)$. By using Lemma 9.2.3, we have for every $u \in L$ and $n \in \mathbb{N}$,

$$\begin{aligned}
&||\tilde{A}^{n+1}(i(u)) - \tilde{A}^n(i(u))|| \\
&\qquad \leq ||\tilde{A}^{n+1}(i(u)) - (n+1)b - (\tilde{A}^n(i(u)) - nb) + b|| \\
&\qquad \leq 2(||i(u)|| + \kappa) + ||b|| \leq 2(\mathrm{Dist}(0, i(L)) + \kappa) + ||b||.
\end{aligned}$$

Next, choose a positive integer N_0 such that $\mathrm{Dist}(0, \mathrm{conv}(\bigcup_{t \geq N_0} \tilde{A}^t(i(L)))) = H > 0$ and $N_0||b|| > \mathrm{Dist}(0, i(L)) + \kappa + ||b|| := M$. We claim there is a positive constant C such that if $u, v \in L, k \geq N_0$ and $j \geq 0$, then

$$||A^{k+j+1}(v) - A^{k+j}(u)|| \leq C \frac{1}{(k+j+1)^2}.$$

In order to see this, note that

$$\begin{aligned}
&||A^{k+j+1}(v) - A^{k+j}(u)|| \\
&\quad \leq ||i(\tilde{A}^{k+j+1}(i(v))) - i(\tilde{A}^{k+j+1}(i(u)))|| \\
&\qquad + ||i(\tilde{A}^{k+j+1}(i(u))) - i(\tilde{A}^{k+j}(i(u)))|| \\
&\quad \leq \sup\{||i'(w)|| : w \in [\tilde{A}^{k+j+1}(i(v)), \tilde{A}^{k+j+1}(i(u))]\}|| \\
&\qquad \times \tilde{A}^{k+j+1} i(v) - \tilde{A}^{k+j+1} i(u)|| \\
&\qquad + \sup\{||i'(w)|| : w \in [\tilde{A}^{k+j}(i(u)), \tilde{A}^{k+j+1}(i(u))]\}|| \\
&\qquad \times \tilde{A}^{k+j+1} i(u) - \tilde{A}^{k+j} i(u)|| \\
&\quad \leq \mathrm{diam}(i(L)) \sup\{||w||^{-2} : w \in [\tilde{A}^{k+j+1}(i(v)), \tilde{A}^{k+j+1}(i(u))]\} \\
&\qquad + 2M \sup\{||w||^{-2} : w \in [\tilde{A}^{k+j}(i(u)), \tilde{A}^{k+j+1}(i(u))]\}.
\end{aligned}$$

Now, if $w \in [\tilde{A}^{k+j+1}(i(v)), \tilde{A}^{k+j+1}(i(u))]$, then by Lemma 9.2.3, $||w - (k+j+1)b|| \leq \mathrm{Dist}(0, i(L)) + \kappa$ and $||w|| \geq (k+j+1)[||b|| - (\mathrm{Dist}(0, i(L)) + \kappa)/N_0]$. Also, since $||\tilde{A}^{k+j}(i(u)) - (k+j+1)b|| \leq ||i(u)|| + \kappa + ||b||$, if $w \in [\tilde{A}^{k+j}(i(u)), \tilde{A}^{k+j+1}(i(u))]$, then $||w - (k+j+1)b|| \leq \mathrm{Dist}(0, i(L)) + \kappa + ||b||$ and $||w|| \geq (k+j+1)[||b|| - (\mathrm{Dist}(0, i(L)) + \kappa + ||b||)/N_0] \geq (k+j+1)[||b|| - M/N_0]$. Combining these inequalities establishes our claim.

Therefore, if $N_0 \leq k \leq n$ we have

$$\begin{aligned}
\mathrm{Dist}(A^k(L), A^n(L)) &\leq \sum_{j=0}^{n-k-1} \mathrm{Dist}\big(A^{k+j+1}(L), A^{k+j}(L)\big) \\
&\leq \sum_{j=0}^{n-k} C(k+j)^{-2} \leq C_{L,2}(k^{-1} - (n+1)^{-1})
\end{aligned}$$

for some constant $C_{L,2} \geq 1$. Clearly, increasing $C_{L,2}$ appropriately, we see that the last inequality is also true for all $1 \leq k \leq n$. The proof of the first part of our lemma is thus complete. □

Lemma 9.2.7 *For every compactum $L \subset I\!R^d \setminus \{0\}$ there exist a constant $C_{L,3} \geq 1$ and an integer $q \geq 0$ such that for all $k \geq 1$ and all $n \geq k+q$*

$$\mathrm{dist}(A^k(L), A^n(L)) \geq C_{L,3}\big(k^{-1} - n^{-1}\big)$$

and

$$\mathrm{dist}(A^n(L), 0) \geq C_{L,3}n^{-1}.$$

Proof. First, notice that it follows from Lemma 9.2.3 that if $w, z \in i(L)$ and $k, n \in N$, then

$$(n-k)||b|| - 2(\mathrm{Dist}(0, i(L)) + \kappa) \leq ||\tilde{A}^n(w) - \tilde{A}^k(z)||.$$

Therefore, there is a positive integer q_0 such that if $n - k \geq q_0$, then $||\tilde{A}^n(w) - \tilde{A}^k(z)|| \geq (1/2)||b||(n-k)$. Let N_0 be as in the proof of Lemma 9.2.6 and $M_{i(L)}$ be as in Corollary 9.2.4. Let $k, n \geq N_1 = \max\{N_0, M_{i(L)}\}$. Consider two arbitrary points $z, w \in i(L)$ and parametrize the line segment γ joining $\tilde{A}^k(z)$ and $\tilde{A}^n(w)$ as

$$\gamma(t) = \tilde{A}^k(z) + t(\tilde{A}^n(w) - \tilde{A}^k(z)), \ t \in [0,1].$$

The curve $i(\gamma)$ is a subarc of either a circle or a line; let $l(i(\gamma))$ be its length. We have

$$\begin{aligned}
&l(i(\gamma)) \\
&= \int_0^1 ||(i\circ\gamma)'(t)||dt = \int_0^1 ||i'(\gamma(t))|| \cdot ||\gamma'(t)||dt \\
&= ||\tilde{A}^n(w) - \tilde{A}^k(z)|| \int_0^1 ||\gamma(t)||^{-2}dt \\
&= ||\tilde{A}^n(w) - \tilde{A}^k(z)|| \int_0^1 ||\tilde{A}^k(z) + t(\tilde{A}^n(w) - \tilde{A}^k(z))||^{-2}dt \\
&\geq ||\tilde{A}^n(w) - \tilde{A}^k(z)|| \int_0^1 \big(||\tilde{A}^k(z)|| + t||\tilde{A}^n(w) - \tilde{A}^k(z)||\big)^{-2}dt
\end{aligned}$$

$$
\begin{aligned}
&= ||\tilde{A}^n(w) - \tilde{A}^k(z)|| \cdot ||\tilde{A}^n(w) - \tilde{A}^k(z)||^{-1} \\
&\times \int_{||\tilde{A}^k(z)||}^{||\tilde{A}^k(z)||+||\tilde{A}^n(w)-\tilde{A}^k(z)||} u^{-2} du \\
&= ||\tilde{A}^k(z)||^{-1} - (||\tilde{A}^k(z)|| + ||\tilde{A}^n(w) - \tilde{A}^k(z)||) \\
&= \frac{||\tilde{A}^n(w) - \tilde{A}^k(z)||}{||\tilde{A}^k(z)|| \cdot (||\tilde{A}^k(z)|| + ||\tilde{A}^n(w) - \tilde{A}^k(z)||)}.
\end{aligned} \tag{9.1}
$$

We have

$$
||\tilde{A}^k(z)|| + ||\tilde{A}^n(w) - \tilde{A}^k(z)|| \leq B_{i(L)}k + C_{i(L),1}\left(\frac{1}{k} - \frac{1}{n+1}\right).
$$

So, there is a constant U such that $||\tilde{A}^k(z)|| + ||\tilde{A}^n(w) - \tilde{A}^k(z)|| \leq Un$. In view of Corollary 9.2.4, there is a constant Q_0 such that

$$
l(i(\gamma)) \geq Q_0 \frac{||\tilde{A}^n(w) - \tilde{A}^k(z)||}{kn}.
$$

Thus, there is a constant Q such that if $k \geq N_1$ and $n \geq k + q_0$, then

$$
l(i(\gamma)) \geq Q(k^{-1} - n^{-1}). \tag{9.2}
$$

If $i(\gamma)$ is a line segment, then

$$
||A^n(i(w)) - A^k(i(z))|| = l(i(\gamma)) \geq Q(k^{-1} - n^{-1}). \tag{9.3}
$$

If, however, $i(\gamma)$ is an arc of a circle, then consider the ray

$$
g(t) = \tilde{A}^k(z) + t(\tilde{A}^n(w) - \tilde{A}^k(z)), \ t \in (-\infty, 0].
$$

Proceeding exactly as in the formula (9.2) and using the estimate $||g(t)|| \leq ||\tilde{A}^k(z)|| - t||\tilde{A}^n(w) - \tilde{A}^k(z)||$, we get

$$
l(i(g)) \geq \int_{||\tilde{A}^k(z)||}^{\infty} u^{-2} du = ||\tilde{A}^k(z)||^{-1}.
$$

And applying Corollary 9.2.4 we get $l(i(g)) \geq B_{i(L)}^{-1} k^{-1} \geq B_{i(L)}(k^{-1} - n^{-1})$. Therefore, invoking (9.2), we deduce that both arcs joining the points $A^k(i(z))$ and $A^n(i(w))$ on the circle $i(\{\tilde{A}^k(z) + t(\tilde{A}^n(w) - \tilde{A}^k(z)) : t \in I\!R \cup \{\infty\}\})$ have the $\geq \min\{B_{i(L)}, Q\}(k^{-1} - n^{-1})$. Thus, also taking into account (9.3), we see there is a constant P_0 such that if $k, n \geq N_1$ and $n - k \geq q_0$, then

$$
\mathrm{dist}(A^k(L), A^n(L)) \geq P_0(k^{-1} - n^{-1}).
$$

Since 0 is not an element of $\bigcup_{j=1}^{N_1} A^j(L)$, and since it follows from Lemma 9.2.3 that $A^k(L) \to 0$ as $k \to \infty$, there is a constant $C_{L,3}$

such that the first part of the conclusion of the lemma holds. Applying the proved part of the lemma, we conclude that

$$\begin{aligned}\operatorname{dist}(A^n(L),0) &= \lim_{k\to\infty} \operatorname{dist}(A^n(L),A^k(L)) \\ &\geq \lim_{k\to\infty} C_{L,3}(n^{-1}-k^{-1}) = C_{L,3}n^{-1}.\end{aligned}$$

□

We end this section by proving the following result concerning general parabolic IFSs in dimension $d \geq 3$. The first is a straightforward consequence of Lemma 9.2.3. First, let us note that Lemma 9.2.6 shows that a conformal parabolic map in $I\!R^d$, $d \geq 3$ has a unique fixed point.

Proposition 9.2.8 *If $\{\phi_i : X \to X\}_{i\in I}$ is an at least three-dimensional parabolic conformal IFS (I is allowed to be infinite), then x_i, the only fixed point of a parabolic map ϕ_i, belongs to ∂X.*

Proof. In view of Lemma 9.2.3, for every $R > 0$ large enough and every $n \geq 1$, the set $\tilde{\phi}_i(\{z : ||z|| > R\})$ is not contained in $\{z : ||z|| > R\}$. Consequently, for every neighborhood U of x_i, the set $\phi_i^n(U)$ does not converge to x_i. Since however $\lim_{n\to\infty} \phi_i^n(X) = x_i$, the point x_i cannot belong to $\operatorname{Int}(X)$. □

9.3 The plane case, $d = 2$

We call a holomorphic map ϕ, defined around a point $\omega \in I\!\!C$, simple parabolic if $\phi(\omega) = \omega$, $\phi'(\omega) = 1$ and ϕ is not the identity map. Then on a sufficiently small neighborhood of ω, the map ϕ has the following Taylor series expansion:

$$\phi(z) = z + a(z-\omega)^{p+1} + b(z-\omega)^{p+2} + \cdots$$

with some integer $p \geq 1$ and $a \in I\!\!C \setminus \{0\}$. Being in the circle of ideas related to Fatou's flower theorem (see [Al] for extended historical information), we now want to analyze qualitatively and especially quantitatively the behavior of ϕ in a sufficiently small neighborhood of the parabolic point ω. Let us recall that the rays coming out from ω and forming the set

$$\{z : a(z-\omega)^p < 0\}$$

are called attracting directions and the rays forming the set

$$\{z : a(z-\omega)^p > 0\}$$

are called repelling directions. Fix an attractive direction, say $A = \omega + \sqrt[p]{-a^{-1}(0,\infty)}$, where $\sqrt[p]{\cdot}$ is a holomorphic branch of the pth radical defined on $\mathbb{C} \setminus a^{-1}(0,\infty)$. In order to simplify our analysis let us change the system of coordinates with the help of the affine map $\rho(z) = \sqrt[p]{-a^{-1}} + \omega$. We then get

$$\phi_0(z) = \rho^{-1} \circ \phi \circ \rho(z) = z - z^{p+1} + b\sqrt[p]{-a^{-1}}z^{p+2} + \cdots$$

and $\rho^{-1}(A) = (0,\infty)$ is an attractive direction for ϕ_0. We want to analyze the behavior of ϕ_0 on an appropriate neighborhood of $(0,\epsilon)$, for $\epsilon > 0$ sufficiently small. In order to do it, similarly as in the previous section, we conjugate ϕ_0 on $\mathbb{C} \setminus (-\infty, 0]$ to a map defined "near" infinity. Precisely, we consider $\sqrt[p]{\cdot}$, the holomorphic branch of the pth radical defined on $\mathbb{C} \setminus (-\infty, 0]$ and leaving the point 1 fixed. Then we define the map

$$H(z) = \frac{1}{\sqrt[p]{z}}$$

and consider the conjugate map

$$\tilde{\phi} = H^{-1} \circ \phi_0 \circ H.$$

Straightforward calculations show that

$$\tilde{\phi}(z) = z + 1 + O(|z|^{-\frac{1}{p}}) \tag{9.4}$$

and

$$\tilde{\phi}'(z) = 1 + O(|z|^{-\frac{p+1}{p}}). \tag{9.5}$$

Given now a point $x \in (0,\infty)$ and $\alpha \in (0,\pi)$, let

$$S(x,\alpha) = \{z : -\alpha < \arg(z - x) < \alpha\}.$$

The formula (9.4) shows that for every $\alpha \in (0,\pi)$ there exists $x(\alpha) \in (0,\infty)$ such that for every $x \geq x(\alpha)$

$$\overline{\tilde{\phi}(S(x,\alpha))} \subset S\left(x + \frac{1}{2}, \alpha\right), \tag{9.6}$$

$$|z| \geq B^p \tag{9.7}$$

and

$$\mathrm{Re}(\tilde{\phi}(z)) \geq \mathrm{Re}(z) + \frac{1}{2} \tag{9.8}$$

for all $z \in S(x, \alpha)$, where B is the constant responsible for $O(|z|^{-\frac{1}{p}})$ in (9.4). The following lemma immediately follows from (9.7), (9.4) and (9.8) by a straightforward induction.

Lemma 9.3.1 *For every compactum $F \subset S(x(\alpha), \alpha)$ there exists a constant $C_F \geq 1$ such that for every $z \in F$ and every $n \geq 1$*

$$C_F^{-1} n \leq |\tilde{\phi}^n(z)| \leq C_F n.$$

Using a straightforward induction, one gets from (9.4) and Lemma 9.3.1 that

$$\tilde{\phi}^n(z) = z + n + O\big(\max\{n^{1-\frac{1}{p}}, \log n\}\big) \tag{9.9}$$

and

$$\tilde{\phi}^n(z) = \tilde{\phi}^k(z) + (n-k) + O\big(|n^{1-\frac{1}{p}} - k^{1-\frac{1}{p}}|\big), \tag{9.10}$$

where the constant involved in "O" depends only on F and ϕ_0. Using Lemma 9.3.1 and (9.5) we shall prove the following.

Lemma 9.3.2 *For every compactum $F \subset S(x(\alpha), \alpha)$ there exists a constant $D_F \geq 1$ such that for every $z \in F$ and every $n \geq 1$*

$$D_F^{-1} \leq |(\tilde{\phi}^n)'(z)| \leq D_F.$$

Proof. For every $z \in S(x(\alpha), \alpha)$ let $g(z) = \tilde{\phi}'(z) - 1$. By the chain rule, we have for every $z \in S(x(\alpha), \alpha)$ and every $n \geq 1$

$$(\tilde{\phi}^n)'(z) = \prod_{j=0}^{n-1} \phi'(\tilde{\phi}^j(z)) = \tilde{\phi}'(z) \prod_{j=1}^{n-1} \big(1 + g(\tilde{\phi}^j(z))\big).$$

Using (9.5) and and the right-hand side of of Lemma 9.3.1, we get for every $z \in F$ and every $j \geq 1$ that

$$|g(\tilde{\phi}^j(z))| = O\big(|\tilde{\phi}^j(z)|^{-\frac{p+1}{p}}\big) \leq C_F^{-\frac{p+1}{p}} O\big(j^{-\frac{p+1}{p}}\big).$$

Since the series $\sum_{j=1}^{\infty} j^{-\frac{p+1}{p}}$ converges, the proof is complete. □

For every $x \in (0, \infty)$ and $\alpha \in (0, \pi)$ let

$$S_0(x, \alpha) = H(S(x, \alpha))$$

and

$$S_\phi^A(x, \alpha) = \rho \circ H(S(x, \alpha)) = \rho(S_0(x, \alpha)).$$

The regions $S_0(x,\alpha)$ and $S_\phi^A(x,\alpha)$ look like flower petals containing symmetrically a part of the ray $(0,\infty)$ and the ray $A = \omega + \sqrt[p]{-a^{-1}(0,\infty)}$ respectively and form with these rays two "angles" of measure α/π at the points 0 and ω respectively. We recall from the previous section that $\mathrm{conv}(M)$ denotes the convex hull of the set M. Combining Lemma 9.3.1 and Lemma 9.3.2 we deduce the following.

Lemma 9.3.3 *For every $\alpha \in (0,\pi/2)$ and for every compactum $F \subset S(x(\alpha),\alpha)$ there exists a constant $C_F \geq 1$ such that for every $n \geq 1$*

$$C_F^{-1} n \leq \mathrm{dist}(0, \mathrm{conv}(\tilde{\phi}^n(F))) \leq \mathrm{Dist}(0, \mathrm{conv}(\tilde{\phi}^n(F))) \leq C_F n.$$

Let us now use the properties of the map $\tilde{\phi}$ and establish useful facts about the map ϕ.

Lemma 9.3.4 *For every compactum $L \subset S_\phi^A(x,\alpha)$ there exists a constant $C_L \geq 1$ such that for every $z \in L$ and every $n \geq 1$*

$$C_L^{-1} n^{-\frac{p+1}{p}} \leq |(\phi^n)'(z)|, \ \mathrm{diam}(\phi_n(L)) \leq C_L n^{-\frac{p+1}{p}}.$$

Proof. It of course suffices to prove this lemma for ϕ replaced by ϕ_0. Since $H^{-1}(L)$ is a compact subset of $S(x(\alpha),\alpha)$ and since $H'(z) = -\frac{1}{p} z^{-\frac{p+1}{p}}$, using the chain rule along with Lemma 9.3.1, Lemma 9.3.2, and (9.7), we deduce that for every $z \in L$ and every $n \geq 1$

$$\begin{aligned}
|(\phi_0^n)'(z)| &= |(H \circ \tilde{\phi}^n \circ H^{-1})'(z)| \\
&= |H'(\tilde{\phi}^n(H^{-1}(z))| \cdot |(\tilde{\phi}^n)'(H^{-1}(z))| \cdot |(H^{-1})'(z)| \\
&= \frac{1}{p} |\tilde{\phi}^n(H^{-1}(z))|^{-\frac{p+1}{p}} |(\tilde{\phi}^n)'(H^{-1}(z))| p |z|^{-(p+1)} \\
&\leq D_{H^{-1}(L)}^{\frac{p+1}{p}} C_{H^{-1}(L)} (\mathrm{dist}(0, H^{-1}(L)))^{-(p+1)} n^{-\frac{p+1}{p}}
\end{aligned}$$

and

$$|(\phi_0^n)'(z)| \leq D_{H^{-1}(L)}^{-\frac{p+1}{p}} C_{H^{-1}(L)}^{-1} \mathrm{Dist}(0. H^{-1}(L))^{-(p+1)} n^{-\frac{p+1}{p}}.$$

□

Lemma 9.3.5 *For every compactum $L \subset S_\phi^A(x,\alpha)$ there exists a constant $C_{L,1} \geq 1$ such that for all $k,n \geq 1$*

$$\mathrm{Dist}(\phi^k(L), \phi^n(L)) \leq C_{L,1} \left| \min(k,n)^{-\frac{1}{p}} - (\max(k,n)+1)^{-\frac{1}{p}} \right|$$

and

$$\mathrm{Dist}(\phi^n(L),\omega) \le C_{L,1}n^{-\frac{1}{p}}.$$

Proof. It suffices again to prove this lemma for ϕ replaced by ϕ_0. Let us prove the first inequality. Without loss of generality we may assume that $n \ge k$. Since $H^{-1}(L)$ and $\mathrm{conv}(H^{-1}(L))$ are compact subsets of $S(x(\alpha),\alpha)$, using (9.4), Lemma 9.3.3, Lemma 9.3.1, and Lemma 9.3.2, we can estimate for every $j \ge 0$ and all $z, \xi \in L$ as follows.

$$\begin{aligned}
&|\phi_0^{k+j+1}(\xi) - \phi_0^{k+j}(z)| \\
&\quad \le |\phi_0^{k+j+1}(\xi) - \phi_0^{k+j+1}(z)| + |\phi_0^{k+j+1}(z) - \phi_0^{k+j}(z)| \\
&\quad \le \sup\{|H'(w)| : w \in \mathrm{conv}\big(\tilde\phi^{k+j+1}(H^{-1}(L))\big)\} \\
&\quad \times \mathrm{diam}\big(\mathrm{conv}\big(\tilde\phi^{k+j+1}(H^{-1}(L))\big)\big) \\
&\quad + \big(1 + B|\tilde\phi^{k+j}(H^{-1}(z))|^{-\frac{1}{p}}\big)\sup\{|H'(w)| : w \in [\tilde\phi^{k+j}(H^{-1}(z)), \\
&\quad \times \tilde\phi^{k+j+1}(H^{-1}(z))]\} \\
&\quad \le \frac{1}{p}\sup\{|w|^{-\frac{p+1}{p}} : w \in \mathrm{conv}\big(\tilde\phi^{k+j+1}(H^{-1}(L))\big)\} \\
&\quad \times \mathrm{diam}(\tilde\phi^{k+j+1}(H^{-1}(L))) \\
&\quad + \frac{2}{p}\sup\{|w|^{-\frac{p+1}{p}} : w \in [\tilde\phi^{k+j}(H^{-1}(z)), \tilde\phi^{k+j+1}(H^{-1}(z))]\} \\
&\quad \le \frac{1}{p}D_{H^{-1}(L)}C_{H^{-1}(L)}\mathrm{diam}(H^{-1}(L))(k+j+1)^{-\frac{p+1}{p}} \\
&\quad + \frac{2}{p}\big(|\tilde\phi^{k+j+1}(H^{-1}(z))| - |\tilde\phi^{k+j}(H^{-1}(z))|\big)^{-\frac{p+1}{p}} \\
&\quad \le \frac{1}{p}D_{H^{-1}(L)}C_{H^{-1}(L)}\mathrm{diam}(H^{-1}(L))(k+j+1)^{-\frac{p+1}{p}} \\
&\quad + \frac{2}{p}\Big(C_{H^{-1}(L)}(k+j+1) - B\big(|\tilde\phi^{k+j}(H^{-1}(z))|^{-\frac{1}{p}} + 1\big)\Big)^{-\frac{p+1}{p}} \\
&\quad \le \frac{1}{p}D_{H^{-1}(L)}C_{H^{-1}(L)}\mathrm{diam}(H^{-1}(L))(k+j+1)^{-\frac{p+1}{p}} \\
&\quad + \frac{2}{p}\Big(C_{H^{-1}(L)}(k+j+1) - B\big(C_{H^{-1}(L)}^{\frac{1}{p}}(k+j)^{-\frac{1}{p}} + 1\big)\Big) \\
&\quad \le \frac{1}{p}D_{H^{-1}(L)}C_{H^{-1}(L)}\mathrm{diam}(H^{-1}(L))(k+j+1)^{-\frac{p+1}{p}} \\
&\quad + \frac{4}{p}C_{H^{-1}(L)}^{\frac{p+1}{p}}(k+j+1)^{-\frac{p+1}{p}} \\
&\quad = \frac{1}{p}\big(D_{H^{-1}(L)}C_{H^{-1}(L)}\mathrm{diam}(H^{-1}(L)) + 4C_{H^{-1}(L)}^{\frac{p+1}{p}}\big)(k+j+1)^{-\frac{p+1}{p}}
\end{aligned}$$

where the last inequality has been written assuming that $k \geq 1$ is large enough, say $k \geq q$, and B is the constant coming from (9.4). Denote the constant appearing in the last line of the above formula by C'_L. Using also Lemma 9.3.4 we then get

$$\begin{aligned}\mathrm{Dist}(\phi_0^k(L), \phi_0^n(L)) &\leq \sum_{j=0}^{n-k-1} \mathrm{Dist}(\phi_0^{k+j}(L), \phi_0^{k+j+1}(L)) \\ &\quad + \sum_{j=0}^{n-k} \mathrm{diam}(\phi_0^{k+j}(L)) \leq \sum_{j=0}^{n-k} C'_L(k+j)^{-\frac{p+1}{p}} \\ &= C_{L,1}(k^{-\frac{1}{p}} - (n+1)^{-\frac{1}{p}})\end{aligned}$$

for some constant $C_{L,1} \geq 1$. Clearly, increasing the constant $C_{L,1}$ appropriately, we see that the last inequality is also true for all $1 \leq k \leq q$. The proof of the first part of Lemma 9.3.5 is thus complete. The second part is a straightforward consequence of the first one. Indeed, it follows from (9.6) that $\phi^k(L)$ converges to ω if $k \to \infty$. Hence, applying the first part of the lemma, we get

$$\begin{aligned}\mathrm{Dist}(\phi^n(L), \omega) &= \lim_{k\to\infty} \mathrm{Dist}(\phi^n(L), \phi^k(L)) \\ &\leq \lim_{k\to\infty} C_{L,1}(n^{-\frac{1}{p}} - (k+1)^{-\frac{1}{p}}) = C_{L,1} n^{-\frac{1}{p}}.\end{aligned}$$

□

Lemma 9.3.6 *For every compactum $L \subset S_\phi^A(x, \alpha)$ there exist a constant $C_{L,2} \leq 1$ and an integer $q \geq 0$ such that for all $k \geq 1$ and $n \geq k + q$,*

$$\mathrm{dist}(\phi^k(L), \phi^n(L)) \geq C_{L,2}|n^{-\frac{1}{p}} - k^{-\frac{1}{p}}|$$

and

$$\mathrm{dist}(\phi^n(L), \omega) \geq C_{L,2} n^{-\frac{1}{p}}.$$

Proof. It suffices of course to prove this lemma with ϕ replaced by ϕ_0. Consider two arbitrary points $z, \xi \in H^{-1}(L)$ and the line segment γ joining $\tilde{\phi}^k(z)$ and $\tilde{\phi}^n(\xi)$. Parametrize it as

$$\gamma(t) = \tilde{\phi}^k(z) + t(\tilde{\phi}^n(\xi) - \tilde{\phi}^k(z)),\ t \in [0, 1].$$

Let $l(H(\gamma))$ be the length of the curve (a subarc of either a circle or a line) $H(\gamma)$. We have

$$\begin{aligned}
l(H(\gamma)) &= \int_0^1 |(H\circ\gamma)'(t)|dt = \int_0^1 |H'(\gamma(t))|\cdot|\gamma'(t)|dt \\
&= |\tilde\phi^n(\xi)-\tilde\phi^k(z)|\int_0^1 |H'(\gamma(t))|dt \\
&= \frac{1}{p}|\tilde\phi^n(\xi)-\tilde\phi^k(z)|\int_0^1 |\gamma(t)|^{-\frac{p+1}{p}}dt \\
&= \frac{1}{p}|\tilde\phi^n(\xi)-\tilde\phi^k(z)|\int_0^1 \left(\tilde\phi^k(z)+t(\tilde\phi^n(\xi)-\tilde\phi^k(z))\right)^{-\frac{p+1}{p}}dt \\
&\ge \frac{1}{p}|\tilde\phi^n(\xi)-\tilde\phi^k(z)|\int_0^1 \left(|\tilde\phi^k(z)|+t|\tilde\phi^n(\xi)-\tilde\phi^k(z)|\right)^{-\frac{p+1}{p}}dt \\
&= \frac{1}{p}\int_{|\tilde\phi^k(z)|}^{|\tilde\phi^k(z)|+|\tilde\phi^n(\xi)-\tilde\phi^k(z)|} u^{-\frac{p+1}{p}}du \\
&= \left(|\tilde\phi^k(z)|^{-\frac{1}{p}} - \left(|\tilde\phi^k(z)|+|\tilde\phi^n(\xi)-\tilde\phi^k(z)|\right)^{-\frac{1}{p}}\right) \\
&= \frac{\left(|\tilde\phi^k(z)|+|\tilde\phi^n(\xi)-\tilde\phi^k(z)|\right)^{\frac{1}{p}} - |\tilde\phi^k(z)|^{\frac{1}{p}}}{|\tilde\phi^k(z)|^{\frac{1}{p}}\left(|\tilde\phi^k(z)|+|\tilde\phi^n(\xi)-\tilde\phi^k(z)|\right)^{\frac{1}{p}}} \\
&\ge C_{H^{-1}(L)}^{-\frac{1}{p}}(3C_{H^{-1}(L)})^{-\frac{1}{p}}\frac{\left(|\tilde\phi^k(z)|+|\tilde\phi^n(\xi)-\tilde\phi^k(z)|\right)^{\frac{1}{p}} - |\tilde\phi^k(z)|^{\frac{1}{p}}}{k^{\frac{1}{p}}n^{\frac{1}{p}}},
\end{aligned} \tag{9.11}$$

where the last inequality has been written due to Lemma 9.3.1. By the mean value theorem there exists $\eta\in[|\tilde\phi^k(z)|, |\tilde\phi^k(z)|+|\tilde\phi^n(\xi)-\tilde\phi^k(z)|]$ such that

$$\begin{aligned}
&\left(|\tilde\phi^k(z)|+|\tilde\phi^n(\xi)-\tilde\phi^k(z)|\right)^{\frac{1}{p}} - |\tilde\phi^k(z)|^{\frac{1}{p}} = \\
&\qquad = \frac{1}{p}|\tilde\phi^n(\xi)-\tilde\phi^k(z)|\eta^{\frac{1-p}{p}} \\
&\qquad \ge \frac{1}{p}|\tilde\phi^n(\xi)-\tilde\phi^k(z)|\left(|\tilde\phi^k(z)|+|\tilde\phi^n(\xi)-\tilde\phi^k(z)|\right)^{\frac{1-p}{p}} \\
&\qquad \ge \frac{1}{p}(3C_{H^{-1}(L)})^{\frac{1-p}{p}}|\tilde\phi^n(\xi)-\tilde\phi^k(z)|.
\end{aligned} \tag{9.12}$$

Now, in view of (3.6), $\tilde\phi^n(\xi)-\tilde\phi^k(z) = \xi - z + O(\max\{n^{1-\frac{1}{p}}, \log n\})$. Hence

$$\begin{aligned}
|\tilde\phi^n(\xi)-\tilde\phi^k(z)| &\ge \mathrm{diam}(H^{-1}(L)) + (n-k) - O\left(\max\{n^{1-\frac{1}{p}}, \log n\}\right) \\
&\ge \frac{1}{2}(n-k)
\end{aligned}$$

if only $n-k$ is large enough, say $n-k \geq q$. Using this, (9.11) and (9.12), if $n \geq k+q$, then

$$l(H(\gamma)) \geq \frac{1}{2p}(3C_{H^{-1}(L)})^{\frac{1-p}{p}} \frac{(n-k)n^{1-\frac{1}{p}}}{k^{\frac{1}{p}}n^{\frac{1}{p}}}. \tag{9.13}$$

Since $t \leq t^{\frac{1}{p}}$ for $t \in [0,1]$, we get $1-t \geq 1-t^{\frac{1}{p}}$ for these t, and consequently $1-\frac{k}{n} \geq 1-\left(\frac{k}{n}\right)^{\frac{1}{p}}$ or $\frac{n-k}{n} \geq 1-\left(\frac{k}{n}\right)^{\frac{1}{p}}$. Multiplying this last inequality by $n^{\frac{1}{p}}$, we get $(n-k)n^{\frac{1-p}{p}} \geq n^{\frac{1}{p}} - k^{\frac{1}{p}}$. Combining this and (9.13) yields

$$l(H(\gamma)) \geq \frac{1}{2p}(3C_{H^{-1}(L)})^{\frac{1-p}{p}}\left(k^{-\frac{1}{p}} - n^{-\frac{1}{p}}\right). \tag{9.14}$$

If $H(\gamma)$ is a line segment, then

$$|\phi_0^k(H(z)) - \phi_0^n(H(\xi))| = l(H(\gamma)) \geq \frac{1}{2p}(3C_{H^{-1}(L)})^{\frac{1-p}{p}}\left(k^{-\frac{1}{p}} - n^{-\frac{1}{p}}\right). \tag{9.15}$$

If however $H(\gamma)$ is an arc of a circle, then consider the curve

$$g(t) = \tilde{\phi}^k(z) + t(\tilde{\phi}^n(\xi) - \tilde{\phi}^k(z)),\ t \in (-\infty, 0].$$

Proceeding exactly as in the formula (9.11) with the estimate $|g(t)| \leq |\tilde{\phi}^k(z)| - t|(\tilde{\phi}^n(\xi) - \tilde{\phi}^k(z)|$, we get

$$l(H(\gamma)) \geq \frac{1}{p}\int_{|\tilde{\phi}^k(z)|}^{\infty} u^{-\frac{p+1}{p}}\, du = |\tilde{\phi}^k(z)|^{-\frac{1}{p}}.$$

Applying now Lemma 9.3.1 this gives

$$l(H(\gamma)) \geq (C_{H^{-1}(L)})^{-\frac{1}{p}} k^{-\frac{1}{p}} \geq (C_{H^{-1}(L)})^{-\frac{1}{p}}\left(k^{-\frac{1}{p}} - n^{-\frac{1}{p}}\right).$$

Therefore, invoking (9.14), we deduce that both arcs joining the points $\phi_0^k(H(z))$ and $\phi_0^n(H(z))$ on the circle $H(\{\tilde{\phi}^k(z) + t(\tilde{\phi}^n(\xi) - \tilde{\phi}^k(z)) : t \in \mathbb{R} \cup \{\infty\})$ have length $\geq C(\left(k^{-\frac{1}{p}} - n^{-\frac{1}{p}}\right)$, where $C = \min\left\{\frac{1}{2p}(3C_{H^{-1}(L)})^{\frac{1-p}{p}}, C_{H^{-1}(L)})^{-\frac{1}{p}}\right\}$. Hence $|\phi_0^k(H(z)) - \phi_0^n(H(\xi))| \geq \frac{C}{\pi}\left(k^{-\frac{1}{p}} - n^{-\frac{1}{p}}\right)$. This and (9.15) imply that

$$\operatorname{dist}\left(\phi_0^k(H(z)), \phi_0^n(H(\xi))\right) \geq \frac{C}{\pi}\left(k^{-\frac{1}{p}} - n^{-\frac{1}{p}}\right)$$

and the proof of the first part of our lemma is complete. Since it follows from (9.6) that $\phi^k(L)$ converges to ω if $k \to \infty$, applying the proved part of the lemma, we conclude that

$$\begin{aligned}\operatorname{dist}(\phi^n(L),\omega) &= \lim_{k\to\infty} \operatorname{dist}(\phi^n(L),\phi^k(L)) \\ &\geq \lim_{k\to\infty} C_{L,2}\big(n^{-\frac{1}{p}} - k^{-\frac{1}{p}}\big) = C_{L,2}n^{-\frac{1}{p}}.\end{aligned}$$

□

Remark 9.3.7 *We would like to remark that all statements proved in this section about the map ϕ continue to be true if we replace the assumption $L \subset S_\phi^A(x(\alpha),\alpha)$ by the assumption $\phi^j(L) \subset S_\phi^A(x(\alpha),\alpha)$ for some $j \geq 0$.*

Lemma 9.3.8 *If $L \subset \mathbb{C} \setminus \omega$ is a compactum and $\lim_{n\to\infty} \phi^n(L) = \omega$, then there exists an attracting direction A such that for every $\alpha \in (0,\pi)$, $\phi^n(L) \subset S_\phi^A(x(\alpha),\alpha)$ for every $n \geq 0$ large enough.*

Proof. First notice that due to (9.6), if $\phi^k(L) \subset S_\phi^A(x(\alpha),\alpha)$, then $\phi^n(L) \subset S_\phi^A(x(\alpha),\alpha)$ for all $n \geq k$. Suppose now on the contrary that our lemma is not true. Since the set of attracting directions is finite, there thus exist $\beta \in (0,\pi)$ and such that for every $n \geq k$

$$\phi^n(L) \cap \bigcup_{i=1}^{p} S_\phi^{A_i^+}(x(\beta),\beta) = \emptyset, \tag{9.16}$$

where $\{A_1^+, A_2^+, \dots, A_p^+\}$ is the set of all attracting directions for ϕ at ω. Taking now $\gamma \in (\pi-\beta,\pi)$ we see that the union

$$\bigcup_{i=1}^{p} S_\phi^{A_i^+}(x(\beta),\beta) \cup \bigcup_{i=1}^{p} S_{\phi^{-1}}^{A_i^-}(x(\gamma),\gamma)$$

(A_i^- being attracting directions for ϕ^{-1}) forms a deleted neighborhood of ω. Along with (9.16) this implies that $\phi^n(L) \subset S_{\phi^{-1}}^{A_i^-}(x(\gamma),\gamma)$ for some $i \in \{1,2,\dots,p\}$ and all $n \geq k$. But since, by (9.6), $\lim_{n\to\infty} \phi^{-n}$ $(S_{\phi^{-1}}^{A_i^-}(x(\gamma),\gamma)) = \omega$, we conclude that $L = \lim_{n\to\infty} \phi^{-n}(\phi^(L)) = \omega$. This contradiction finishes the proof. □

We end this section with a result concerning parabolic IFSs in dimension $d = 2$.

Proposition 9.3.9 *If $S = \{\phi_i : X \to X\}_{i\in I}$ is a parabolic IFS and $d = 2$, then the fixed point of each parabolic element ϕ_i belongs to the boundary of X. In addition, the derivative of each parabolic element evaluated at the corresponding parabolic fixed point is a root of unity.*

Proof. Suppose that $i \in I$ is a parabolic index and that the corresponding fixed point x_i is in $\mathrm{Int}(X)$. Let C_i be the component of $\mathrm{Int}(X)$ containing x_i. So, C_i is an open connected subset of $\mathbb{C}$ missing at least three points, since X is a compact subset of $\mathbb{C}$. Therefore, by the uniformization theorem, there exists a holomorphic covering map $R: D \to C_i$ sending 0 to x_i, where $D = \{z \in \mathbb{C} : |z| < 1\}$ is the open unit disk in $\mathbb{C}$. Since $\phi_i(x_i) = x_i$, $\phi_i(C_i) \subset C_i$. Considering, if necessary, the second iterate of ϕ_i we may assume that ϕ_i is holomorphic. Hence, all its lifts to D i.e. satisfying the equality $\phi_i \circ R = R \circ \psi$) are holomorphic. Take $\psi : D \to D$, the lift fixing the point 0. Then $\psi'(0) = \phi_i'(x_i)$, whence $|\psi'(0)| = 1$. Therefore, in view of Schwarz's lemma, $\psi : D \to D$ is a rotation with center at 0. In particular

$$\phi_i(C_i) = \phi_i \circ R(D) = R \circ \psi(D) = R(D) = C_i.$$

This contradicts condition (4) from Section 8.1. Finally, suppose i is a parabolic index. If $\phi_i'(x_i)$ were not a root of unity, then the images of finitely many iterates of ϕ_i of an open cone witnessing the cone condition at x_i would cover a punctured neighborhood of X. This contradicts the fact the the boundary of X has no isolated points. □

9.4 Proofs of the main theorems

In order to apply the results of Sections 8.2 and 8.3 we need the following. Recall that for each parabolic index i, x_i is the unique fixed point of the map ϕ_i.

Proposition 9.4.1 *If $\{\phi_i : X \to X\}_{i \in I}$ is a parabolic IFS (I is allowed to be infinite), then for every parabolic index $i \in I$ and every $j \in I \setminus \{i\}$, we have $x_i \notin \phi_j(X)$.*

Proof. Suppose on the contrary that $x_i \in \phi_j(X)$ for some parabolic index $i \in I$ and some $j \in I \setminus \{i\}$. Then by the cone condition and conformality of ϕ_j, the set $\phi_j(X)$ contains a central cone with positive measure and vertex x_i. On the other hand, since ϕ_i is conformal, $X \setminus \phi_i(X)$ contains no central cone with positive measure and vertex x_i. This is a contradiction since, by the open set condition, $\mathrm{Int}(\phi_i(X)) \cap \mathrm{Int}(\phi_j(X)) = \emptyset$. □

Consider a parabolic IFS, $S = \{\phi_i : X \to X\}_{i \in I}$. If S is two-dimensional, then dealing with the family of second iterates $S^2 = \{\phi_{ij} : i, j \in I\}$, instead of S, we may assume that all the parabolic maps are holomorphic.

Also, from Proposition 9.3.9 the derivative of each parabolic element evaluated at the corresponding parabolic fixed point is a root of unity. Therefore, for some appropriate positive integer q, the derivative of each parabolic element of S^q evaluated at the corresponding parabolic fixed point is equal to 1. Thus, without loss of generality, we may assume that in case $d = 2$, all the parabolic elements of S are simple parabolic mappings in the sense of Section 8.3. Grouping together now the results of Sections 8.2 and 8.3, we deduce that for any given $d \geq 2$, there exists a constant $Q \geq 1$ and an integer $q \geq 0$ such that for every parabolic index $i \in I$ there exists an integer $p_i \geq 1$ such that for every $j \in I \setminus \{i\}$ and all $n, k \geq 1$ we have

$$Q^{-1}n^{-\frac{p_i+1}{p_i}} \leq \inf_X\{||\phi'_{i^nj}(x)||\}, ||\phi'_{i^nj}||, \operatorname{diam}(\phi_{i^nj}(X)) \leq Qn^{-\frac{p_i+1}{p_i}}, \tag{9.17}$$

$$Q^{-1}n^{-\frac{1}{p_i}} \leq \operatorname{dist}(x_i, \phi_{i^nj}(X)) \leq \operatorname{Dist}(x_i, \phi_{i^nj}(X)) \leq Qn^{-\frac{1}{p_i}}, \tag{9.18}$$

$$\operatorname{Dist}(\phi_{i^nj}(X), \phi_{i^kj}(X)) \leq Q\left|\min\{k,n\}^{-\frac{1}{p_i}} - (\max\{k,n\}+1)^{-\frac{1}{p_i}}\right| \tag{9.19}$$

and, furthermore, if $|n-k| \geq q$, then

$$\operatorname{dist}(\phi_{i^nj}(X), \phi_{i^kj}(X)) \geq Q|n^{-\frac{1}{p_i}} - k^{-\frac{1}{p_i}}|. \tag{9.20}$$

We also need the following.

Theorem 9.4.2 *If $\{\phi_i : X \to X\}_{i\in I}$ is a parabolic IFS (I is allowed to be infinite), then*

$$\operatorname{HD}(J_S) > \max\left\{\frac{p_i}{p_i+1} : i \text{ is parabolic}\right\},$$

where p_i is the integer indicated in (9.20).

Proof. Using (9.17), if we take t slightly larger than $\frac{p_i}{p_i+1}$, then $\psi(t)$ can be made as large as we like. Since $\mathrm{P}^*(t) \geq -t\log K + \log\psi(t)$, $\mathrm{P}^*(t) > 0$. Therefore, $h = \operatorname{HD}(J_{S^*}) > \frac{p_i}{p_i+1}$. It therefore immediately follows from Lemma 8.4.3 that

$$\operatorname{HD}(J_S) = \operatorname{HD}(J_{S^*}) > \max\left\{\frac{p_i}{p_i+1} : i \ \text{ is parabolic}\right\}.$$

□

If in addition S is finite, then we conclude from (9.17) that

$$\theta_{S^*} = \max\left\{ \frac{p_i}{p_i + 1} : i \ \text{ is parabolic} \right\}$$

and $\psi(\theta_{S^*}) = \infty$. This means that the system S^* is hereditarily regular and we have proved Theorem 9.1.2.

Lemma 9.4.3 *For every parabolic index $i \in I$, there exists an open cone $C_i \subset X$ with vertex x_i and such that $x_i \in \overline{J \cap C_i}$.*

Proof. In case $d \geq 3$ this is an immediate consequence of Lemma 9.2.3. In case $d \geq 3$ this is an immediate consequence of (9.9) and Lemma 9.3.8. □

In view of Theorem 9.1.3, in order to prove Theorem 9.1.4 it suffices to demonstrate the following four lemmas assuming the finite parabolic system S satisfies the strong open set condition.

Lemma 9.4.4 *If $h < 1$, then $\mathcal{H}^h(J) = 0$.*

Lemma 9.4.5 *If $h \leq 1$, then $\mathcal{P}^h(J) < \infty$.*

Lemma 9.4.6 *If $h > 1$, then $\mathcal{P}^h(J) = \infty$.*

Lemma 9.4.7 *If $h \geq 1$, then $\mathcal{H}^h(J) > 0$.*

Proof of Lemma 9.4.4. Let $i \in I$ be a parabolic index. Fix $j \in I \setminus \{i\}$. Since $\phi_{i^n j}(X) \subset B(x_i, r)$ if and only if $\mathrm{Dist}(x_i, \phi_{i^n j}(X)) < r$, it follows from (9.18) that if $Qn^{-\frac{1}{p_i}} < r$, then $\phi_{i^n j}(X) \subset B(x_i, r)$. Hence using (9.17) and the conformality of m, we get

$$\begin{aligned}
r^{-h} m(B(x_i, r)) &\geq r^{-h} \sum_{n: Qn^{-\frac{1}{p_i}} < r} m(\phi_{i^n j}(X)) \\
&\geq r^{-h} \sum_{n > (Qr^{-1})^{p_i}} Q^{-h} n^{-\frac{p_i+1}{p_i} h} \\
&\geq Q^{-h} (\mathrm{const})\, r^{-h} (Q^{p_i} r^{-p_i})^{1 - \frac{p_i+1}{p_i} h} \\
&\geq (\mathrm{const})\, r^{-h} r^{-p_i + (p_i+1)h} \\
&= (\mathrm{const})\, r^{p_i(h-1)}.
\end{aligned}$$

Since $h < 1$, this implies that $\lim_{r\to 0} r^{-h}m(B(x_i,r)) = \infty$. By Proposition 9.1.1, $x_i \in S^*(\infty)$. It therefore follows immediately from Theorem 4.5.3 that $\mathcal{H}^h(J_S) = \mathcal{H}^h(J_{S^*}) = 0$. □

Proof of Lemma 9.4.5. Fix a parabolic index $i \in I$, $j \in I\setminus\{i\}$, $n \geq 1$ and fix r, $2\text{diam}(\phi_{i^n j}(X)) < r \leq 1$. Take an arbitrary point $x \in \phi_{i^n j}(X)$. It follows from (9.19) and the inequality $r > 2\text{diam}(\phi_{i^n j}(X))$ that if $k \leq n$ and $\overline{Q}\big(k^{-\frac{1}{p_i}} - n^{-\frac{1}{p_i}}\big) < r$, where we take an appropriate constant $\overline{Q} \geq Q$, then $B(x,r) \supset \phi_{i^k j}(X)$. Hence, using (9.17) andTheorem 9.4.2 and letting $E(x)$ denote the greatest integer in x, we get

$$\begin{aligned}
m(B(x,r)) &\geq \sum_{k=E\left(\left(\overline{Q}^{-1}r+n^{-\frac{1}{p_i}}\right)^{-p_i}\right)+1}^{n} m(\phi_{i^k j}(X)) \\
&\geq \sum_{k=E\left(\left(\overline{Q}^{-1}r+n^{-\frac{1}{p_i}}\right)^{-p_i}\right)+1}^{n} \overline{Q}^{-h} k^{-\frac{p_i+1}{p_i}h} \\
&\geq (\text{const})\left(\left(\overline{Q}^{-1}r+n^{-\frac{1}{p_i}}\right)^{-p_i\left(1-\frac{p_i+1}{p_i}h\right)} - n^{1-\frac{p_i+1}{p_i}h}\right) \\
&\geq (\text{const})\left(\left(\overline{Q}^{-1}r+n^{-\frac{1}{p_i}}\right)^{(p_i+1)h-p_i} - n^{-\frac{1}{p_i}((p_i+1)h-p_i)}\right).
\end{aligned} \tag{9.21}$$

It follows from the mean value theorem that there exists some η with $n^{-\frac{1}{p_i}} \leq \eta \leq \overline{Q}^{-1}r + n^{-\frac{1}{p_i}}$ such that

$$\begin{aligned}
&\left(\overline{Q}^{-1}r+n^{-\frac{1}{p_i}}\right)^{(p_i+1)h-p_i} - n^{-\frac{1}{p_i}((p_i+1)h-p_i)} \\
&= ((p_i+1)h-p_i)(\overline{Q}^{-1}r)\eta^{(p_i+1)h-p_i-1} \\
&= ((p_i+1)h-p_i)\overline{Q}^{-1}r\eta^{(p_i+1)(h-1)} \\
&\geq ((p_i+1)h-p_i)\overline{Q}^{-1}r\left(\overline{Q}^{-1}r+n^{-\frac{1}{p_i}}\right)^{(p_i+1)(h-1)}.
\end{aligned} \tag{9.22}$$

But, by our constraints on r and by (9.17), $n^{-\frac{1}{p_i}} \leq Q^{\frac{1}{p_i+1}}\text{diam}^{\frac{1}{p_i+1}}(\phi_{i^n j}(X)) \leq (1/2)Q^{\frac{3}{p_i+1}}r^{\frac{1}{p_i+1}}$. Thus, combining this, (9.22) and (9.21), we get

$$
\begin{aligned}
m(B(x,r)) &\geq (\text{const})\, r \left(\overline{Q}^{-1} r + n^{-\frac{1}{p_i}}\right)^{(p_i+1)(h-1)} \\
&\geq (\text{const})\, r \left(r^{\frac{1}{p_i+1}}\right)^{(p_i+1)(h-1)} = (\text{const})\, r^h.
\end{aligned}
$$

Therefore, the proof follows by applying Theorem 4.5.5 with $\xi = 1$ and $\gamma = 1$. □

Proof of Lemma 9.4.6. Fix a parabolic index $i \in I$. Since the system is finite, by applying Proposition 9.4.1, there is some $R > 0$ such that if $0 < r < R$, then $B(x_i, r)$ does not intersect $\phi_j(X)$, for any $j \neq i$. Fix such a radius r. Using (9.18) and (9.17), we derive

$$
\begin{aligned}
r^{-h} m(B(x_i,r)) &\leq r^{-h} \sum_{j \neq i} \sum_{n: Q^{-1} n^{-\frac{1}{p_i}} < r} m(\phi_{i^n j}(X)) \\
&\leq r^{-h} \sum_{j \neq i} \sum_{n > (Qr)^{-p_i}} Q^h ||\phi'_{i^n j}||^h \\
&\leq Q^h r^{-h} \sum_{j \neq i} \sum_{n > (Qr)^{-p_i}} n^{-\frac{p_i+1}{p_i} h} \\
&\leq (\text{const}) \# I Q^h \left(\frac{p_i+1}{p_i} h - 1\right) r^{-h} (Qr)^{(-p_i)\left(1 - \frac{p_i+1}{p_i} h\right)} \\
&= (\text{const})\, r^{-h + (p_i+1)h - p_i} = (\text{const})\, r^{p_i(h-1)}.
\end{aligned}
$$

Since $h > 1$, this implies that $\lim_{r \to 0} r^{-h} m(B(x_i, r)) = 0$. Applying Theorem 4.5.5 along with Lemma 9.4.3 and Proposition 9.1.1, we conclude that $\mathcal{P}^h(J) = \infty$. □

Proof of Lemma 9.4.7. Fix a parabolic index $i \in I$, $j \in I \setminus \{i\}$, $n \geq \max\{2q, q+1\}$ and $x \in \phi_{i^n j}(X)$. Given $1 \geq r > \text{diam}(\phi_{i^n j}(X))$ and using (9.17) twice we obtain

$$
\begin{aligned}
\Sigma_1 := \sum_{a \neq i} \sum_{k=n-q}^{n+q} m(\phi_{i^k a}(X)) &\leq \sum_{a \neq i} \sum_{k=n-q}^{n+q} ||\phi'_{i^k a}||^h \\
\leq \sum_{a \neq i} \sum_{k=n-q}^{n+q} Q^h k^{-\frac{p_i+1}{p_i} h} &\leq \# I Q^h 2q (n-q)^{-\frac{p_i+1}{p_i} h}
\end{aligned} \tag{9.23}
$$

$$= 2\#IqQ^h \left(\frac{n}{n-q}\right)^{\frac{p_i+1}{p_i}h} n^{-\frac{p_i+1}{p_i}h}$$

$$\leq 2qQ^h \#I 2^{\frac{p_i+1}{p_i}h} Q \text{diam}^h(\phi_{i^n j}(X)) \leq 2qQ^{h+1} 2^{\frac{p_i+1}{p_i}h} \#I r^h.$$

Put $l = E\left(\left(n^{-\frac{1}{p_i}} - Qr\right)^{-p_i}\right) + 1$ if $Qr < n^{-\frac{1}{p_i}}$ and $l = \infty$ otherwise. Using (9.17) we get

$$\Sigma_2 := \sum_{a \neq i} \sum_{k: |n^{-\frac{1}{p_i}} - k^{-\frac{1}{p_i}}| < Qr} m(\phi_{i^k a}(X))$$

$$\leq \sum_{a \neq i} \sum_{k=E\left(\left(Qr + n^{-\frac{1}{p_i}}\right)^{-p_i}\right)}^{l} Q^h k^{-\frac{p_i+1}{p_i}h}$$

$$\leq \#IQ^h \sum_{k=E\left(\left(Qr - n^{-\frac{1}{p_i}}\right)^{-p_i}\right)}^{l} k^{-\frac{p_i+1}{p_i}h}.$$

Suppose first that $Qr < n^{-\frac{1}{p_i}}$. Then

$$\Sigma_2 \leq \#IQ^h \left(\frac{p_i+1}{p_i}h - 1\right) \left(\left(Qr + n^{-\frac{1}{p_i}}\right)^{-p_i+(p_i+1)h} - \left(n^{-\frac{1}{p_i}} - Qr\right)^{-p_i+(p_i+1)h}\right).$$

It follows now from the mean value theorem that there exists $\eta \in [n^{-\frac{1}{p_i}} - Qr, n^{-\frac{1}{p_i}} + Qr]$ such that

$$\left(Qr + n^{-\frac{1}{p_i}}\right)^{-p_i+(p_i+1)h} - \left(n^{-\frac{1}{p_i}} - Qr\right)^{-p_i+(p_i+1)h} = ((p_i+1)h - p_i) 2Qr\eta^{(p_i+1)(h-1)}.$$

Since by (9.17)

$$n^{-\frac{1}{p_i}} \leq Q^{\frac{1}{p_i}} \text{diam}(\phi_{i^n j}(X))^{\frac{1}{p_i}} \leq Q^{\frac{1}{p_i}} r^{\frac{1}{p_i}},$$

we therefore find

$$\Sigma_2 \leq (\text{const})\, r\eta^{(p_i+1)(h-1)} \leq (\text{const})\, r \left(Qr + n^{-\frac{1}{p_i}}\right)^{(p_i+1)(h-1)}$$
$$\leq (\text{const})\, r \left(Q^{\frac{1}{p_i}} r^{\frac{1}{p_i}} + Qr\right)^{(p_i+1)(h-1)} \leq (\text{const})\, r \left(r^{\frac{1}{p_i}}\right)^{(p_i+1)(h-1)}$$
$$= (\text{const})\, r^h. \tag{9.24}$$

Suppose in turn that $Qr \geq n^{-\frac{1}{p_i}}$. Then

$$\begin{aligned}
\Sigma_2 &\leq Q^h \# I \sum_{k=E\left(\left(Qr+n^{-\frac{1}{p_i}}\right)^{-p_i}\right)}^{\infty} k^{-\frac{p_i+1}{p_i}h} \\
&\leq Q^h \# I \left(\frac{p_i+1}{p_i}h - 1\right)\left(Qr + n^{-\frac{1}{p_i}}\right)^{-p_i\left(1-\frac{p_i+1}{p_i}h\right)} \\
&= Q^h \# I \left(\frac{p_i+1}{p_i}h - 1\right)\left(Qr + n^{-\frac{1}{p_i}}\right)^{(p_i+1)h-p_i} \leq (\text{const})\, r^{(p_i+1)h-p_i} \\
&= (\text{const})\, r^h r^{p_i(h-1)} \leq (\text{const})\, r^h.
\end{aligned} \tag{9.25}$$

Since, by (9.20), $m(B(x,r)) \leq \Sigma_1 + \Sigma_2$, it follows from (9.23)–(9.25) that $m(B(x,r) \leq (\text{const})\, r^h$. Finally, applying Theorem 4.5.3 completes the proof. □

Proof of Theorem 9.1.5. It is a straightforward consequence of formulae (9.18), (9.19) and (9.20) that for every parabolic index $i \in I$, every $j \in I \setminus \{i\}$ and every $x \in X$, $\overline{\text{BD}}\left(\{\phi_{i^n j}(x)\}_{n\geq 1}\right) = \frac{1}{\frac{1}{p_i}+1} = \frac{p_i}{p_i+1}$. Hence, it follows from Theorem 4.2.18 along with Theorem 4.2.14 that $\overline{\text{BD}}(J) = \dim_{\text{H}}(J)$. □

Appendix 1
Ergodic theory

If $(X, \mathcal{F}, \mu)$ is a probability space, then a map $T : X \to X$ is called measure-preserving (or the measure μ is called T-invariant) if for every $A \in \mathcal{F}$, $T^{-1}(A) \in \mathcal{F}$ and $\mu(T^{-1}(A)) = \mu(A)$. A set $A \in \mathcal{F}$ is called T-invariant if $T^{-1}(A) = A$. The measure-preserving map $T : X \to X$ is said to be ergodic if the measure μ of every T-invariant set A is either 0 or 1. The most significant property of ergodic measure-preserving mappings is contained in the following.

Theorem A1.0.8 *(Birkhoff's Ergodic Theorem) If $T : X \to X$ is an ergodic measure-preserving endomorphism of a probability space $(X, \mathcal{F}, \mu)$ and $g : X \to I\!R$ is an integrable function then*

$$\lim_{n\to\infty} \frac{1}{n} S_n g(x) = \int f d\mu \quad \textit{for } \mu\textit{-a.e. } x \in X.$$

Let α be a countable partition of the set X by elements in $\mathcal{F}$. For every $1 \le n \le \infty$ let

$$\alpha^n = \{A_1 \cap A_2 \cap \cdots \cap A_n : A_i \in \alpha \ \text{ for all } \ i = 1, 2 \ldots, n\}.$$

Of course α^n is again a countable partition of X. The partition α is called generating if after subtracting from X a set of measure zero, all the elements of α^∞ become singletons. By $\mathrm{H}_\mu(a)$, the entropy of the partition α, we mean the following quantity.

$$\mathrm{H}_\mu(\alpha) = -\sum_{A\in\alpha} \mu(A) \log(\mu(A)).$$

It can be proved (see e.g. [Par], [Wa], [KH], or [PU]) that the limit

$$\mathrm{h}_\mu(T,\alpha) = \lim_{n\to\infty} \frac{1}{n}\mathrm{H}_\mu(\alpha^n)$$

exists and $\mathrm{h}_\mu(T)$, the entropy of the endomorphism T, is defined to be the supremum of the numbers $\mathrm{h}_\mu(T,\alpha)$ taken over all countable partitions with finite entropy. Two basic facts we use in this book to deal with the concept of entropy are the following.

Theorem A1.0.9 *(Kolmogorov–Sinai) If α is a generating partition with finite entropy, then $\mathrm{h}_\mu(T) = \mathrm{h}_\mu(T,\alpha)$.*

Given $x \in X$ let $\alpha(x)$ be the only element of the partition α containing x.

Theorem A1.0.10 *(Shannon–McMillan–Breiman) If α is a generating partition with finite entropy, then*

$$\lim_{n\to\infty} \frac{-1}{n} \log \mu(\alpha^n(x)) = \mathrm{h}_\mu(T) \quad \textit{for } \mu\textit{-a.e.} \quad x \in X.$$

Appendix 2
Geometric measure theory

We start with Hausdorff measures and dimensions. Let $g : [0, \infty) \to [0, \infty)$ be a non-decreasing function continuous at 0, positive on $(0, \infty)$ and such that $g(0) = 0$. Let (X, ρ) be a metric space. For every $\delta > 0$ define

$$\mathrm{H}_g^\delta(A) = \inf\Big\{\sum_{i=1}^{\infty} g(\mathrm{diam}(U_i))\Big\}$$

where the infimum is taken over all countable covers $\{U_i : i = 1, 2, \ldots\}$ of A of with the diameter of each U_i not exceeding δ. The following limit

$$\mathrm{H}_g(A) = \lim_{\delta \to 0} \Lambda_\phi^\delta(A) = \sup_{\delta > 0} \Lambda_\phi^\delta(A)$$

exists, but may be infinite, since $\mathrm{H}_g^\delta(A)$ increases as δ decreases. Since all the functions H_g^δ are outer measures, H_g is an outer measure. Moreover, H_g turns is a metric outer measure and therefore all Borel subsets of X are H_g-measurable. A particular role is played by functions g of the form $t \mapsto t^\alpha$, $t, \alpha > 0$ and in this case the corresponding outer measure H_g is denoted by H_α. It is easy to see that there exists a unique value, $\mathrm{HD}(A)$, called the Hausdorff dimension of A, such that

$$\mathrm{H}_t(A) = \begin{cases} \infty & \text{if } 0 \le t < \mathrm{HD}(A) \\ 0 & \text{if } \mathrm{HD}(A) < t < \infty. \end{cases}$$

Note that like Hausdorff measures, Hausdorff dimension is consequently an intrinsic property of a subset of a given metric space in that in only depends on the metric restricted to the subset. Hausdorff dimension is an increasing function with respect to inclusion of sets and is σ-stable, meaning that the following is true.

Theorem A2.0.11 *If* $\{A_n\}_{n\geq 1}$ *is a countable family of subsets of* X *then*

$$\mathrm{HD}(\bigcup_n A_n) = \sup_n\{\mathrm{HD}(A_n)\}.$$

Passing to packing measures and dimensions, a collection $\{(x_i, r_i) : i \in I\}$ is a packing of $A \subset X$ if and only if for any pair $i \neq j$

$$\rho(x_i, x_j) \geq r_i + r_j$$

and each $x_i \in A$. This property is not generally equivalent to the requirement that all the balls $B(x_i, r_i)$ are mutually disjoint. It is obviously so if X is a Euclidean space. For every $A \subset X$ and every $r > 0$ let

$$\Pi_g^{*r}(A) = \sup\{\sum_{i=1}^{\infty} g(r_i)\}$$

where the supremum is taken over all packings $\{(x_i, r_i) : i = 1, 2, \ldots\}$ of A of radius not exceeding r, where the radius of a packing is $\sup\{r_i : i \in I\}$. Let

$$\Pi_g^*(A) = \lim_{r\to 0} \Pi_g^{*r}(A) = \inf_{r>0} \Pi_g^{*r}(A)$$

The limit exists since $\Pi_\phi^{*r}(A)$ decreases as r decreases. Π_g^* need not be an outer measure. In order to construct an outer measure, called the packing measure associated with the function g, we put

$$\Pi_\phi(A) = \inf\{\sum \Pi_\phi^*(A_i)\}.$$

It is this two stage process involved in the definition of packing measure that makes it more complicated to deal with (cf. [MM]). Now, in exactly the same way as Hausdorff dimension HD, one can define the packing* dimension PD* and packing dimension PD using respectively $\Pi_t^*(A)$ and $\Pi_t(A)$ instead of $\mathrm{H}_t(A)$. One has monotonicity and σ-stability for the packing dimension also.

Some basic sufficient conditions for finiteness and positivity of Hausdorff and packing measures are described as follows. Let ν be a Borel probability measure on X and let $t \geq 0$ be a real number. Define the function $\rho = \rho_t(\nu) : X \times (0, \infty) \to (0, \infty)$ by

$$\rho(x, r) = \frac{\nu(B(x, r))}{r^t}.$$

The following two theorems are for our aims some key facts from geometric measure theory.

Theorem A2.0.12 *Assume that X is a compact subspace of a d-dimensional Euclidean space. Then for every $t \geq 0$ there exist constants $h_1(t)$ and $h_2(t)$ with the following properties.*

(1) *If A is a Borel subset of X and $C > 0$ is a constant such that for all (but countably many) $x \in A$*

$$\limsup_{r \to 0} \rho(x, r) \geq C^{-1},$$

then for every Borel subset $E \subset A$ we have $\mathrm{H}_t(E) \leq h_1(t) C \nu(E)$ and, in particular, $\mathrm{H}_t(A) < \infty$.

(2) *If A is a Borel subset of X and $C > 0$ is a constant such that for all $x \in A$*

$$\limsup_{r \to 0} \rho(x, r) \leq C^{-1},$$

then for every Borel subset $E \subset A$ we have $\mathrm{H}_t(E) \geq C h_2(t) \nu(E)$.

Theorem A2.0.13 *Assume that X is a compact subspace of a d-dimenbreak sional Euclidean space. Then for every $t \geq 0$ there exist constants $p_1(t)$ and $p_2(t)$ with the following properties.*

(1) *If A is a Borel subset of X and $C > 0$ is a constant such that for all $x \in A$*

$$\liminf_{r \to 0} \rho(x, r) \leq C^{-1},$$

then for every Borel subset $E \subset A$ we have $\Pi_t(E) \geq C p_1(t) \nu(E)$.

(2) *If A is a Borel subset of X and $C > 0$ is a constant such that for all $x \in A$*

$$\liminf_{r \to 0} \rho(x, r) \geq C^{-1},$$

then for every Borel subset $E \subset A$ we have $\Pi_t(E) \leq p_2(t) C \nu(E)$ and, consequently, $\Pi_t(A) < \infty$.

(1') *If ν is non–atomic then (1) holds under the weaker assumption that the hypothesis of part (1) is satisfied on the complement of a countable set.*

Passing to ball-counting dimensions, for every $r > 0$ consider the family of all collections $\{B(x_i, r)\}$ which cover A and are centered at A, meaning that all x_i are in A. Put $N(A, r) = \infty$ if this family is empty. Otherwise define $N(A, r)$ to be the minimum of all cardinalities of elements of this family. The lower ball-counting dimension and upper ball-counting dimension of A are defined respectively by

$$\underline{\mathrm{BD}}(A) = \liminf_{r \to 0} \frac{\log N(A,r)}{-\log r} \text{ and } \overline{\mathrm{BD}}(A) = \limsup_{r \to 0} \frac{\log N(A,r)}{-\log r}.$$

If $\underline{\mathrm{BD}}(A) = \overline{\mathrm{BD}}(A)$, the common value is called simply the ball-counting dimension of A and is denoted by $\mathrm{BD}(A)$. In the literature the names box-counting dimension, Minkowski dimension and capacity are also frequently used for the ball-counting dimension. The basic relation between the dimensions we have introduced is provided by the following.

Theorem A2.0.14 *For every set $A \subset X$*

$$\begin{aligned}\mathrm{HD}(A) &\le \min\{\mathrm{PD}(A), \underline{\mathrm{BD}}(A)\} \\ &\le \max\{\mathrm{PD}(A), \underline{\mathrm{BD}}(A)\} \le \overline{\mathrm{BD}}(A) = \mathrm{PD}^*(A).\end{aligned}$$

We finish this section with the following definition.

Definition A2.0.15 *Let μ be a Borel measure on (X, ρ). Then the Hausdorff dimension* $\mathrm{HD}(\mu)$ *of the measure μ is defined as*

$$\mathrm{HD}(\mu) = \inf\{\mathrm{HD}(Y) : \mu(X \setminus Y) = 0\}$$

and an analogous definition can be formulated for packing dimension.

An extended exposition of the material contained in this section can be found for example in [Fa2] or [Ma1]. An appropriate tool useful for calculating Hausdorff dimensions of measures is provided by the following.

Theorem A2.0.16 *Suppose that μ is a Borel probability measure on $\mathbb{R}^n$, $n \ge 1$.*

(a) If there exists θ_1 such that for μ-a.e. $x \in \mathbb{R}^n$

$$\liminf_{r \to 0} \frac{\log \mu(B(x,r))}{\log r} \ge \theta_1$$

then $\mathrm{HD}(\mu) \ge \theta_1$.

(b) If there exists θ_2 such that for μ-a.e. $x \in \mathbb{R}^n$

$$\liminf_{r\to 0} \frac{\log \mu(B(x,r))}{\log r} \leq \theta_2$$

then $\mathrm{HD}(\mu) \leq \theta_2$.

Glossary of Notation

$\mathrm{AI}(\tilde{m})$	set of densities of absolutely continuous invariant probability measures, 39
α^n	standard partition into cylinders of length n, 10
T_zQ	approximate tangent k-plane, 167
CGDMS	abbreviation for conformal graph directed Markov system, 71
$\chi_\mu(\sigma)$	characteristic Lyapunov exponent, 90
$\mathrm{Con}(x, \alpha, u)$	central cone with vertex x, measure α and the axis parallel to the vector U, 69
$\lambda A \circ i_{a,r} + b$	form of any conformal map in $\mathbb{R}^d$ with $d \geq 3$, 62
d_α	metric on coding space, 4
$\mathrm{Dist}(A, B)$	$\sup\{\|\|a - b\|\| : (a, b) \in A \times B\}$, 240
$\mathrm{dist}(A, B)$	$\inf\{\|\|a - b\|\| : (a, b) \in A \times B\}$, 240
E_F^∞	admissible words with entries form the set F, 5
E_A^*	finite admissible words on an alphabet A, 1
E_A^∞	infinite admissible words on an alphabet A, 1
E_A^n	admissible words of length n on an alphabet A, 1
$\mathrm{H}_{\tilde{\mu}}(\beta)$	entropy of μ over the partition β, 10
E_n^ω	set of words of length n whose last entry connects to the first entry of ω, 13
$E^{\mathbb{Z}}$	two-sided shift space, 119
$\mathcal{F}in(I)$	family of all finite subsets of I, 81
$\mathcal{F}in(F)$	finite set of the one-parameter family generated by F, 124
$\mathcal{F}in(q)$	$\inf\{t : \mathcal{L}_{G_{q,t}}(\mathbb{1}) < \infty\}$, 124
$\mathrm{Fix}(\mathcal{L}_0)$	set of integrable functions fixed by $\mathcal{L}_0$, 39
$f_\mu(\alpha)$	Hausdorff dimension of the level set $K_\mu(\alpha)$, 126
$F(S)$	finiteness interval, 78
$G_{q,t}$	family of functions $\{g_{q,t}^{(i)} := qf^{(i)} + t\log\|\phi_i'\|\}$ appearing in multifractal analysis, 123
$G(d, k)$	Grassmannian manifold, 167
$\mathcal{H}_\alpha$	functions in $\mathcal{H}_0$ with finite α variation, 32
$\mathcal{H}_\alpha^s$	summable functions in $\mathcal{H}_\alpha$, 32

$\mathrm{HD}(\nu)$	Hausdorff dimension of the measure ν, 90
$\mathcal{H}_0$	bounded continuous complex valued functions on E^∞, 32
$\mathcal{H}_\beta^0$	functions in $\mathcal{H}_\beta$ with integral 0, 36
$\mathcal{H}_\beta^{0,1}$	unit ball in $\mathcal{H}_\beta^0$, 36
$\mathcal{H}_0^0$	functions in $\mathcal{H}_0$ with integral 0, 36
inversion, 62	
$i_{a,r}$	inversion with respect to the sphere centered at the point a and with radius r, 62
I_*	the alphabet of the hyperbolic system associated with a parabolic system, 223
J^*	limit set generated by the hyperbolic system associated with a parabolic system, 224
J_v	the part of the limit set coded by words starting with vertex v, 2
$\mathcal{K}_\beta$	complex Hölder continuous functions of order β on E^∞ , 43
$\mathcal{K}_\beta^s$	summable functions in $\mathcal{K}_\beta$, 43
$K_\mu(\alpha)$	α-level set of the measure μ, 126
$J = J_S$	the limit set, 2
$\mathcal{L}_0$	normalized Perron–Frobenius operator, 29
$N_r(E)$	minimum number of balls of radius $\leq r$ covering a se t E., 83
μ_F	S-invariant version of m_F, 61
$\gamma_{d,k}$	natural measure on the Grassmannian manifold $G(d,k)$, 167
$\|\|g\|\|_\alpha$	norm on $\mathcal{H}_\alpha$, 32
$\overline{\mathrm{OD}}(S)$	maximum of upper box-counting dimensions of first level orbits of points in X, 85
$\|\omega\|$	length of the word ω, 1
$[\omega]_m^n$	cylinder generated by ω with coordinates between m and n, 119
$\omega\|_n$	the word ω restricted to its first n entries, 1
Ω_*	the set of orbits of parabolic points, 222
$\omega \wedge \tau$	the longest initial block common to ω and τ, 4
$\mathrm{osc}(f)$	the sup of the variation of f over basic cylinders, 8
$\overline{D}_\mu(x)$	upper local cylindrical dimension of the measure μ at the point x, 126
$\overline{d}_\mu(x)$	upper local dimension of the measure μ at the point x, 126
$\overline{\mu}_F$	Rokhlin's natural extension of the invariant measure $\tilde{\mu}_F$, 119
$\mathcal{L}_f(g)$	Perron–Frobenius operator on space of bounded continuous functions, 27
$\mathcal{L}_f^n(g)$	nth composition of the Perron–Frobenius operator on space of continuous functions, 27
$\mathcal{L}_f^*$	conjugate Perron–Frobenius operator acting on dual space of continuous functions, 27
$P_F(f)$	the topological pressure of f over the subsytem determined by F, 7
ϕ_ω	the map coded by the word ω, 2

π the coding map from the coding space to the limit set, 2
$\mathrm{P}(F)$ pressure of the family F, 54
$I\!P I\!R^d$ $d-1$ dimensional projective space, 162
$\psi^*(t)$ ψ function of the hyperbolic system associated with a parabolic system, 232

R lower bound on sums of iterates of $1\!\!1$, 29

$S(\infty)$ the asymptotic boundary of the system, 2
$S_v(\infty)$ the asymptotic boundary from the vertex v, 2
$S_n f$ the nth partial orbit sum of f, 7
$S_\omega(F)$ iterated sum of the family F, 55
$S(\omega, K)$ preparatory quantitity needed to define scaling functions, 176
S^* hyperbolic system associated with a parabolic system, 223
$S^w(\{\omega_n\}_0^\infty, i)$ weaker scaling function, 177

$T(g)$ bounding constant, 26
$\theta(F)$ infimum of $\mathcal{F}in(F)$, 124
$\theta(q)$ infimum of $\mathcal{F}in(q)$, 124
$\theta(S)$ finiteness parameter, 78
$T(F)$ bounded distortion constant, 56
$T(q)$ temperature function associated with the family $G_{q,t}$, 125
TD topological dimension, 165
$\boldsymbol{T}: \mathcal{H}_0 \to \mathcal{H}_0$ weighted normalized operator, 36

$u_n(\omega)$ balancing function for the nth iterate of $\boldsymbol{T}$, 36
$\underline{D}_\mu(x)$ lower local cylindrical dimension of the measure μ at the point x, 126
$\underline{d}_\mu(x)$ lower local dimension of the measure μ at the point x, 126

$V_\alpha(f)$ total variation of order α, 19
$V_{\alpha,n}(f)$ variation of order α on cylinder sets of length n, 19

ζ distinguished potential funcion associated with a conformal graph directed Markov system, 90
$Z_n(F,f)$ nth partition function, 6
$Z_n(t)$ partition function, 78

Bibliography

[Al] D. S. Alexander, *A History of Complex Dynamics*, Vieweg 1994.

[Be] T. Bedford, Hausdorff dimension and box dimension in self-similar sets, *Proc. Topology and Measure V*, Ernst-Moritz-Arndt Universität Greisfwald (1988).

[BP] Benedetti, Petronio, *Lectures on Hyperbolic Geometry*, Springer-Verlag, New York, 1991.

[B1] R. Bowen, *Equlibrium States and the Ergodic Theory of Anosov Diffeomorphisms*, Lect. Notes in Math. 470 (1975), Springer Verlag.

[B2] R. Bowen, Hausdorff dimension of quasi-circles. *Publ. IHES* **50** (1980), 11–25.

[CG] L. Carleson, T. W. Gamelin, *Complex Dynamics*, Springer (1993).

[CM] R. Cawley, R. D. Mauldin, Multifractal decompositions of Moran fractals, *Adv. Math.* **92** (1992), 196–236.

[De] A. Denjoy, Sur une priopriété des séries trigonométriques, *Verlag v.d.G.V. der Wis-en Natuur. Afd.*, 30 Oct. (1920).

[DKU] M. Denker, G. Keller, M. Urbański, On the uniqueness of equilibrium states for piecewise monotone maps, *Studia Math.* **97** (1990), 27–36.

[DU1] M. Denker, M. Urbański, Relating Hausdorff measures and harmonic measures on parabolic Jordan curves, *Journal für die Reine und Angewandte Mathematik*, **450** (1994), 181–201.

[DU2] M. Denker, M. Urbański, Hausdorff and conformal measures on

Julia sets with a rationally indifferent periodic point, *J. London Math. Soc.* **43** (1991), 107–118.

[DU3] M. Denker, M. Urbański, Geometric measures for parabolic rational maps, *Ergod. Th. Dynam. Sys.* **12** (1992), 53–66.

[E] G. A. Edgar, *Measure, Topology and Fractal Geometry*, Springer, New York, 1990.

[EM] G. A. Edgar, R. D. Mauldin, Multifractal decompositions of digraph recursive fractals, *Proc. London Math. Soc.* **65** (1992), 604–628.

[EH] S. Eilenburg, O. G. Harrold,Jr., Continua of finite linear measure, *Amer. J. Math.*, **65** (1943), 137–146.

[Fa1] K. J. Falconer, *The Geometry of Fractal Sets*, Cambridge University Press, 1986.

[Fa2] K. J. Falconer, *Fractal Geometry*, J. Wiley, New York, 1990.

[Fa3] K. J. Falconer, *Techniques in Fractal Geometry*, J. Wiley, New York, 1997.

[Fe] H. Federer, Dimension and measure, *Trans. Amer. Math. Soc.*, **62** (1947), 536–547.

[FU] A. Fisher, M. Urbański, On invariant line fields, *Bull. London Math. Soc.* **32**, no. 5 (2000), 555–570.

[FP] U. Frisch, G. Parisi, On the singularity structure of fully developed turbulence. appendix to U. Frisch, Fully developed turbulence and intermittency, in "Turbulence and predictability in geophysical fluid dynamcis and climate dynamics," (*Proc. Int. Sch. Phys. "Enrico Fermi," Course LXXXVIII*), 84–88, North-Holland, Amsterdam, 1985.

[GM] R.J. Gardner, R. D. Mauldin, On the Hausdorff dimension of a set of continued fractions, *Illinois J. Math*, 27 (1983), 334–345.

[GL] S. Graf and H. Luschgy, *Foundations of Quantization for Probability Distributions*, Lecture Notes in Mathematics 1730, Springer, Berlin, 2000.

[GMW] S. Graf, R. D. Mauldin and S. C. Williams, The exact Hausdorff dimeninsion in random recursive constructions, *Memiors Amer. Math. Soc.* **381**, 1988.

[Gr] P. Grassberger, Generalized dimension of strange attractors. *Phys. Lett. A* **97** (1983), 346–349.

[GP] P. Grassberger, I. Procaccia, Characterization of strange attractors, *Phys. Lett.* **50** (1983), 346–349.

[Gu] B. M. Gurevich, Shift entropy and Markov measures in the path space of a denumerable graph, *Dokl. Akad. Nauk. SSSR*, **192** (1970); English trans. in *Soviet Math. Dokl.*, **11** (1970), 744–747.

[GS] B. M. Gurewich, S.V. Savchenko, Thermodynamic formalism for countable symbolic Markov chains, *Russ. Math. Surv.* **53:2** (1998), 245–344.

[HMU] P. Hanus, R. D. Mauldin, M. Urbański, Thermodynamic formalism and multifractal analysis of conformal infinite iterated function systems, *Acta Math. Hungarica* **96** (2002), 27–98.

[HU] P. Hanus, M. Urbański, A new class of positive recurrent functions, *Contemp. Math. Series Amer. Math. Soc.* **246** (1999), 123–136.

[HU1] P. Hanus, M. Urbański, Rigidity of infinite iterated function systems, *Real Anal. Ex.*, **24** 1998/9, 275–288.

[Ha] T. C. Halsey, M. H. Jensen, L. P. Kadanoff, I. Procaccia B. J. Shraiman, Fractal measures and their singularities: The characterization of strange sets, *Phys. Rev. A* **33** (1986), 1141–1151.

[HeU], S. Heinemann, M. Urbański, Hausdorff dimension estimates for infinite conformal iterated function systems, *Nonlinearity*, **15** (2002), 727–734.

[Hu] J. E. Hutchinson, Fractals and self-similarity, *Indiana Math. J.* **30** (1981), 713–747.

[IL] I. A. Ibragimov, Y. V. Linnik, *Independent and Stationary Sequences of Random Variables*, Wolters-Noordhoff Publ; Groningen 1971.

[JJM] E. Järvenpää, M. Järvenpää, R. D. Mauldin, Deterministic and random aspects of porosities, *Discrete Continuous Dynam. Sys.* **8** (2002), 121–136.

[Ka] T. Kato, *Perturbation Theory for Linear Operators*, Springer (1995).

[KH] A. Katok, B. Hasselblatt, *Introduction to the Modern Theory of Dynamical Systems*, Cambridge University Press, 1995.

[KS], M. Keesböhmer, B. Stratmann, *A multifractal formalism for growth rates and applications to geometrically finite Kleinian groups*, Preprint 2001.

[LM] L. Lindsay and R. D. Mauldin, Quantization dimension for conformal iterated function systems, *Nonlinearity* **15** (2002), 189–199.

[Lj] V. Ljubich, Entropy properties of rational endomorphisms of the Riemann sphere, *Ergod. Th. Dynam. Sys.* **3** (1983), 351–385.

[Man] B. B. Mandelbrot, Intermittent turbulence in self-similar cascades: divergence of high moments and dimension of the carrier, *J. Fluid Mech.* **62** (1974), 331–358.

[MM] A. Manning, H. McCluskey, Hausdorff dimension for horseshoes, *Ergod. Th. Dynam. Sys.* **3** (1983), 251–260.

[Ma1] P. Mattila, *Geometry of Sets and Measures in Euclidean Spaces, Fractals and Rectifiability*, Cambridge University Press, 1995.

[Ma2] P. Mattila, Distribution of sets and measures along planes, *J. London Math. Soc.* (2) **38** (1988), 125–132.

[Ma3] P. Mattila, On the structure of self-similar fractals, *Ann. Acad. Sci. Fenn. Ser. A I Math.* **7** (1982), 189–192.

[MM] P. Mattila, R. D. Mauldin, Measure and dimension functions: measurability and densities, *Math. Proc. Camb. Phil. Soc.* **121** (1997), 81–100.

[MMU] R. D. Mauldin, V. Mayer, M. Urbański, Rigidity of connected limit sets of conformal IFS, *Michigan Math. J.* **49** (2001), 451–458.

[MPU] R. D. Mauldin, F. Przytycki, M. Urbański, Rigidity of conformal iterated function systems, *Compositio Math.* **129** (2001), 273–299.

[MU1] R. D. Mauldin, M. Urbański, Dimensions and measures in infinite iterated function systems, *Proc. London Math. Soc.* (3) **73** (1996), 105–154.

[MU2] R. D. Mauldin, M. Urbański, Conformal iterated function systems with applications to the geometry of continued fractions, *Trans. Amer. Math. Soc.* **351** (1999), 4995–5025.

[MU3] R. D. Mauldin, M. Urbański, Gibbs states on the symbolic space over an infinite alphabet, *Israel J. Math.* **125** (2001), 93–130.

[MU4] R. D. Mauldin, M. Urbański, On the uniqueness of the density for the invariant measure in an infinite hyperbolic iterated function system, *Periodica Math. Hung.* **37** (1998), 47–53.

[MU5] R. D. Mauldin, M. Urbański, Jordan curves as repellors, *Pacific J. Math.* **166** (1994), 85–97.

[MU6] R. D. Mauldin, M. Urbański, The doubling property of conformal measures of infinite iterated function systems, *J. Number Th., to appear.*

[MU7] R. D. Mauldin, M. Urbański, Parabolic iterated function systems, *Ergod. Th. Dynam. Sys.* **20** (2000), 1423–1447.

[MU8] R. D. Mauldin, M. Urbański, Dimension and measures for a curvlinear Sierpinski gasket or Apollonian packings, *Adv. Math.* **136** (1998), 26–38.

[MU9] R. D. Mauldin, M. Urbański, Fractal measures for parabolic IFS, *Adv. Math.* **168** (2002), 225–253.

[MW1] R.D. Mauldin, S.C. Williams, Random recursive constructions: asymptotic geometric and topological properties, *Trans. Amer. Math. Soc.* **259** (1986), 325–346.

[MW2] R. D. Mauldin, S. C. Williams, Hausdorff dimension in graph directed constructions, *Trans. Amer. Soc.* **309** (1988), 811–829.

[MyW] V. Mayer, M. Urbański, Finer geometric rigidity of limit sets of conformal IFS, *Proc. Amer. Math. Soc*, to appear

[Mo] P.A.P. Moran, Additive functions of intervals and Hausdorff measure, *Proc. Cambridge, Philos. Soc.* 42 (1946), 15–23.

[Ne] Z. Nehari, *Conformal Mapping*, Dover Publications, 1975.

[O1] L. Olsen, A multifractal formalism, *Adv. Math.* **116** (1995), 82–196.

[O2] L. Olsen, *Random Geometrically Graph Directed Self-similar Multifractals*, Pitman Research notes in Math. Series, Vol. 307, Longman Scientific, 1994.

[O3] L. Olsen, Multifractal geometry, Proceedings Fractal Geometry and Stochastics II, Greifswald, Germany Aug. 1998. Progress in Probability, Vol. 46 (Editors C. Bandt, S. Graf, and M. Zähle), 1–37. Birkhäuser, 2000.

[O4] L. Olsen, Multifractal analysis of divergence points of deformed measure theoretical Birkhoff averages, preprint.

[Or] D. Orstein, *Ergodic Theory, Randomness and Dynamical Systems*, Yale University Press, New Haven and London 1974.

[Par] W. Parry, *Entropy and Generators in Ergodic Theory*, W. A. Benjamin, Inc., New York, 1969.

[Pat] S. J. Patterson, The limit set of a Fuchsian group, *Acta Math.* **136** (1976), 241–273.

[Pat1] N. Patzschke, Self-conformal multifractal measures, *Adv. Appl. Math.* **19** (1997), 486–513.

[Pat2] N. Patzschke, Self-conformal multifractal measures and self-similar random thermodynamics, Preprint 2001.

[Pe] Y. B. Pesin, *Dimension Theory in Dynamical Systems*, Chicago 1997.

[PP] Y. Pesin, B. Pitskel, Topological pressure and variational principle for non-compact sets, *Funct. Anal. Appl.* **18** (1985), 307–318.

[PW] Y. B. Pesin, H. Weiss, The multifractal analysis of equilibrium measures for conformal expanding maps and Markov Moran geometric constructions, *J. Sta. Phys.* **86** (1997), 233–275.

[Po1] C. Pommerenke, Uniformly perfect sets and the Poincaré metric, *Arch. Math.* **32** (1979), 192–199.

[Po2] C. Pommerenke, *Boundary Behaviour of Conformal Maps*, Springer 1992.

[PS] W. Philipp, W. Stout, Almost sure invariance principles for partial sums of weakly dependent random variables, *Memoirs Amer. Math. Soc.* **161** (2), (1975)

[Pr] F. Przytycki, On holomorphic perturbations of $z \to z^n$, *Bull. Polish Acad. Sci. Math.* **24** (1986), 127–132.

[PT] F. Przytycki, F. Tangerman, Cantor sets in the line: Scaling functions of the shift map, Preprint 1992.

[PU1] F. Przytycki, M. Urbański, *Fractals in the Plane - Ergodic Theory Methods*, to be published by the Cambridge University Press, available on Urbański's webpage http://www.math.unt.edu/~urbanski.

[PUZ] F. Przytycki, M. Urbański, A. Zdunik, Harmonic, Gibbs and Hausdorff measures on repellers for holomorphic maps I, *Ann. Math.* **130** (1989), 1–40.

[Ra] D. Rand, the singularity spectrumf(α) for cookie-cutters, *Ergod. Th. Dynam. Sys.* **9** (1989), 527–541.

[Ru] D. Ruelle, *Thermodynamic Formalism*, Encyclopedia of Mathematics and its Applications. vol. 5, Addison-Wesley 1978.

[Sal] A. Salli, On the Minkowski dimension of strongly porous fractals in $I\!R^n$, *Proc. London Math. Soc.* (3) **62** (1991), 353–372.

[Sar] O. Sarig, Theormodynamic formalism for countable Markov shifts, *Ergod. Th. Dynam. Sys.* **19** (1999), 1565–1593.

[Sc] H. Schaefer, *Banach Lattices and Positive Operators*, Springer, New York, 1974.

[SU] B. Stratmann, M. Urbański, Real Analyticity of Topological Pressure, for parabolically semihyperbolic generalized polynomial-like maps, Preprint 2001, *Indagationes Math.*, to appear.

[Su1] D. Sullivan, Seminar on conformal and hyperbolic geometry by D. P. Sullivan (Notes by M. Baker and J. Seade), Preprint IHES (1982).

[Su2] D. Sullivan, Differentiable structures on fractal-like sets, determined by intrinsic scaling functions on dual Cantor sets, The Mathematical Heritage of Herman Weyl, *Proc. Symp. Pure Math.* **48** (1988).

[Su3] D. Sullivan, Quasiconformal homeomorphisms in dynamics, topology, and geometry, Proc. International Congress of Mathematicians, A.M.S. (1986), 1216–1228.

[Su4] D. Sullivan, Entropy, Hausdorff measures old and new, and limit sets of geometrically finite Kleinian groups, *Acta Math.* **153** (1984), 259–277.

[U1] M. Urbański, Hausdorff measures versus equilibrium states of conformal infinite iterated function systems, *Periodica Math. Hung.* **37** (1998), 153–205.

[U2] M. Urbański, Porosity in conformal infinite iterated function systems, *J. Numb. Th.* **88** No. 2 (2001), 283–312.

[U3] M. Urbański, On the Hausdorff dimension of a Julia set with a rationally indifferent periodic point, *Studia Math.* **97** (1991), 167–188.

[U4] M. Urbański, Parabolic Cantor sets, *Fund. Math.* **151** (1996), 241–277.

[U5] M. Urbański, Rigidity of multi-dimensional conformal iterated function systems, *Nonlinearity* **14** (2001), 1593–1610.

[U6] M. Urbański, Rational functions with no recurrent critical points, *Ergod. Th. Dynam. Sys.* **14** (1994), 391–414.

[U7] M. Urbański, Thermodynamic formalism, topological pressure and escape rates for non-recurrent conformal dynamics, Preprint 2000, Fund. Math., to appear.

[UV] M. Urbański, A. Volberg, A rigidity theorem in complex dynamics, in Fractal Geometry and Stochastics, *Progress in Probability* **37**, Birkhäuser, 1995.

[UZd] M. Urbański, A. Zdunik, Hausdorff dimension of harmonic measure for self-conformal sets, *Adv. Math.* **171** (2002), 1–58.

[UZi] M. Urbański, M. Zinsmeister, The geometry of hyperbolic Julia-Lavaurs sets, *Indagationes Math.* **12** (2001) 273–292.

[Wa] P. Walters, *An Introduction to Ergodic Theory*, Springer 1982.

[Wh] G.T. Whyburn, Concerning continua or finite degree and local separating points, *Amer. J. Math.* **57** (1935), 11–18.

[Zaj] L. Zajicek, Porosity and σ-porosity, *Real Anal. Exchange* **13** (1987/8), 314–350.

[Zar] A. Zargaryan, Variational principles for topological pressure in the case of Markov chain with a countable number of states, *Math. Notes Acad. Sci. USSR* **40** (1986), 921–928.

[Z1] A. Zdunik, Parabolic orbifolds and the dimension of the maximal measure for rational maps, *Invent. Math.* **99** (1990), 627–649.

[Z2] A. Zdunik, Harmonic measure versus Hausdorff measures on repellors for holomorphic maps, *Trans. Amer. Math. Soc.* **326** (1991), 633–652.

Index

F-conformal, 57
σ-finite, 225
k-rectifiable, 166
t-conformal, 77

finitely primitive, 5
strongly connected, 5

absolutely regular along a filtration, 40
acceptable function, 8
almost sure invariance principle, 41
almost-periodic, 145
amalgamated function, 56
angle between linear hyperspaces, 196
Apollonian packing, 234
approximate tangent k-plane, 166
associated iterated function system, 107

bi-Lipschitz, 174
Bloch function, 63
bounded distortion lemma, 26

central limit theorem, 41
characteristic Lyapunov exponent, 90
coding map π, 2
coefficients of absolute regularity, 40
cofinite subsystem, 87
cohomologous, 19
Cone property, 72
conformal mapping of $I\!R^d$, 62
conformal affine contractions, 182
conformal affine hoeomorphism, 62
conformal parabolic iterated function systems, 209
conformal-like, 58
Continued fractions, 137
cookie-cutter Cantor set, 176

dynamical rigidity, 170

entropy, 10
entropy with respect to a partition, 10
equilibrium state, 22
essentially affine, 182
expanding maps, 136
exponential decay of correlations, 41

finitely irreducible, 5
finitely supported, 9

generalized Hausdorff measures, 109
geometric rigidity, 160
Gibbs state, 12
graph directed system(GDS), 3
Graph Directed Markov System(GDMS), 1
Grassmannian manifold, 167
guling, 136

Hölder continuous with exponent $\alpha > 0$, 19
Hölder family of functions, 54
hereditarily regular, 87

invariant Gibbs state, 12
Ionescu-Tulcea and Marinescu inequality, 32
irreducible, 5
irregular, 88
iterated function system(IFS), 3

Jacobian, 152

Kleinian groups of Schottky type, 136
Koebe's Distortion Theorem, 63

law of the iterated logarithm, 41
Legendre transform, 126
Legendre transform pair, 127
limit set, 2

line field, 191
Liouville's theorem, 63
lower class, 42

metric d_α, 4
Minkowski, 83
Montel's theorem, 150
Multifractal Analysis, 123
multifractal formalism, 123

normalized Perron–Frobenius operator, 29

Open set condition, 71
orbit sum, 7

partition function, 6
Perron–Frobenius operator, 27
porosity, 103
porous, 103
potential, 22
primitive, 5
purely k-unrectifiable, 166

Radon-Nikodym derivative, 144
Rate of approximation of the Hausdorff dimension, 153
real-analytic extension, 148
Refined Geometry, 109
refined volume lemma, 110
regular, 78
restrictions, 136
Rigidity in Dimension $d \geq 3$, 195
Rokhlin canonical system of measures, 119
Rokhlin's natural extension, 119

scaling function, 176
standard partition, 10
strong open set condition(SOSC), 96
strong separation condition, 58
strong tangent, 162
stronger scaling function, 177
strongly connected, 49
strongly regular, 87
summable, 27

tangent, 162
topological dimension, 165
topological pressure, 7
topologically mixing, 5
two-sided shift space, 119

uniformly perfect, 156
upper class, 42

variational principle, 9
Volume lemma, 90

weak-Bernoulli, 40
weaker scaling function, 177
weighted normalized Perron–Frobenius operator, 36